Chemical Derivatization in Analytical Chemistry

Volume 1: Chromatography

MODERN ANALYTICAL CHEMISTRY

Series Editor: David Hercules
University of Pittsburgh

Chemical Derivatization in Analytical Chemistry

Volume 1: Chromatography

Edited by

R. W. Frei

Free University
Amsterdam, The Netherlands

and

J. F. Lawrence

Food Research Division of the Department
of National Health and Welfare
Ottawa, Ontario, Canada

PLENUM PRESS • NEW YORK AND LONDON

Library of Congress Cataloging in Publication Data

Main entry under title:

Chemical derivatization in analytical chemistry.

    (Modern analytical chemistry)
    Bibliography: v. 1, p.
    Includes index.
    Contents: v. 1. Chromatography
    1. Gas chromatography. I. Frei, R. W. (Roland W.) II. Lawrence, James F. III.
Series.
QD79.C45C48   543'.0896   81-5901
ISBN 0-306-40608-X   AACR2

© 1981 Plenum Press, New York
A Division of Plenum Publishing Corporation
233 Spring Street, New York, N.Y. 10013

Printed in the United States of America

22824

Contributors

W. P. Cochrane, Laboratory Services Division, Agriculture Canada, Ottawa, Canada

R. W. Frei, Department of Analytical Chemistry, Free University, Amsterdam, The Netherlands

W. C. Kossa, Applied Science Division, Milton Roy Company, State College, Pennsylvania

L. A. Sternson, Department of Pharmaceutical Chemistry, The University of Kansas, Lawrence, Kansas

Preface

The first volume in this series is devoted to derivatization techniques in chromatography, for very obvious reasons. In gas chromatography (GC) chemical derivatization as an aid to expand the usefulness of the technique has been known for more than a decade and has become an established approach.

The first chapter deals to a great extent with derivatization for the purpose of making compounds amenable to GC. Although the discussion concentrates on pesticides, some generally valid conclusions can be drawn from this chapter. Chemistry will not be limited to the separation—it can also have a pronounced impact on the sample cleanup, another topic covered in Chapter 1.

Since the introduction of coupled GC–mass spectroscopy (GC–MS), a very powerful tool, derivatization techniques have taken still another direction—taking into consideration chromatographic as well as mass spectrometric improvement of the compounds of interest. Cyclic boronates are discussed as derivatization reagents for this purpose in the second chapter.

Chemical derivatization in liquid chromatography (LC) is a somewhat younger branch. In principle, one differentiates between prechromatographic and postchromatographic (precolumn and postcolumn) techniques. The former has now gained a status comparable to derivatization in GC. The third chapter deals with this aspect. The various possibilities for reacting different groups of compounds are critically discussed and a few examples are given, usually from the pharmaceutical area; but the chapter is by no means restricted to pharmaceuticals. The treatment of the subject in Chapter 3 permits extrapolation into any analytical area in need of derivatization procedures. Also discussed are optical isomers and ion pair formation.

The fourth and final chapter deals with postcolumn reaction detectors in LC. This is no doubt one of the newest lines in the field of derivatization, but it is currently undergoing very rapid development. This chapter gives a comprehensive survey of the state-of-the-art of reaction detectors, including theoretical aspects, a discussion of tubular reactors with segmented and nonsegmented streams, and bed reactors. The technical aspects are grouped according to detection modes (fluorescence, uv–visible) and information on application possibilities are given. Finally, a critical assessment of development potential and trends (i.e., coupling to other detection modes, column switching, automation, etc.) is presented.

The selection of areas and the manner and level of treatment of fundamental and applied aspects should render this book of interest to analytical chemists and investigators in many fields where chromatography is used, e.g., in pharmaceutical, environmental, medicinal, agricultural, and biochemical disciplines.

R.W. Frei

J.F. Lawrence

Contents

1. Chemical Derivatization in Pesticide Analysis
W. P. Cochrane

2. Cyclic Boronates as Derivatives for the Gas Chromatography–Mass Spectrometry of Bifunctional Organic Compounds
Walter C. Kossa

3. General Aspects of Precolumn Derivatization with Emphasis on Pharmaceutical Analysis
Larry A. Sternson

4. Reaction Detectors in Liquid Chromatography
R. W. Frei

Chapter 1

Chemical Derivatization in Pesticide Analysis

W. P. Cochrane

1. INTRODUCTION

Many approaches have been used in the identification and determination of pesticides at both the macro- (formulations) and micro- (residues) levels in various substrates. A number of the more commonly used methods of analysis are shown in Table 1 and the range of techniques used gives an indication of the difficulties encountered by the pesticide analyst. Traditionally, chemical derivatization has played an integral part in the analysis of pesticides since the widespread use of the organochlorine (OC) insecticides and herbicides in the 1940s. In 1955 Gunther and Blinn published *Analysis of Insecticides and Acaracides*, which gave detailed formulation and residue procedures for about 90 inorganic and organic compounds used in current pest control practices.[1] Of the 15 or so organochlorine insecticides discussed, many of the formulation procedures employed a total chlorine method while others utilized specific reactions to produce colored compounds. While DDT formulations could be determined via total chlorine, DDT residues were analyzed by the Schechter–Haller method involving nitration to a tetranitro derivative which produced a colored compound on

W. P. Cochrane • Laboratory Services Division, Food Production and Inspection Branch, Agriculture Canada, Ottawa, Ontario, Canada K1A OC5.

TABLE 1. Methods Used in the Identification of Pesticides

---

1. Instrumental methods

    Spectral techniques — infrared, uv, visible, mass, NMR
    Paper and thin-layer chromatography
    Gas–liquid chromatography — multiple column and detector systems
    Neutron activation analysis

2. Microchemical methods

    Chemical and photochemical conversion of pesticides into derivatives

3. Biological assay methods

    Enzymatic
    Immunological
    Phytotoxicity, etc.

4. Partition methods

    Liquid–liquid — p values
    Liquid–solid — column chromatography

---

treatment with alcoholic sodium methoxide. An alternate method was the dehydrochlorination of DDT to DDE and uv quantitation which was similar to the recommended residue method for methoxychlor. Again for the 15 organophosphorus (OP) insecticides included, colorimetry or spectrophotometry was generally the method of choice although enzymatic methods were also recommended. Diazinon, described at that time as a promising new insecticide, was hydrolyzed with alcoholic KOH to 2-isopropyl-4-methyl-6-hydroxypyrimidone which was quantitated by uv at 272 nm. No carbamate or herbicidal compounds were covered in this early book; these pesticides together with plant growth regulators and food additives were subsequently included in the multivolume series started by Zweig in 1964.[2] Again different chemical reactions were used for formulation and residue analysis. For example, the recommended formulation method for carbaryl was alkaline hydrolysis to 1-naphthol and methylamine with volumetric determination of the methylamine. For carbaryl residues the 1-naphthol produced was coupled with p-nitrobenzenediazonium fluoroborate then quantitated at 590 nm. By today's standards these methods appear rather crude. However, with the introduction of gas chromatography (GC) in the 1950s many organic compounds were qualitatively and quantitatively analyzed intact by this procedure without further alteration. The application of GC to pesticide formulation and residue analysis was not successfully used before 1960 primarily due to the lack of the selectivity of the then available detection systems. The microcoulometric[3] and ³H-electron capture detectors[4] were both reported in 1960–1961 and became commercially available within the following few years. These detectors were quickly followed by the alkali flame ionization (1964), electrolytic conductivity (1965), and flame photometric (1966) detectors which displayed varying degrees of selectivity and sensitivity to N-, P-, S-, or halogen-containing compounds. They were immediately utilized in the pharmaceutical and pesticide fields of residue analysis and to verify product integrity. By 1967, the GC operating parameters for approximately 50 pesticides had been established.[5] In addition, derivatization procedures to enhance sensitivity or impart selectivity were incorporated into many methods for pesticide analysis. Since GC is a "multiple-detection" end method for quantitative analysis the limitations of the EC (electron capture) and other detectors quickly became apparent. Even after the application of the more common cleanup techniques, EC–GC interferences occurred not only from peak overlap of the various pesticides themselves (Figure 1) but also from interferring coextractives which originate from the sample being analyzed, as well as extraneous contamination from solvents or other materials used in the method. With the increased use of capillary column GC in pesticide analysis in the 1970s earlier separation problems were easily solved. For example, in the chlordane field it

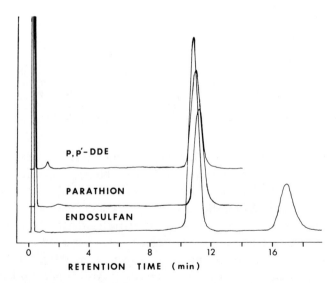

p, p'-DDE

PARATHION

ENDOSULFAN

RETENTION TIME (min)

Fig. 1. Chromatograms of p,p'-DDE, parathion, and endosulfan I on 4% SE-30/6% QF-1 on chromosorb W at 190°C with electron-capture detection showing potential for misidentification.

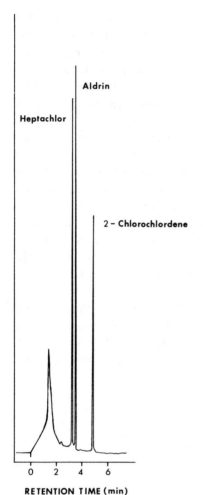

Aldrin

Heptachlor

2 - Chlorochlordene

Fig. 2. Separation of heptachlor and 2-chlorochlordene on a 25 m × 0.25 mm i.d. OV-17 WCOT at 175°C with electron capture detection; injection time 25 sec; solvent effect with purified dodecane.

0   2   4   6

RETENTION TIME (min)

had been previously stated in 1971 that "no suitable GLC column had been obtained that will successfully separate the two compounds" heptachlor and its isomer 2-chlorochlordene.[6] However, as seen in Figure 2, a capillary OV-101 column will resolve these two isomeric compounds.

The identity of one large area of "interfering coextractives" observed by residue chemists involved in multiresidue analysis was solved by the late 1960s. The interfering compounds were shown to be the polychlorinated biphenyl (PCB) class of industrial chemicals.[7] Also implicated have been the polychlorinated terphenyls (PCT), naphthalenes, terpenes, and chlorinated paraffins. Even in the late 1960s there was ample evidence that a significant proportion of the then-current residue data was not based on adequate analytical information since cases of "mistaken identity" were not infrequent. This, of course, led to much research into ways and means of either separating the PCBs and other interferences from the organochlorine residues or chemically altering one or other to eliminate coelution on GC and improve quantitation. Therefore, chemical derivatization techniques for confirmation of residue identity of all classes of pesticides has received considerable attention from the late 1960s to the present.[6,8,9]

It should be stressed that by far the best method of confirmation of pesticide identity is by mass spectrometry (MS). The GC–MS system has become the single most powerful tool for analysis of trace organic contaminants in biological and environmental samples, primarily owing to the development of the computerized measurement of mass spectra.[10] With the use of an interactive data system it is possible to get a complete spectrum on 100 pg of 2,4,5-T methylester depending upon type of system used. Increased sensitivity is possible, e.g., 20 pg of 2,4,5-T methyl ester can be identified by multiple ion monitoring of the six most intense peaks in the spectrum. However, the use of a GC–MS data system is often limited by availability, cost, or sample type (with or without cleanup and residue level). Once the nature or identity of a particular interference is known, the residue analyst is then in a much better position to deal with it—for example, chemical derivatization to alter its retention time or by proper choice of cleanup procedure.

From the above introduction it can be seen that two different types of chemical derivatization techniques have been mentioned. There is the chemical derivatization of a pesticide as a prerequisite of the method of analysis, for example, the esterification of chlorophenoxy acid herbicides prior to EC–GC analysis. Since this derivatization step is part of the actual method it must meet all the requirements associated with a practical, viable analytical procedure, namely, reproducibility, good recovery, freedom from interferences, and accuracy. In the case of a confirmatory test the demands are less severe in that the main criteria are speed, ease of operation, and acceptable yields. In addition, there are two other uses of chemical and photochemical reactions in pesticide analysis. On-column derivatization occurs in the heated injection port or column of the GC. The most commonly used conversion to date has been the alkylation of pesticides or their derivatives which contain acidic NH or OH functional groups. When an OP insecticide in alcohol is injected onto a precolumn containing NaOH-coated glass beads, the P–O–C bond is cleaved with formation of the appropriate ester, e.g., the methyl ester if methanol is used. Although good yields are obtained, the method characterizes only the P-containing portion of the molecule and as such this type of derivatization technique is better suited as a screening method. Similarly, on-column transesterification of the *N*-methyl carbamates results in the formation of methyl-*N*-methyl carbamate, which indicates the original compound was a carbamate but not the identity of the pesticide. Finally, there is that class of chemical reactions that are used to clean up or remove specific interferences rather than form identifiable derivatives. This type of chemical cleanup approach has a long history of use in pesticide analysis. For example, treatment of a wildlife sample extract containing cointerfering mirex and PCB residues can be treated with a 1:1 mixture of concd H_2SO_4–fuming HNO_3 to "eliminate" PCB interference in the quantative analysis of mirex and its derivatives (Figure 3). Essentially, nitration of the PCBs occurs and their GC retention times are sufficiently long to effect a separation between them and the mirex constituents.

The aim of this review is to survey the application of chemical and photochemical derivatization techniques that have been used in the analysis of pesticides. This includes a survey of the four general areas discussed above, namely, chemical derivatization as used in (a) the analytical method, (b) confirmation of identity, (c) screening procedures, and (d) cleanup reactions.

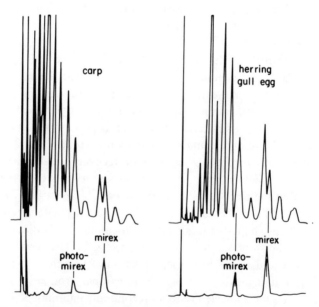

Fig. 3. Chromatograms of carp and herring gull egg extracts, from Lake Ontario, before (top) and after (bottom) nitration. Column 180 cm × 2 mm i.d., 1% SP-2100 on 80-100 mesh Supelcoport at 180°C.

TABLE 2. Chemical Derivatization of Pesticides: Advantages and Reaction Criteria

---

A. Advantages in GC analysis
   1. Increase in volatility
   2. Increase in thermal stability
   3. Improve peak shape and/or separation behaviors
   4. Increase in sensitivity of detection
   5. Impart selectivity

B. Advantages in cleanup
   1. Increase or decrease in polarity, e.g., for column cleanup
   2. Improve extractability, e.g., extractive derivatization

C. Criteria for derivative formation
   1. Must be facile and rapid
   2. Preparation should require the minimum of manipulation
   3. Reaction should give a good reproducible and/or quantitative yield of a specific derivative
   4. Derivative should be stable
   5. Short (acceptable) GC retention time of the derivative
   6. Absence of interference from sample background, other pesticides solvents, or reagents used in the procedure
   7. Sensitivity at the appropriate level of analysis, e.g., mg/kg at the residue level

---

Some advantages to be gained by chemical derivatization are summarized in Table 2 together with the criteria that should be met once the use of derivative formation has been decided.

2. ORGANOCHLORINE INSECTICIDES

Since the beginning of the 1960s the majority of the early chemical methods of analysis described for the OC insecticides[1,2] have been replaced by quantitative GC analysis of the intact pesticide, the EC being the detector of choice for multiresidue methods and the flame ionization detector (FID) or an AFID for formulation analysis. With the increased use of the element-selective detectors, a corresponding increase in shortened methods of residue analysis occurred. Here the final result is obtained by GC after extraction but without prior cleanup of the extract. Owing to the volume of analytical work being done in monitoring environmental and agricultural samples as

TABLE 3. Commonly Used Methods of Confirmation of Pesticide
Residue Identity

---

1. Thin-layer chromatography

    Retention times on different absorbents or using different mobile phases
    Specific visualizing reagents, e.g., $AgNO_3$ for OC insecticides or chlorinesterase
    inhibition for OP insecticides, etc.

2. Gas chromatography

    Retention times on different columns
    Element-selective detectors
    On-column reactions

3. Partition coefficients (p values)

    Partition coefficients using immiscible pairs of solvents

4. Microscale chemical and photochemical reactions

    GC retention times before and after reaction
    Reaction to eliminate one or more of the cointerfering compounds

---

well as the wide range of food products, the need for confirming the identity
of a suspect pesticide took on new importance.

Table 3 gives the most commonly used techniques currently employed
in the confirmation of pesticide residue identity. The most frequently used
method of confirmation is by retention time comparison on two or three
columns of different polarity. At best this method is barely adequate, as
pointed out by Elgar[11] who demonstrated, using published data, that the
correspondence between retention times on different GC phases which gave
good resolution of individual pesticides is highly significant. That is to say,
running the same sample extract on different phases tells the analyst little
more than running it on one phase only except for phases differing markedly
on polarity, e.g., nonpolar DC-200 complimented by the highly polar DEGS
phase. Elgar extended his examples to show that rerunning a sample on
paper or thin-layer chromatography under different conditions or obtaining
p values from different pairs of immiscible solvents again added little to the
positive identification of a particular pesticide. Therefore, complimentary
techniques must use different physical effects to achieve separation. The
use of GC, which depends on vapor pressure, combined with TLC, which
depends on partition, would give extra support to a tentative identification
by GC. In conclusions it was suggested by Elgar,[11] as it had been by others[12,13] in the early 1960s, that the change in retention time after a chemical

reaction, together with the retention time of the parent compound, would provide confirmation of the identity of a pesticide. Since the mid-1960s chemical derivatization techniques have been extensively used for the confirmation of OC, OP, carbamates, herbicides and fungicides, and their metabolites. In most instances they are fairly simple routine tests that can be performed on the final extract used for GC analysis. The ideal situation would be a collection of reagents in which each individual chemical was specific for one pesticide or metabolite and unreactive towards others. This would make the confirmation of a pesticide, even in the presence of artifacts, etc., extremely easy. In practice this situation is far from true and most general reactions have not only an effect on the pesticide in question but on one or more other compounds that may be present. Also, in many instances, the plant extract or sample background is also cause for further concern. In particular, the separation or elimination of the interfering PCBs in OC residue procedures has been the subject of a large body of literature, which will be discussed under cleanup procedures.

2.1. The DDT Group

The most intensely studied methods used in the pesticide residue analysis of DDT and related compounds have been confirmatory techniques using alkali treatment for dehydrochlorination and saponification of fats as part of the cleanup procedure. Using EC–GC for the determination of DDT in butter and some vegetables Klein et al.[13,14] drew early attention to the need for confirming the presence of any DDT residue found. Confirmation was required because of the presence of an unidentified substance (or artifact) which had the same retention time as p,p'-DDT on the nonpolar SF-96 column used.[13] After a 30-min reflux of 2% NaOH in ethanol the residues of o,p'-DDT, p,p'-DDT, p,p'-DDD, and perthane were converted to their respective olefins:

$$(p\text{-Cl}-C_6H_4)_2-CH-CCl_3 \quad \xrightarrow{\text{alkali}} \quad (Cl-C_6H_4)_2-C{=}CCl_2$$

p,p'-DDT mp 109°C $\qquad\qquad\qquad$ DDE mp 88°C

$$(p\text{-Cl}-C_6H_4)_2-CH-CHCl_2 \quad \xrightarrow{-\text{HCl}} \quad (Cl-C_6H_4)_2-C{=}CHCl$$

p,p'-DDD mp 110°C $\qquad\qquad\qquad$ DDMU mp 65°C

$$(p\text{-}C_2H_5-C_6H_4)_2-CH-CCl_3 \quad \longrightarrow \quad (p\text{-}C_2H_5-C_6H_4)_2C{=}CCl_2$$

perthane mp 60°C $\qquad\qquad\qquad$ mp 32°C

Even using this simple technique the SF-96 column alone failed to separate o,p'-DDT, DDD, and perthane as well as their respective olefins.

Investigation showed base-line separation could be achieved using a more polar mixed phase SF-96 and polyester GC column even though the parent compounds perthane and *o,p'*-DDT as well as *p,p'*-DDT and DDD overlapped, respectively. Interestingly, this mixed SF-96/polyester column resolved malathion and heptachlor epoxide which overlapped on the 4% SE-30/6% QF-1 column used extensively in later residue work (Figure 1).

Subsequent publications optimized reaction parameters, simplified the handling techniques, extended the use to other OC insecticides–sample substrate combinations and investigated the resolution characteristics of various GC column stationary phases. A summary is given in Table 4. Dehydrochlorination with alcoholic NaOH or KOH has been carried out under reflux for 30, 15, 12 min, or at room temperature for 5 min with or without preliminary evaporation of the original solvent used for extraction. In some instances where reflux has been used, it has performed the dual function of dehydrochlorination as well as saponification of any fat present.[13,18] One of simplest procedures is to shake the 1-ml final sample extract (normally in *n*-hexane) with 1 ml 5% sodium methylate (NaOMe) in methanol for 1 min, allow to separate and reinject the upper hexane layer.[20] Using 2% ethanolic KOH, Young and Burke[18] optimized reaction parameters to obtain quantitative recovery of the various olefin derivatives, investigated its effect on various other OC insecticides, and detailed a procedure for routine application on residue analysis. Similarly, Krause[19] reported the effect of alcoholic KOH on 46 miscellaneous pesticides and listed their retention time data, before and after reaction, on three different GC columns. In most instances the limit of detection for the DDT-type insecticides is 0.02 ppm of parent compound based on a 10-g sample except for methoxychlor, which is 0.1 ppm. This higher level results from its longer retention time and lesser EC response. In most samples, *p,p'*-DDT is detected together with its metabolite *p,p'*-DDE, which is unaffected by alkali treatment, and therefore must be removed by column cleanup or TLC prior to confirmation of the parent compound. The 2% KOH treatment does not affect the PCBs/HCB, aldrin, dieldrin, endrin *p,p'*-DDE, and other olefins to any great extent and only partially derivatizes heptachlor epoxide, heptachlor, and mirex. The isomers of BHC form trichlorobenzenes which appear with the solvent front, while the endosulfan isomers and endosulfan sulfate are derivatized to compounds with shorter retention times. Dicofol is also completely derivatized but only a 65% yield of *p,p'*-dichlorobenzophenone was recovered[18]:

$$(p\text{-Cl}—C_6H_4)_2—CCl_3 \xrightarrow{\text{KOH}} (p\text{-Cl}—C_6H_4)_2—C{=}O$$

difocol mp 79°C   DBP mp 146°C

TABLE 4. Confirmatory Tests for the DDT Group of Insecticides

Reaction type	Reagent	Substrate(s)	GC column	Reference
Dehydrochlorination	Ethanolic NaOH	Butter, vegetables	2.5% SE-96	13, 14
	Alcoholic KOH	Animal viscera		15
		Soil, vegetable	5% DC-11 or 5% QF-1	16, 17
		Butterfat, kale	10% DC-200/15% QF-1	18
		Spinach, peas, butterfat	3% DEGS, 10% DC-200 or 10% DC-200/15% QF-1	19
	NaOMe	Vegetables, wheat, potatoes	4% SE-30/6% QF-1	20
	1,5-diazabicyclo-5.4.0-undec-5-ene	—		21
	Alkaline precolumn	Carrot, onion	10% DC-200 or 10% DC-200/15% QF-1	22
	KOH- or t-BuOK-impregnated alumina	Fish, mud, lake water	4% OV-101/6% OV-210	23
Oxidation (DDE and other olefins)	CrO$_3$/acetic acid	Soil, vegetables	5% DC-11 or 5% QF-1	16, 21
	Tetrabutylammonium permanganate	—	4% SF-96	24
Dechlorination	CrCl$_2$	Forage crops, cereals	4% DC-11/6% QF-1	27, 30
uv photolysis	Various wavelengths	Soil, water, plants	10% DC-200/15% QF-1	33–39
Esterification (DDA only)	BCl$_3$/2-chloroethanol or BCl$_3$/MeOH	Urine	195% QF-1/1.5% OV-17	40

A minor peak with retention time at 1.71 relative to aldrin was observed on a 10% DC-200 column. When stronger reagents are used, such as potassium *tert*-butoxide (*t*-BuOK),[25] a number of the cyclodiene insecticides of the chlordane series can also be confirmed, e.g., *cis*- and *trans*-chlordane, nonachlor, heptachlor, and heptachlor epoxide (see Section 2.2). Although the above reactions were all carried out in solution, other techniques have been used routinely. Miller and Wells[22] utilized a dual-injection-port–single-column setup in which one side was used as a normal heated injection port for quantitative residue analysis while the other port contained an inert support coated with NaOH and KOH as an alkaline precolumn for interpretation of residue chromatograms. Retention time data on a DC-200 and mixed DC-200/QF-1 columns for 32 commonly used pesticides was reported, as well as the ability of the alkaline precolumn to clean up crop material not removed by conventional column chromatography. The major drawback to the single-column setup is that eventually some alkaline material deposits on the column and thus interferes with normal quantitative work. A much better arrangement is to use two similar columns to obtain before and after chromatograms. This precolumn approach has the distinct advantage of easy operation and speed once it is set up. Another approach has been to carry out the derivatization on alumina microcolumns impregnated with 10% KOH or 20% *t*-BuOK with reaction taking place at 70–75°C for 1 hr or overnight at room temperature.[23] The DDT-type compounds plus *cis*-chlordane were completely reacted overnight using either the Al_2O_3/*t*-BuOK or Al_2O_3/KOH column. The advantage of the latter column over the former was that the KOH column yielded the respective hydroxy derivatives of heptachlor and heptachlor epoxide, especially at 70–75°C for 1 hr, while none were observed with the Al_2O_3/*t*-BuOK column. Again perthane, *o,p'*-DDD and *o,p'*-DDT were difficult to confirm since their dehydrochlorinated products were not sufficiently resolved on the mixed OV-101/OV-210 column when present together. In such instances, chromic acid[16,21] or tetrabutylammonium permanganate[24] oxidation of these olefins and DDE to give their respective chlorobenzophenones are follow-up reactions that can be utilized since the phenones have different retention times. Conversion of DDT, DDE, DDD, and DDMU to DBP can be accomplished in a single step.[24] However, the EC–GC responses of these phenones are relatively low; therefore an appreciable level of parent residue, e.g., *o,p'*-DDT, is required. Since the dehydrochlorinated or dechlorinated derivatives can initially appear as metabolites, the possible sources of interference during CrO_3 oxidation of DDT-type extracts are many.

Dechlorination with chromous chloride ($CrCl_2$) is another reaction that has been extensively studied for confirmation of identity in the OC field,[26–30] certain organophosphorus insecticides,[31] and more recently for NO$_2$-containing herbicides and fungicides.[32] CrCl$_2$ reacts preferentially with the p,p'-isomer of DDT to give, in 45 min, predominantly p,p'-DDD,[27] which on prolonged contact with CrCl$_2$ converts to mainly *trans-p,p'*-dichlorostilbene with DDNU and DDMU as secondary products. Therefore, the complexity of the resulting chromatogram is time dependant. Interestingly, methoxychlor and p,p'-DDE did not react, while o,p'-DDT reacted slowly:

$$(p\text{-}ClC_6H_4)_2\text{---}CH\text{---}CCl_3 \xrightarrow{\ CrCl_2\ } (p\text{-}ClC_6H_4)_2\text{---}CH\text{---}CHCl_2$$

DDT   DDD

$$\xrightarrow{\ CrCl_2\ } p\text{-}ClC_6H_4\text{---}CH\text{=}CH\text{---}p\text{-}ClC_6H_4$$

DCS (*trans* isomer)

+

$$(p\text{-}ClC_6H_4)_2C\text{=}CH_2$$

DDNU

+

$$(p\text{-}ClC_6H_4)_2C\text{=}CHCl$$

DDMU

Also, p,p'-DDE and *trans*-DCS have identical retention times on a 4% DC-11/6% QF-1 column. The major drawback to the usefulness of CrCl$_2$ is that each of the DDT-type compounds (and also cyclodienes, Section 2.2) which react give rise to a definite pattern consisting of, in most cases, two or more GC peaks. Therefore, no more than two pesticide residues should be confirmed at one time.

A similar situation exists in the use of photolysis as a means of OC identification.[33–39] Whether uv photolysis of the DDT-type compounds is carried out in solution[33,34,36–39] or in the solid phase,[35] "fingerprint" patterns are obtained. Wavelengths used have varied from 253.7 to 340 nm, which in turn has necessitated the determination of the optimum period of irradiation required to yield the most characteristic degradation pattern for a given insecticide. Using hexane as solvent, irradiation at 253.7 nm was found quite destructive to the pesticides and caused large tailing solvent peaks probably due to chlorination of the solvent.[37] In the solid phase, irradiation at 253.7 nm resulted in shorter reaction times, e.g., 30 sec, for p,p'-DDE and slightly simpler patterns.[35] The major uv degradation product for the p,p' isomers of DDT, DDD, and DDE is p,p'-dichlorobenzophenone.[34] After 2 hr irradiation of p,p'-DDE at 280–320 nm in hexane three major photoproducts were identified[37]:

$$(p\text{-ClC}_6\text{H}_4)_2\text{—C}{=}\text{CCl}_2 \xrightarrow[\text{2 hr}]{\text{uv}} (p\text{-Cl—C}_6\text{H}_4)_2\text{—C}{=}\text{O}$$

p,p'-DDE   DBP

$+$

$(p\text{-Cl—C}_6\text{H}_4)_2\text{—C}{=}\text{CHCl}$

DDMU

$+$

$(p\text{-Cl—C}_6\text{H}_4)(\text{Cl}_2\text{—C}_6\text{H}_3)\text{—C}{=}\text{CHCl}$

DBP was quite stable to further photodegradation. However, the presence or absence of dissolved oxygen has a significant effect on the yield of DBP. Saturation of the hexane solution containing the pesticide with oxygen prior to irradiation maximizes the yield of DBP.[37] Since the uv photolysis of p,p'-DDT and p,p'-DDE has been carried under different irradiation conditions and fingerprinting affected on various GC columns, they are shown in Figure 4 for comparison purposes. In each instance of irradiation in hexane the major DBP photoproduct appeared at relative retention times (RRT) of 0.303–0.36 to p,p'-DDT and RRT of 0.55–0.66 to DDE. From the findings of Leavitt et al.[37] it would appear the DDE degradation peaks

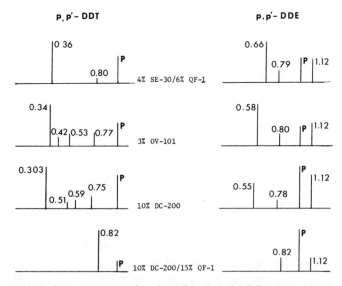

Fig. 4. uv degradation patterns of p,p'-DDT and p,p'-DDE presented as bar graphs of peak heights versus retention time. Each degradation peak has been normalized to the parent compound (P) and retention times are relative to P. The top three bar graphs were obtained from irradiation in hexane[33,34,38] and the bottom graphs from solid phase irradiation.[35]

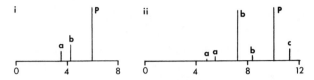

Fig. 5. (*i*) Arochlor 1254 peak 2 plus aldrin (P) irradiated for 75 sec: (a) PCB degradation peak; (b) aldrin degradation peak. (*ii*) Arochlor 1254 peak 5 plus *p,p'*-DDE (P) irradiated for 30 sec: (a) PCB degradation peaks; (b) PCB and *p,p'*-DDE degradation peaks; (c) *p,p'*-DDE degradation peak.

at RRT 0.78–0.80 and 1.12 are DDMU and 2-(*p*-chlorophenyl)-2-(dichlorophenyl)-1-chloroethylene, respectively. Interestingly no DBP is formed during uv solid phase photolysis. In this case dechlorination is one of the main degradation routes, with DDT giving DDD (i.e., RRT of 0.82 to DDT) and DDE forming DDMU (RRT of 0.82 to DDE) (Figure 3). In the application of uv photolysis to PCB–insecticide containing samples Hannen et al.[36] collected each peak (single compound or mixture) following GC separation, irradiated and rechromatographed, and noted any changes. In most cases, the DDT family had overlapping degradation peaks with the photoproducts from their corresponding coeluting PCB isomer. For example, *o,p'*-DDD, *p,p'*-DDD, and *o,p'*-DDT did not have any characteristic peaks which did not coincide with a PCB degradation peak, while *o,p'*-DDE, *p,p'*-DDE, and *p,p'*-DDT had only one characteristic degradation peak each which is separated from the rest. The degradation pattern for a mixture of *p,p*-DDE and PCB peak 5 is presented in Figure 5. Therefore, uv photolysis provides an alternative technique for the confirmation of DDT-type insecticide identity but quantitative analysis by this method is necessarily complicated, especially in the presence of PCBs. Also, depending upon irradiation time, incomplete conversion occurs together with poor EC response of the derivatives.

Finally, the GC analysis of DDA [bis(*p*-chlorophenyl)acetic acid] can be accomplished after esterification with diazomethane or BF_3–MeOH. However, the formation of the 2-chloroethanol ester has two distinct advantages[40]:

(a) The 2-chloroethanol ester is 3.7 times more EC responsive than the methyl ester.

(b) Its longer retention time on a OV-17/QF-1 column separates it from the other OCs which might interfere with the analysis. Since no single-packed GC column has been found that will separate all 18 DDT compounds and derivatives–metabolites, Table 5 lists their retention times relative to aldrin on five different columns of varying polarity.[40,41]

TABLE 5. Relative Retention Times of DDT-type Compounds[a] on Various GC Stationary Phases

Compound	Relative retention time (aldrin = 1.00)				
	10% DC-200	5% SE-30/5% QF-1	10% DC-200/ 15% QF-1	5% QF-1	3% DEGS
DDM	0.62	0.70	—	0.91	—
DDNU	0.78	0.81	—	0.91	—
DBP	0.99	1.28	1.30	2.26	4.15
DBH	0.99	1.28	—	2.26	—
o,p'-DDMU	1.13	—	1.19	—	2.83
DDOH	1.25	2.44	—	3.43	—
p,p'-DDMU	1.45	1.58	1.48	1.83	4.02
Perthane olefin	1.47	—	1.31	—	2.00
o,p'-DDE	1.48	—	1.47	—	3.24
DDMS	1.66	2.00	—	2.96	—
p,p'-DDE	1.89	2.00	1.83	2.30	4.32
o,p'-DDD (TDE)	1.89	2.18	2.04	2.78	8.20
Perthane	2.13	—	2.02	—	5.51
p,p'-DDD (TDE)	2.43	2.70	2.70	4.09	15.10
o,p'-DDT	2.57	2.92	2.48	2.96	6.20
Methoxychlor olefin	2.68	—	2.93	—	15.6
p,p'-DDT	3.24	3.65	2.28	4.48	12.00
Methoxychlor	4.16	—	4.67	—	37.9

[a] Dicofol (Kethane) consistently breaks down to p,p'-dichlorobenzophenone (DBP), while DDA has to be esterified prior to EC–GC (see text).

2.2. Cyclodiene Insecticides

2.2.1. Aldrin–Endosulfan Group

The most commonly utilized reactions for the confirmation of aldrin, dieldrin, endrin, and endosulfan are shown in Table 6. Addition to the unhindered double bond of aldrin has been accomplished using Cl_2, Br_2, and tert-butyl hypochlorite (t-BuOCl) to distinguish "aldrinlike" peaks in a long list of sample types, such as soils,[83] sediments,[84] turnips,[16,46] onions,[46] carrots,[46] sewage sludges,[85] etc. The identity of one of the interferences has been proven to be sulfur.[83] With electron capture detection and a DC-200 column, sulfur displays three peaks with retention times (relative to aldrin, i.e., RRTA) of 0.23, 0.55, and 1.13.[18] Sulfur interference has been solved in various ways, namely, the use of metallic mercury[88] or

TABLE 6. Confirmatory Tests Used in the Identification of the Aldrin–Endosulfan Group of Insecticides

Insecticide	Reaction used	Limit of detection (ppm)	Interfering pesticides	References
Aldrin	Addition			
	Cl$_2$	0.03	p,p-DDD	15
	Br$_2$	0.04	—	6, 43
	t-BuOCl	0.04	—	44
	Epoxidation			
	Peracids	0.01	Dieldrin	45, 46
	CrO$_3$	0.01	Dieldrin	16
	Dechlorination			
	CrCl$_2$	0.01	2-chlorochlordene	28
	uv photolysis	0.008	Dieldrin	33–35, 39
Dieldrin	Halohydrin formation	0.01	Endrin	15, 47–49, 62
	Rearrangement	0.04	Endrin	25, 49–52
	uv photolysis	0.002		33, 37, 57
Endrin	Rearrangement	0.04	Dieldrin	25, 28, 29, 39, 58, 60, 66
	Dechlorination	0.04	—	25, 28, 29, 58, 59
Endrin photoproducts	Rearrangement	0.02	Endrin	61
Endosulfan	Reduction	0.02	Trans-chlordane	62
	Ethanolic alkali	0.1 ppb		63
	Acetylation	0.005 (fish) 0.003 (water)	—	62, 64, 65, 66
	uv photolysis	—	—	39, 67

copper ribbon[89] to convert the sulfur to the corresponding sulfides, refluxing with alkali[18] various GC columns to resolve sulfur from the residues of interest,[90,91] and derivatization of aldrin (Table 6). By far the easiest procedures are the addition of Cl$_2$ to aldrin to produce dichloroaldrin and aldrin oxidation to dieldrin. The majority of other OC insecticides are unaffected by chlorination; however, dichloroaldrin may coincide with p,p'-DDD, o,p'-DDT, or endrin on some GC columns.[6,15] When the oxidation reaction is utilized, prior removal of dieldrin is mandatory. If CrO$_3$ is

TABLE 7. Evaluation of Confirmatory Reactions Performed on 500 ng
Dieldrin

Reagent	R_A values[a]		Identity of derivatives	References
	OV-17/QF-1 column	SE 30/QF-1 column		
1. Aqueous HBr solution	8.3	5.8	Aldrin bromohydrin	15
2. HBr/Ac₂O reagent	8.3	5.8	Aldrin bromohydrin	15
	8.1	7.4	Aldrin bromoacetate	
3. HBr in acetic acid (30–32%)	8.1	7.4	Aldrin bromoacetate	49
4. ZnCl₂/HCl	6.0	4.6	Aldrin chlorohydrin	47
5. BCl₃/2-chloroethanol	6.0	4.6	Aldrin chlorohydrin	48
6. Concd H₂SO₄	9.5	7.3	Dieldrin ketone[b]	25
7. BF₃/diethyl ether	9.5	7.1	Dieldrin ketone[b]	51
8. Ac₂O/H₂SO₄	12.0	Not observed	Dieldrin cross-linked acetate	25,52

[a] Retention time relative to aldrin.
[b] The term *dieldrin* is used to denote the starting material and not a specific structure.

used as reagent, p,p'-DDE will form p,p'-dichlorobenzophenone and hence eliminate any subsequent dieldrin/p,p'-DDE peak overlap experienced on some columns.[15] However, with peracids as oxidants p,p'-DDE, if present, remains intact.

Since a considerable number of chemical tests had been proposed for use in confirming dieldrin residues, Maybury and Cochrane[49] compared eight of the more commonly used reagents on several substrates. Table 7 outlines the reagents used and the identity and relative retention times of the derivatives formed. It was found that aqueous HBr[15] and BCl₃/2-chloroethanol[48] were especially useful because of their application and sensitivity. A lower level of 0.003-ppm dieldrin could be confirmed in a 10-g dry sample, such as an animal feed, and a 0.001-ppm level for a 25-g wet sample. The ZnCl₂/HCl [47] reagent was a practical alternative to BCl₃/2-chloroethanol since each produced aldrin-*trans*-chlorohydrin. The other reagents tried (Table 7) either resulted in multiple derivatives or products with lengthy retention times. Using HCl/acetic anhydride as reagent, Musial *et al.*[66] were able to confirm the presence of 5–10-ppb dieldrin in river silts. Here two derivatives were obtained with relative retention times to dieldrin of 2.0 and 2.3, respectively, on a 10% OV-1 column. The exact structures of the derivatives obtained from these acid catalyzed reactions of dieldrin have been the subject of a number of publications.[50-54] The H_2SO_4/Ac_2O reaction was first reported by Baker and Skerrett[52] as a first step in a colorimetric method for dieldrin residues. Their experiments indicated that a *gem*-diacetate was found instead of a *vic*-diacetate. Subsequently, other workers showed that at least four distinct products could be obtained in this reaction.[50,53] One minor product, a *cis*-diacetate,[53] was correlated with the known *exo-cis*-diol, while another minor product was identified as a monoacetoxy derivative of photoisodrin[54] (Structure II, Figure 6). The major H_2SO_4/Ac_2O product was, in fact, a "half-cage" *gem*-diacetate (structure III), which is probably the compound actually involved in the original colorimetric method for dieldrin.[51] The formation of these products can be explained *via* a Wagner–Meerwein-type rearrangement to give a substituted isodrin carbonium (structure I) ion.[54] It was also suggested[54] that the $BF_3/MeOH$ results in a dimethyl acetal, which was also obtained by warming the *gem*-diacetate in 6 N HCl/MeOH. Originally Skerrett and

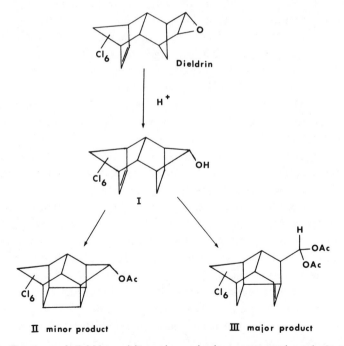

Fig. 6. Structure of dieldrin and its major and minor rearranged products obtained on reaction with a H_2SO_4/Ac_2O mixture.

Baker[51] suggested that BF_3–ether might form a "dieldrin ketone".[11] Dieldrin also undergoes skeletal rearrangements on uv photolysis to form a number of products. When dieldrin is subjected to uv light (253.7 nm) in the laboratory photodieldrin is formed as the major product and pentachlorodieldrin as one of three minor products.[55] However, pentachlorodieldrin was not formed when plants sprayed with dieldrin were exposed to sunlight.[56] Subsequent investigations identified one of the other minor products as photoaldrin chlorohydrin.[57] The retention times relative to aldrin of these three photoproducts on a mixed QF-1/DC-200 column are 1.86 (pentachlorodieldrin), 7.5 (photodieldrin), and 11.6 (photoaldrin chlorohydrin) with dieldrin appearing at 2.35. Erney[39] used a 2-min uv photolysis (254-nm) reaction time to confirm the identity of dieldrin in iso-octane solutions. The photoproduct obtained had a 0.78 retention time relative to dieldrin on a 10% OV-101 column.

The rearrangement of endrin to the half-cage endrin pentacyclic ketone is well known and is easily accomplished thermally,[25] photochemically,[6,18,39] and chemically[16,25] using mineral acids or Lewis acids such as BF_3. This particular reaction has been applied to a wide range of sample types (Table 6), with H_2SO_4 being the preferred reagent.[25,60] Reductive monodechlorination of endrin with chromous chloride ($CrCl_2$) [28,58] or Zn/acetic acid[58] is also another approach that has been used in its confirmation. The final product is a half-cage pentachloropentacyclic ketone[29] which is formed via the intermediate hexachloroendrin ketone.[59] Endrin can easily be confirmed at the 0.01-ppm level in a 10-g sample.[30] The aqueous $CrCl_2$ reaction, normally carried out at 55–60°C, is also an alternative to the potassium t-BuOK/t-BuOH reagent[61] and to confirm the two endrin photoproducts, since monodechlorination occurs at the gem-dichloro groups, respectively. A number of similar approaches have been utilized in the confirmation of the α and β isomers of endosulfan. Dilute ethanolic KOH converts the α and β isomers to the same internal ether,[63] which was found to be three times less responsive to electron capture detection than the endosulfan diacetate formed from the H_2SO_4/acetic anhydride reaction.[64] The use of H_2SO_4-impregnated alumina for the confirmation of endosulfan in water samples yields endosulfan ether. Here 3 ppt of the parent compounds can routinely be confirmed in a 2-l water extract, and 5 ppb in a 10-g fish or sediment extract.[65] A 2–3-min photolysis of α-endosulfan yields one major and two minor peaks, while the β isomers give one major and three minor peaks, and these characteristic GC fingerprint patterns[67] or the major peaks alone[39] could be used for routine confirmation of identity.

2.2.2. Chlordane/Mirex Group

Many chemical derivatization techniques have been devised for the confirmation of identity of the residues of the chlordane group. These are listed in Table 8 together with those used for mirex, Kepone, and Kelvan. Cochrane and Forbes investigated extensively the practical application of $CrCl_2$ to the confirmation of organochlorine pesticide residues.[28] It was found that these pesticides and their epoxy metabolites resulted in two general transformations: (a) reductive deoxygenation of epoxides (e.g., oxychlordane), and (b) replacement of chlorine by hydrogen.

The latter transformation can be further subdivided according to the reactivity of the various chloro structures, namely,

$$-CH=CH-\overset{\displaystyle |}{\underset{\displaystyle |}{C}}Cl \;>\; \overset{\displaystyle Cl}{\underset{\displaystyle Cl}{C}} \;>\; \overset{\displaystyle Cl\;\;Cl}{-\underset{\displaystyle |\;\;\;|}{\overset{\displaystyle |\;\;\;|}{C-C}}-} \;>\; -\overset{\displaystyle |}{CH}-Cl \;>\; -\overset{\displaystyle |}{C}=\overset{\displaystyle |}{C}Cl$$

allylic   geminal   vicinal   secondary   vinylic

Therefore, the allylic chlorine atom of heptachlor is especially susceptible to chemical attack and can be easily replaced by hydrogen using $CrCl_2$[26,28,30] (or acetoxy[68] or hydroxyl[68,25] groups using other reagents). Other dechlorination reagents, such as $NiCl_2$, have proved equally effective.[70] Extension of the $CrCl_2$ reaction to other cyclodiene pesticides with various functional groups in the cyclopentane ring revealed a gradation in reactivity.[28] These were in the following order: 1-keto and 1-hydroxychlordene > 2- and 3-chlorochlordene > *trans*-nonachlor and *trans*-chlordane > *cis*-nonachlor and *cis*-chlordane.

In all, 26 organochlorine insecticides were investigated and the results are shown in Table 9.[28] In most cases a definite derivative pattern consisting of two or more peaks was obtained on $CrCl_2$ reduction. Therefore, like uv photolysis[33,34,35,39] no more than two pesticide residues should be confirmed at the one time. Depending on the pesticide, $CrCl_2$ confirmation could be accomplished at the 0.005–0.01-ppm level in agricultural samples. Since $CrCl_2$ can also reduce a NO_2 group to NH_2, this reagent has been used in the confirmation of certain herbicides (see Section 5.2).

Another very useful and well-studied reaction is the action of strong basic reagents. Using *t*-BuOK/*t*-BuOH, heptachlor produces 1-hydroxychlordene, while heptachlor epoxide undergoes rearrangement to the secondary alcohol, 1-hydroxy-3-chlorochlordene (Figure 7). Further silylation or acetylation produces more EC-responsive derivatives. Also basic dehydrochlorination of the *cis* and *trans* isomers of chlordane and nonachlor

TABLE 8. Confirmatory Tests for the Chlordane–Mirex Group
of Insecticides

Compound	Reaction used	Substrate	References
Heptachlor	Allylic acetylation	Wheat	68, 69
	Allylic hydroxylation	Forage crops, cereals, fish, water	68, 69, 23, 25
	Allylic dechlorination	Crops, cereals	26, 28, 30, 70
	Addition	Cereals	44, 65
	Epoxidation–oxidation	Soil, root crops, corn, cereals	16, 71, 72
	Photolysis	Food, eggs, root crops, soil, water, wheat, fish, oil	33–35, 38, 39
Heptachlor epoxide	Rearrangement	Cereals, carrots, peas, butterfat	44, 69, 25, 23, 66, 18
	Dechlorination	Cereals	28
	uv photolysis	Water, food, plants, eggs	33, 34, 35, 36, 38[a], 39
	$K_2Cr_2O_7$ precolumn	—	12
Octachlor epoxide (oxychlordane)	Dehydrochlorination	Spinach, pears, butterfat	19
	Dechlorination	Cereals	28
cis- and trans-Chlordane	Dehydrochlorination	Cereals, fish, water	44, 69, 25, 23
Technical chlordane	Alkali precolumns	Carrots, onions	12, 22
Trans-nonachlor	Dechlorination	—	28
	Dehydrochlorination	Cereals	73
	Chlorination	Fish	96, 97
Mirex	uv photolysis	Chicken fat, eggs, milk	74–77
	Dechlorination	Water, sediment, fish	78, 79
	Nitration	Water, fish, milk, fat, sediment	80, 81
	Alcoholic KOH	Vegetables, butterfat	18, 19
Kepone	KOH–esterification	—	82
	Perchlorination	Blood, oyster	83, 84
Kelvan	Photolysis of oxidation	Potatoes	85, 86

[a] See Figure 12.

TABLE 9. Reactivity of Some Chlorinated Pesticides, Metabolites, and Derivatives with Chromous Chloride at 60°C

Compound	$R_A{}^a$	R_A of products[b]	Reactivity[c] hr
α-BHC	0.35	0.00	2
β-BHC	0.40	0.00	4
γ-BHC (lindane)	0.44	0.00	(5 min)
Chlordene	0.54	*0.42*, 0.37	16
Heptachlor	0.78	*0.54*, 0.42, 0.37	0.75
2-Chlorochlordene	0.80	*0.68*, 0.59	10
Chlordene epoxide	0.81	0.63, 0.54, 0.42, 0.37	6
α-Dihydroheptachlor	0.93	0.71, 0.65, *0.49*, 0.43	18
l-Ketochlordene	0.93	*0.83*, 0.67	6
l-Keto-2, 3-epoxychlordane	0.98	*0.93*, 0.83, 0.67	(15 min)
Aldrin	1.00	*0.78*, 0.69	24
β-Dihydroheptachlor	1.02	0.81, 0.74, 0.49, 0.43	20
3-Chlorochlordene	1.03	*0.86*, 0.75	10
l-Hydroxychlordene	1.08	0.89, 0.77	6
Heptachlor Expoxide	1.28	*1.02*, 0.91	2
1,2-Dichlorochlordene epoxide	1.30	*1.09*, *0.92*, 0.80	2
Trans-chlordane	1.50	*1.28*, 1.14, *0.42*, 0.37	10
1-Hydroxy-2,3-epoxychlordene	1.52	*1.08*, 0.89	1.25
Endosulfan	1.62/2.32	1.06	8
1,3-Dichlorochlordene epoxide	1.66	*0.91*, *0.83*, 0.68, 0.54	1.5
Cis-chlordane	1.70	*1.37*, 1.29, *0.42*, 0.37	16
Trans-nonachlor	1.80	*1.58*, 1.50, *0.42*, 0.37	10
Dieldrin	1.98	1.68, 1.40, *1.0*, *0.78*, 0.69	6
Endrin	2.22	3.7	0.5
Cis-nonachlor	2.7	*2.31*, 2.21, *0.42*, 0.37	16
Ketoendrin	4.0	3.7	0.5

[a] Relative to aldrin on a 5% SE-30 on Anakrom ABS column at 185°C.
[b] On complete consumption of starting material; major products are in italics.
[c] Average of at least three replicates.

lead to their respective olefin derivatives[69] which appear at shorter GC retention times than the parents. Therefore, it can be seen that the reaction with base can be used for the simultaneous confirmation of DDT-type, chlordane-type, and other OC insecticides (Figure 8). This aspect has been studied by a number of workers using sodium methylate[20] 2% alcoholic KOH,[18,19] *t*-BuOK/*t*-BuOK,[25,44,69] *t*-BuOK-impregnated Al_2O_3 column,[23,60,65] and a GC alkaline precolumn.[22] Depending on conditions used, one[69] or more[19] GC peaks can be obtained. The reaction mechanisms

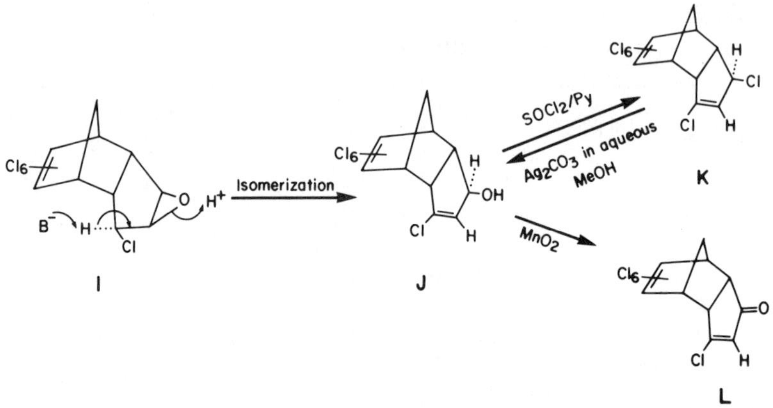

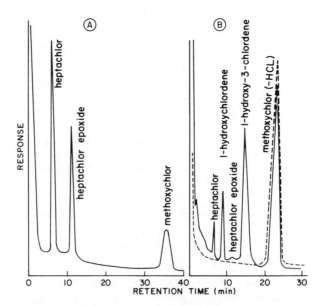

Fig. 7. Base-promoted isomerization of heptachlor epoxide (*I*) to the corresponding allylic alcohol (1-hydroxy-3-chlorochlordene, *J*) with subsequent derivatization by chlorination and oxidation.

and identity of the products from the *t*-BuOH/*t*-BuOK reaction has been investigated in some detail for heptachlor, its epoxide, and *cis*- and *trans*-chlordane.[69] Prior to a study of the effect of base on *cis*- and *trans*-nonachlor their exact stereochemistry had to be confirmed. The *trans* isomer

Fig. 8. (A) Mixture of 200 pg each of heptachlor and its epoxide and 400 pg methoxychlor. (B) 3–5 times the concentration of A after derivatization at 75°C on a 1-in. Al_2O_3/*t*-BuOK microcolumn for 1 hr (——) and on 1-in. Al_2O_3/KOH for 3 hr (- - - -).

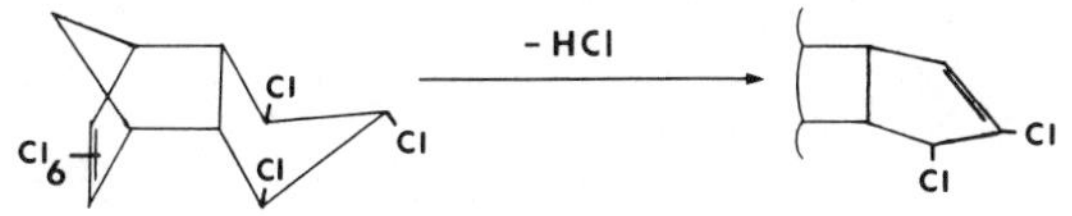

trans - nonachlor 1,2 - dichlorochlordene

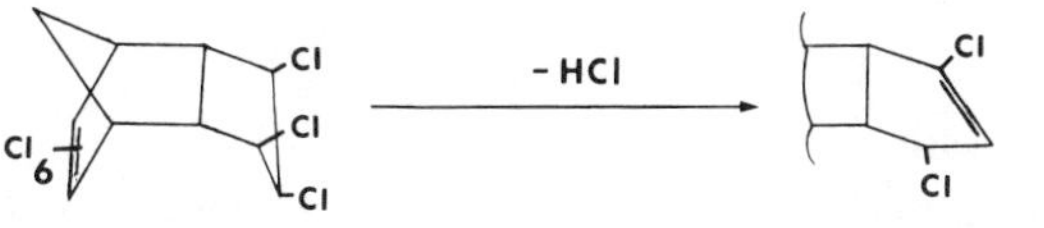

cis - nonachlor 1,3 - dichlorochlordene

Fig. 9. Dehydrochlorination reactions of *cis*- and *trans*-nonachlor.

proved to have a 1-*exo*-2-*endo*, 3-*exo* arrangement of chlorines in the cyclopentane ring, while the *cis* isomer had a 1-*exo*, 2-*exo*, 3-*exo* arrangement.[73] On dehydrochlorination, 1,2- and 1,3-dichlorochlordene were obtained from *trans*- and *cis*-nonachlor, respectively (Figure 9). Subsequently, 1,2-dichlorochlordene was used as precursor in the synthesis[92,93] of 1,2-dichlorochlordene epoxide, which is more commonly referred to as octachlor epoxide[19] or oxychlordane.[92] One major drawback in the early identification of oxychlordane as a distinct metabolite of *cis*- and *trans*-chlordane was the fact that on nonpolar columns such as DC-200, OV-1, SE-30, etc., oxychlordane and heptachlor epoxide were not satisfactorily separated.[94,95] Eventually columns containing 11% OV-17/QF-1[94] or 1.5% OV-17/1.95% OV-210[95] were found to give separation of oxychlordane, heptachlor epoxide, and *cis*- and *trans*-chlordane. The formation of heptachlor epoxide via the action of chromic oxide (CrO_3) [16,72] on heptachlor has proved a very useful confirmatory test for heptachlor as well as a convenient method of synthesizing the heptachlor *exo* epoxide found in biological systems. However, it was noted that at the residue level the CrO_3 oxidation of heptachlor did not appreciably increase the total concentration of any epoxide already present in the extract. Cochrane and Forbes[72] showed that CrO_3 oxidation of heptachlor produced not only its epoxide (structure II, Figure 10, 40–60% yield) but also 1-carboxyendosulfan lactones (structure III, 30–50% yield) with chlorendic acid as a minor product (structure IV, 10–20% yield). Since the double bond of heptachlor is sterically hindered by the allylic chlorine, addition of chlorine normally does not take place unless an initiator, such as antimony pentachloride, is present.

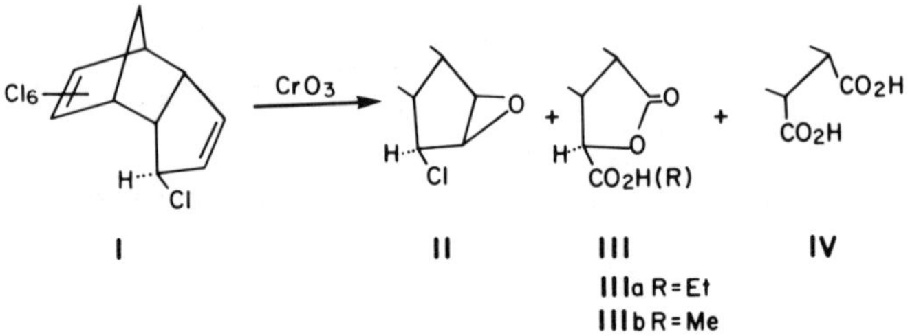

Fig. 10. Reaction of heptachlor epoxide with chromic oxide.

However, chlorine bubbled through a refluxing solution of heptachlor in chloroform for 6 hr produced *cis*- and *trans*-nonachlor, which were subsequently used as reference standards for the GC–MS identification of both nonochlors in goby fish from Tokyo bay.[96,97] When heptachlor is treated with *t*-BuOCl/HOAc a single chloroacetate derivative is formed which has been used for confirmatory purposes as well as a precursor in synthesizing the isomeric, nonbiologically produced heptachlor *endo* epoxide.[44,69,73] This isomeric heptachlor *endo* epoxide also reacts with sodium methoxide to form an unsaturated secondary alcohol by initial nucleophilic attack on the proton in the *trans* position to the *endo* epoxide group, i.e., the proton situated on the C-3a ring junction position.[98]

It should be emphasized that the "chlordane group" comprises a vast array of compounds and metabolites possessing the basic methanoindene skeleton. For example, the complexity of technical chlordane itself has been the subject of a number of publications. Originally, seven crystalline compounds were isolated from technical chlordane, and this number has increased from 11 to 26 to 45 components with the advent of GC and combined GC–MS techniques.[99,100] In addition, isomeric "chlordenes" have been structurally elucidated and found not to contain the cyclodiene (i.e., methanoindene) type of structure (Figure 11).[101–103] Similarly, there has

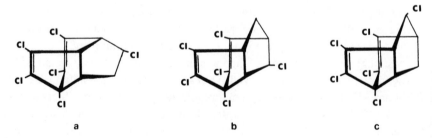

Fig. 11. Structures of the isomeric chlordenes. (a) α-, (b) β-, and (c) γ-chlordene.

been a diversity of *in vivo* and *in vito* metabolic pathways leading to a multitude of "chlordane" epoxides, hydroxylated or oxidatively dechlorinated or dehydrochlorinated metabolites.[104,105] In all of these studies the epoxidation, dehydrochlorination, and dechlorination reactions described above have played an important role in synthesizing metabolites–reference materials and structural elucidation. Photochemistry has also played an important role in the confirmation of identity and the preparation of photoproducts found under environmental conditions. The intermolecular rearrangements that can occur in the "chlordane group" of insecticides have been divided into three classes:

(a) *Photoisomerization of the* ($\pi\sigma \rightarrow 2\sigma$) *Type*. These include the *cis* and *trans* isomers of chlordane and nonachlor, which do not have a double bond present in the cyclopentane ring. Here a new C–C bond is formed to give a "half-cage" photoproduct, e.g., photoheptachlor epoxide.

(b) *Photoisomerization of the* ($2\pi \rightarrow 2\sigma$) *Type*. This represents a four-center cyclo addition between a double bond in the cyclopentene ring and the ClC=CCl group to form two new C–C bonds to give the so-called "bird cage" photoproducts, e.g., photoheptachlor. Similarly, the isomeric "chlordanes" can also undergo ($2\pi \rightarrow 2\sigma$)-type cyclo additions, e.g., photo-α-chlordene.

(c) *Photoreversible and Photoirreversible Hydrogen Transfer Reactions*. Among the group-transfer reactions, a large number of suprafacial transfers of two hydrogen atoms from saturated to unsaturated systems are known, e.g., the reversible irradiation of the diester of chlorendic acid. For nonconcerted processes, it is probable that a radical forms in the first step and this radical can stabilize either by displacing a second hydrogen atom or by forming a new C–C bond, e.g., the irreversible intermolecular photoisomerization of chlordene to form the β- and γ-chlordene isomers.[101]

Another very important photoreaction observed in the chlordane group is dechlorination reactions in various solvents. In this case the didechlorinated product and the two possible monodechlorinated ClC=CCl double-bond isomers (only for asymmetric cyclodienes) are formed. The above photoreactions have been thoroughly reviewed by Parler and Korte,[106] their GC retention indices determined on OV-17,[107,108] and their application in confirmatory test analysis are as listed in Table 8 (see also Figure 12).

Because of the structural similarity between the chlordane photoproducts such as photoheptachlor and the insecticides mirex–Kepone they have also been included in Table 8. Although mirex is thermally stable and is stated[74] to be resistant to chemical degradation by strong acids and bases

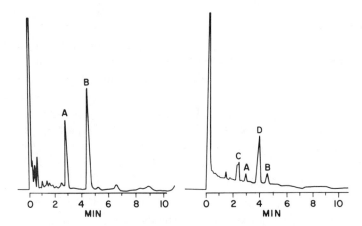

Fig. 12. Gas chromatograms of an extract of a fortified pelletized animal feed sample before (left) and after (right) irradiation: (A) heptachlor; (B) heptachlor epoxide; (C) heptachlor photoderivatives; and (D) heptachlor epoxide photoderivative. The right chromatogram illustrates "problem" products where sample extracts contain uv-absorbing substances that prevent complete conversion of parent pesticides to derivatives in the 30-min exposure period specified.

as well as oxidizing and reducing agents, it does, in fact, react with strong bases and certain reducing agents. Using refluxing 2% alcoholic KOH, Krause reported[19] that mirex ($R_A = 5.43$ on DC-200) formed an unidentified degradation product with $R_A = 0.94$ on DC-200, while Young and Burke[18] obtained a product with $R_A = 1.83$. To date the identity of the alkaline degradation product(s) has not been elucidated. Two other "direct" chemical reactions have been used for confirmatory purposes, namely, dechlorination reactions using $CrCl_2$ or uv photolysis. The major monodechlorinated photoproduct of mirex in hydrocarbon solvents,[74] by sunlight when absorbed on silica[74] or by γ irradiation,[76] is 1,2,3,4,5,5,6,7,9,10,10-undecachloropentacyclo(5.3.0.0^{2,6},0^{3,6},0^{4,8})decane (i.e., photomirex). That is dechlorination occurring at the 8-chloro atom to give 8-monohydromirex. However, when irradiated in sunlight in the presence of aliphatic amines the major products result from dechlorination occurring at one or both *gem*-dichloro groups, i.e., positions 5 and/or 10.[74] Similarly, the use of $CrCl_2$ reductively dechlorinates mirex, and depending upon temperature and reaction conditions, three different major products can be obtained,[78,79] none of which appear to be photomirex. In general, it was found that photomirex (8-monohydromirex) was easier to dechlorinate than mirex and Kepone was more difficult to react than mirex under the same conditions. The major difficulty in analyzing or confirming mirex and photomirex in biological samples are the interfering PCBs.[79] Without prior separation the retention times of mirex and photomirex are similar to that of major heptachlorobiphenyls on most GC columns[75–79] (Figure 3). In mirex–PCB containing human tissue extracts Lewis *et al.*[75] utilized a diethylamine-assisted photolysis procedure at wavelength >280 nm to selectively eliminate the PCB interference. After a 100-min photolysis in hexane containing diethylamine, complete elimination of the interfering heptachlorobiphenyl component was obtained with only a 0–5% loss of mirex. However, other workers[77] have used a 30-min irradiation time and pointed out that although mirex can be identified by GC, using irradiation to limit PCB interference and its presence confirmed by mass spectrometry, quantitative results at low levels (0.01–0.1 ppm) obtained by GC remain questionable, as indicated by the poor agreement between MS and GC data. Other indirect approaches to the PCB–mirex problem have been the perchlorination of the PCBs to decachlorobiphenyl with $SbCl_5$-containing reagents and nitration of the PCBs with a 1:1 mixture of concd. H_2SO_4:fuming HNO_3.[80,81] These will be discussed in more detail later (Section 2.4).

Although Kepone (chlordecone) is an insecticide in its own right it can also arise as an environmental[107] or photolytic[108] degradation product of Mirex. Since Kepone contains a reactive carbonyl group it has been confirmed by conversion back to mirex.[80,83] Using a 4:1 ratio of phosphorous pentachloride to aluminum chloride in CCl_4 at 145°C for 3 hr in a closed tube as perchlorination reagent, Kepone could be confirmed at the 11-ppb level in blood samples.[80] The reaction was found to be quantitative, eight of the more common organochlorine insecticides were shown not to interfer, and prior separation of mirex must be carried out if present in the same samples. Another interesting reaction is the prolonged treatment of Kepone with refluxing KOH pellets in xylene to yield a nonachloro-saturated acid via a Favorskii rearrangement.[82] This particular reaction has not, so far, been applied to residue confirmation. Similarly, Kepone can be converted to the corresponding dihydroderivative via a Wolff–Kishner reduction using basic hydrazine hydrate in diethylene glycol at 200°C for 4 hr.[83] Kelvan is the ethyl levulinate of chlordecone (Kepone) and can be confirmed via photolytic dechlorination[85] or by oxidation back to Kepone.[86] Using the latter reaction a lower level of 5 ppb Kelvan could be detected in potatoes.[86]

2.3. Miscellaneous

In addition to those organochlorine insecticides discussed above, the BHC isomers, HCB, and toxaphene residues also pose confirmatory problems. Traditionally, lindane (γ-BHC) has been confirmed by various separation methods and reliance placed on R_f, R_t, p values, or appearance in the appropriate fraction on column cleanup. However, Zimmerli *et al.*[111] employed a 3.5% ethanolic KOH solution at 80°C to determine the total isomers of BHC in fat samples. Quantitation was based on GLC analysis, of the trichlorobenzenes formed, on DC-200 and DC-11 columns at 80–110°C. Although Zimmerli *et al.*[111] obtained 90–98% recovery of BHC residues as the trichlorobenzenes, Krause[19] recovered only 22% of 1,2,4-trichlorobenzene, the major dehydrochlorination product of the BHC isomers, after treatment with 2% ethanolic KOH solution at 100°C. Using milder conditions, Cochrane and Maybury[112] achieved quantitative conversion of lindane to trichlorobenzenes in a 10-min reaction with sodium methoxide (NaOMe) in methanol at 60°C. Similarly, α-BHC was derivatized in 5 min, while β-BHC was not affected after a 15-min reaction time. Normally a three-peak dehydrochlorination pattern consisting of 1,3,5-, 1,2,4-, and 1,2,3-trichlorobenzene is obtained and their respective ratios are different for α-, β-, and γ-BHC (Figure 13). For adequate GC separation

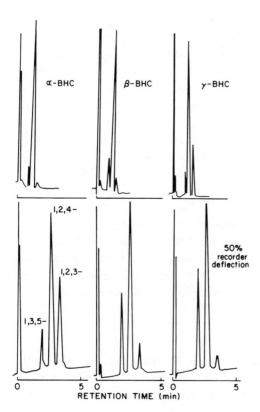

Fig. 13. Dehydrochlorination patterns obtained from α-, β-, and γ-BHC: upper chromatograms, alkaline precolumn treatment (11% OV-17/QF-1 at 150°C); lower chromatograms, sodium methoxide/methanol treatment (6% SE-30/4% QF-1 at 100°C).

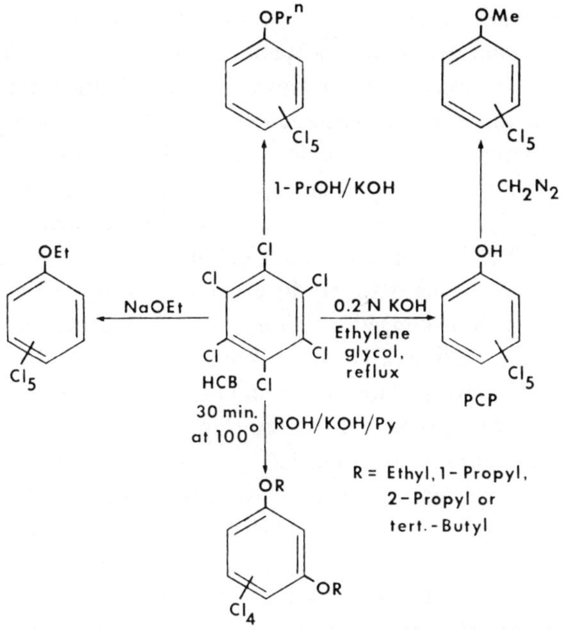

Fig. 14. Various reactions used in the analysis of hexachlorobenzene (HCB).

a column temperature of about $100°C$ is required, since at temperatures of $180–210°C$ the trichlorobenzenes merge with the solvent front. Comparison of the NaOMe solution reaction with an alkaline precolumn technique revealed that the latter saved time in the analysis of cereal, animal feed, and cheese extracts.[112] However, the alkaline precolumn did not eliminate interferences in some meat and fat samples. Here the NaOMe reaction was the preferred technique.[112] Nickel boride (Ni_2B), prepared *in situ* by the reaction of $NaBH_4$ with alcoholic nickel chloride, totally dechlorinates lindane to give benzene as the major reaction product, with cyclohexane and cyclohexene as minor products.[113] Using $NaBH_4$, alone in methanol, produced compounds containing four and five chlorines, none of which appeared to be chlorobenzenes.

Hexachlorobenzene (HCB) is a fungicide which has been used to control Bunt in wheat and it has attracted attention because of its common occurrence in milk, adipose tissue, wild life samples, wheat, human blood, and lake water. A common approach to the chemical confirmation of HCB has been its reaction with basic reagents. As shown in Figure 14 reaction of HCB with KOH in 1-propanol for 10 min at reflux temperature yields the pentachlorophenyl propyl ether.[114] This propyl ether was chosen over the pentachloroanisole (i.e., the methyl ether) since it had a longer GC retention time than HCB, the pentachloroanisole and HCB had similar retention times on the GC columns investigated.[114] Other approaches have been the conversion of HCB to pentachlorophenol, which is then methylated with diazomethane to give pentachloroanisole.[115] In this case the HCB, at levels of 0.005–1 ppm, was initially quantitated on a 4%-SE-30/6%-QF-1 column at 180°C, while confirmation, using the pentachloroanisole, was carried out using a 6% Carbowax 20M column at 150°C. Sodium ethoxide (NaOEt) can also be used to produce a monoethoxy derivative[116] or a diethoxy product on prolonged reaction,[117] which has been used to confirm HCB residues in Harp seal tissues. Similarly, Crist and co-workers[118] confirmed HCB at levels as low as 5 ppb in fatty tissue by the selection and preparation of appropriate derivatives using various alcohols as solvent (Figure 14). For example, 0.3 ppm HCB was confirmed and quantitated in rat adipose tissue containing 160 ppm Arochlor 1016, based on conversion to isopropoxypentachlorobenzene. This reaction has also been applied in the confirmation of HCB down to a lower level of 0.1 ppb in human milk.[118]

Toxaphene, like technical chlordane, is a complex mixture and has been shown by GC–MS to contain at least 177 C_{10} polychlorinated derivatives.[120] Therefore, the detection and quantitation of toxaphene residues is difficult, expecially in the presence of other organochlorine insecticides. Various changes in the toxaphene GC pattern can be brought about by using various precolumn reagents,[12] alkali precolumns,[22] or alkali reactions in alcoholic solutions.[19,25,121] More recently, Gomes[122] determined toxaphene in the presence of other OCs by refluxing with 50% methanolic KOH for 1 hr to produce a pattern displaying three prominent, well-defined GC peaks. With the aid of a small Florisil column a specific "peak" was isolated which was used to identify toxaphene in the presence of chlordane, Strobane, or PCBs.

2.4. Cleanup Techniques Using Chemical Reactions

As stated in the Introduction, interfering substances can be eliminated by utilizing differences in physical and chemical properties. Four general approaches have been used in the analysis of pesticides. These are discussed in the following sections.

2.4.1. Differences in Polarity

This approach includes the various liquid–solid column and thin-layer chromatographic procedures used for the separation of PCBs from OC residues. Florisil[7,123] and silica gel[124,125] have been used on the macro as well as micro scale for the separation of PCBs from DDT and its analogs. Similarly, a combined silica gel–alumina column[126] has performed the simultaneous cleanup–separation of PCB–DDE containing extracts from sea bird eggs and cod liver oil. Charcoal[127] and polyurethanefoam–charcoal[128] column methods have also been used. In these procedures the PCBs and OC insecticides are separated without being chemically altered.

2.4.2. Treatment with Acids

Sulfuric acid has long been used to remove lipid interferences in extracts of biological origin.[1] As examples, Doguchi et al.[129] used a fuming H_2SO_4 treatment to clean up human fat extracts prior to polychlorinated terphenyl analysis, while others have employed a concd H_2SO_4 treatment in the OC residue analysis[96,97,126,130] of substrates ranging from fish to river water. Normally, the sample extract (commonly in hexane) is shaken with the acid or the extract is passed through a celite–H_2SO_4 [131] or alumina–H_2SO_4 [132] column. In the latter case,[132] heptachlor was converted to 1-hydroxychlordene (90% yield) and endrin to its ketone (75% yield) thereby allowing for their simultaneous confirmation in the presence of large quantities of PCBs. The concentration of the fuming H_2SO_4 used is important especially if above 10% (concentration of SO_3), since some polychlorinated terphenyl GC peaks were found to decrease or disappear.[131].

An early chemical means of separating pesticides from PCBs involved treating the cleaned-up extract with concd HNO_3–H_2SO_4 mixture[7] DDT was apparently nitrated to give a compound with a GC retention time of 2hr. However, it was concluded[7,124] that this procedure did not satisfactorily separate PCBs and DDT since the PCBs were also nitrated. However, nitration with a 1:1 fuming HNO_3–concd H_2SO_4 mixture has been utilized in the separation and quantitation of mirex and photomirex (8-hydromirex) in the presence of PCBs and other aromatic compounds.[81] Essentially the PCBs are nitrated and removed on an alumina microcolumn prior to the GC analysis of the unchanged mirex–photomirex. It was stressed that fuming nitric acid is essential and a reaction time of 30 min at 70°C was chosen to ensure essentially complete nitration of PCBs with seven or fewer chlorines per molecule. Lindane recoveries are erratic and HCB recoveries are low (10%) primarily owing to volatilization losses rather than reaction. After a four-laboratory collaborative study using extracts of naturally contaminated substrates (sediment, carp, eel, and gull egg) this nitration procedure was found reliable for the routine determination of mirex and photomirex at levels ≥ 10 ppb in the presence of 1000-fold greater levels of PCBs (Figure 3). Similarly, Holdrinet[80] used the nitration procedure to confirm mirex, and cis- and trans-chlordane, at the 0.02–0.1 ppm level, in water, sediment, sludge, shearwaters, fish, seals, milk, and human fat.

2.4.3. Treatment with Alkali

Like the above sulfuric acid treatment, procedures involving alkali treatment for the dehydrochlorination of certain OC insecticides and the saponification of fats has also had a long history of use in pesticide residue analysis.[6,7,19,20,25] Young and Burke,[25] using butterfat as substrate and refluxing with 2% ethanolic KOH for 15 min, found that 1 ml of reagent would saponify about 50 mg of fat. Also when the weight of fatty substances exceeded about 50 mg, complete dehydrochlorination of Perthane (40 μg) and methoxychlor (4.0 μg) was not achieved. However, p,p'-DDT (8.0 μg) was completely dehydrochlorinated in the presence of 100–120 mg of fat. Krause[19] also investigated the use of 2% ethanolic KOH in the quantitation analysis of Perthane residues in food products. In the process, the effect of ethanolic KOH on 46 miscellaneous pesticides was reported together with retention data, before and after reaction, on 10% DC-200, 10% DC-200/15% QF-1, and 3% DEGS columns.

The stability of the PCBs to alkali, with no change in their GC pattern, is a characteristic that has been readily utilized in the confirmation of identity of this complex mixture in PCB–OC-containing environmental samples. This has been used for the confirmation of PCB residues in fish and fishery products from the Northwest Atlantic,[133] fish from a French river,[134] and bivalves and sediments.[135] Trotter[136] used ethanolic KOH as one step in a procedure to remove DDT interferences in PCB analysis. Initially, DDT and DDD (TDE) were dehydrochlorinated to their respective olefins, which are then oxidized by CrO_3–acetic acid to the more polar dichlorophenzophenone. As a final step the PCBs were eluted from a Florisil microcolumn with petroleum ether. Although the DDT group was not quantitatively recovered, the procedure was successfully applied to the ppb level analysis of PCBs in Lake Michigan chubs.

2.4.4. Special Reactions or Techniques

Even after column cleanup or treatment with acids or alkalis, a specific reagent may still have to be used to remove interferences. For example, sulfur can be removed by alkali,[18] but if this treatment has not been used in the workup it can be removed by mercury,[88] a copper ribbon[89] or column,[135] or Raney nickel.[130] Similarly, chromic acid oxidation has been used as an adjunct to alkali treatment[136] or column separation[137] to convert DDE and related olefins to their corresponding chlorobenzophenones. However, chromic acid treatment has been found to reduce PCB recovery,[138] produce early elating GC peaks from PCBs and polychloroterphenyls (Arochlor 5460),[131] or give incomplete conversion of p,p'-DDE to p,p'-dichlorobenzophenone if fat or oil is present.[137,139] The use of GC precolumns has also been extended from the use of alkali to "gaseous microreactions" using magnesium oxide (MgO) for the quantitative dehydrochlorination of DDT and metabolites in the presence of background levels of PCBs.[140] Dechlorination reagents have also been used in residue analysis to confirm the presence of chlorinated parafins, mirex, etc. Zitko[144] used sodium bis(2-methoxy-ethoxy) aluminum hydride for the confirmation of chlorinated paraffins in lipid-containing extracts of biological samples and in antifouling paint.[141] Chlorinated paraffins containing up to 50% chlorine yielded, after 2 hr reaction at 150°C, unsaturated hydrocarbons and hydroxy compounds, while those containing up to 70% chlorine gave unsaturated hydroxy compounds. Similarly, Dennis et al.[142] found that Arochlor 1254 could be 97% dechlorinated to biphenyl after a 30-min room temperature reaction with excess sodium borohydride containing nickel boride in 2-propanol. While chromous chloride ($CrCl_2$) has no effect on PCBs it can dechlorinate mirex. Lusby and Hill[79] developed a procedure for the determination of mirex in fish containing a PCB–mirex mixture. Reaction with $CrCl_2$ for 45 min at 100°C selectively dechlorinated mirex to products that eluted prior to the Arochlor 1260 pattern on gas chromatography.

As noted in previous sections photochemically assisted dechlorination is also a well-used technique for confirmation and the elimination of interferences. It was observed that when a 25-ppb aqueous solution of Arochlor 1254, in the presence of suspended titanium dioxide, was irradiated at 365 nm for 30 min, no unreacted Arochlor could be detected in solution or absorbed on the surface of TiO_2.[143] The mechanism of reaction is different than that observed in hexane, and apparently biphenyl is not a primary product in PCB–TiO_2 photolysis. However, it was postulated that photodechlorination of PCBs at catalytic surfaces may be a significant pathway for environmental PCB degradation. By using hexane as solvent and irradiation at 280–320 nm, Leavitt et al.[37] were able to analyze certain OCs in the presence of 20–40 times more concentrated amounts of PCBs. The PCBs were degraded to residual peaks having relatively short GC retention times. Under the condition used, the OCs (DDT, dieldrin, and DDE) underwent varying degrees of degradation. The lower limits of practical quantitation were 0.5 ppm for PCBs and 0.05 ppm for individual OCs. It has also been shown that reductive dehalogenation and ring methoxylation were the main photolytic reactions occurring when hexabromo- and hexachlorobiphenyls were irradiated ($\lambda > 286$ nm) in methanol for 0.5–2.0 hr.[144] It was found that the brominated biphenyl reacted about seven times faster than its chloro counterpart.

The selective photolytic reactivity of PCBs and OCs has been used by Hannen et al.[36] as an aid in the identification of OCs in PCB–OC-containing extracts such as salmon and herring oils. While these workers covered primarily aldrin–dieldrin and DDT-type OCs in combination with over-lapping PCB peaks, Mansour and Parlar[145] investigated selective photo-isomerization for the determination of several cyclodiene insecticides in the presence of PCBs. uv irradiation in acetone at wavelengths greater than 290 nm convert most cyclodiones, e.g., heptachlor, to their cross-linked photo-products, which normally have longer GC retention times than the parent compounds. PCBs are not affected under these conditions; however, ir-radiation of the same sample at 230 nm in methanol dechlorinates the PCBs while the cyclodienes remain unchanged. Similarly, the selective photode-chlorination of chlorinated aromatic and unsaturated materials, including DDT and PCBs, was used to eliminate interferences in the analysis of chlo-rinated paraffins at the 1–5-ppm level in fish.[146,147] This technique has also been used to confirm the presence of 0.1 ppm polybrominated biphenyls in cheese[148] and food commodities.[149] The polychlorinated naphthalenes (e.g., Halowax 1014) also interfere with OC residue analysis, with hepta-chlor, aldrin, heptachlor epoxide, p,p'-DDT, etc. having similar GC reten-tion times to major Halowax 1014 peaks when separated on 2% SE-30/2% QF-1.[149] However, using a GC effluent splitting and trapping technique[34] followed by uv irradiation the coeluting OC insecticides could easily be distinguished from the Halowax 1014 peaks since each insecticide gave a photoproduct that was largely resolved from the Halowax degradation products.[149] Also as mentioned above mirex has been confirmed in the presence of PCBs in chicken fat[75] and Canadian human milk[77] using diethylamine-assisted uv irradiation ($\lambda > 280$ nm) to limit PCB interference.

Another approach to the OC–PCB interference problem has been the perchlorination of the PCBs to decachlorobiphenyl with $SbCl_5$-containing reagents. Berg et al.[151] originally reported the conversion of PCBs to a single decachlorobiphenyl peak, which helped to eliminate the problems associated with quantitation of a multiple peak mixture and also provided a chemical confirmatory test. Reaction was carried out using $SbCl_5$ under anhydrous conditions in sealed tubes at elevated temperature (e.g., 5 hr at 170°C) to give 85% yields at the 1-μg level. On a normal 6-ft SE-30 column, decachlorobiphenyl has a retention time of 8.20 and an EC detector response of 1.6 relative to DDE. Others have reported on optimum reaction conditions[152] and the presence of two contaminants in certain batches of SbCl$_5$ which led to erratic recoveries of decachlorobiphenyl.[153] Recently, Berg et al.[154] described a simplified procedure in which perchlorination was achieved in 2 hr at 130°C using a SbCl$_5$–sulfurylchloride–iodine reaction mixture for the detection of 1–40 ng/m^3 PCB in air samples and 2 ppm PCB in combustible municipal refuse. With this reagent mixture p,p'-DDT gave a single GC peak while the p,p'-DDE yielded four peaks which may be used for the simultaneous determination of DDT, DDE in the presence of PCBs. Similarly, Stratton et al.[155] found that the perchlorination method of Armour[154] did not yield quantitative results on PCB species with less than three chlorines in the molecule or for less than 1-μg quantities of PCB when applied to the analysis of ambient airborne PCB. After method modification, quantitative recovery was achieved for the most volatile PCB species over the range 100 ng to 10 μg in ambient air, human blood, and ground water samples. Mes and Davis[156] also observed that the perchlorination method of Armour[152] accounted for only 38% of the PCBs as calculated on the basis of GC results. However, a duplicate human milk sample fortified before the perchlorination step with Arochlor 1260 at 0.2 ppm gave 75.7% recovery, while a single perchlorination of 10 μg Arochlor 1260 gave 110% recovery. It was concluded that these results could indicate the presence of unknown contaminants in the PCB fraction of Canadian human milk. Robbins and Willhite[157] have suggested that quantitation be accomplished by comparison of the perchlorinated PCB from the sample (milk) with a perchlorinated standard of Arochlor 1254.

PCBs are degraded to hydroxylated metabolites by certain organisms and these compounds are chemically similar to chlorophenols.[158] The analysis of these chlorobiphenylols has been carried out via GC–MS of their TMS (trimethylsilyl ether) derivatives on 3% OV-7 programmed from 90 to 265°C at 5°C/min.[158] Also, TMS derivatives were used for both qualitative and quantative aspects of the determination of the hydroxylated metabolites of p-chlorobiphenyl and p,p'-dichlorobiphenyl in the rat.[159] In one instance BSFTA (N,O-bistrimethylsilyltrifluoroacetamide), as silylation reagent, in acetonitrile at 60°C for 15 min was used,[158] while another employed 0.1 ml reagent (hexamethyldisilazane-trimethyldichlorosilane–pyridine, 3:1:9) at room temperature for 30 min.[159] Zitko et al.[160] utilized the corresponding acetates of the chlorobiphenylols for their detection at the μg/g level in shark liver, herring gull eggs, and salmon samples.

3. ORGANOPHOSPHORUS INSECTICIDES

Most organophosphorus (OP) pesticides may be determined directly by GC without derivatization. The most commonly used detectors being the flame photometric (FPD) and the alkali flame ionization detector (AFID), which are selective to P and/or S and have helped in minimizing sample cleanup. However, their specificity is limited, and in the case of the FPD, cross-channel interference in the S mode by P compounds can occur depending on the oxygen: hydrogen ratio used.[161,162] The AFID will respond to N, As, Cl, Br, I, and B in addition to P, the selectivity being controlled by varying parameters such as electrode distance, type of alkali salt, temperature of the flame or the salt bead in the case of the recent electrically heated type "NP detectors".[162] Hence, chemical derivatization methods are still required in the OP area to further characterize residues and metabolites found in environmental–biological samples.

With the recent interest in on-column transesterification reactions on GC there are now four general techniques that can be used for the confirmation of OP insecticides, namely, 1. alkaline hydrolysis and further derivatization of the P moiety produced, 2. hydrolysis followed by derivatization of the alkyl or aryl moiety formed, 3. on-column transesterification, and 4. derivatization of the intact OP insecticide.

Figure 15 illustrates reactions from areas 1, 2, and 4 using parathion as a typical example.

3.1. Derivatization of the P Moiety

St. John and Lisk[163] described a rapid method of quantitative analysis for temephos, chlorpyrifos, dasanit, and azinphos-ethyl in soil, grapes, and cow urine based on their hydrolysis to the corresponding dialkyl phosphate, methylation with diazomethane, and determination of the subsequent trialkyl phosphate. This approach was expanded to 32 OP pesticides and metabolites[164] but conditions of hydrolysis and extraction were not optimized and interference from inorganic phosphate occurred with those OPs which formed trimethyl phosphate. In a series of publications, Shafik, Enos, and co-workers[165–168] resolved these limitations and extended the method to other substrates, primarily human blood and urine. The interference from inorganic phosphates was eliminated by formation of the methyl and ethyl derivatives of the dialkyl phosphates to give limits of detection in the 0.01–0.2-ppm range using a P-mode FPD.[165] Confirmation was achieved by forming the n-propyl derivative. Also use of a 20% Versamid column at

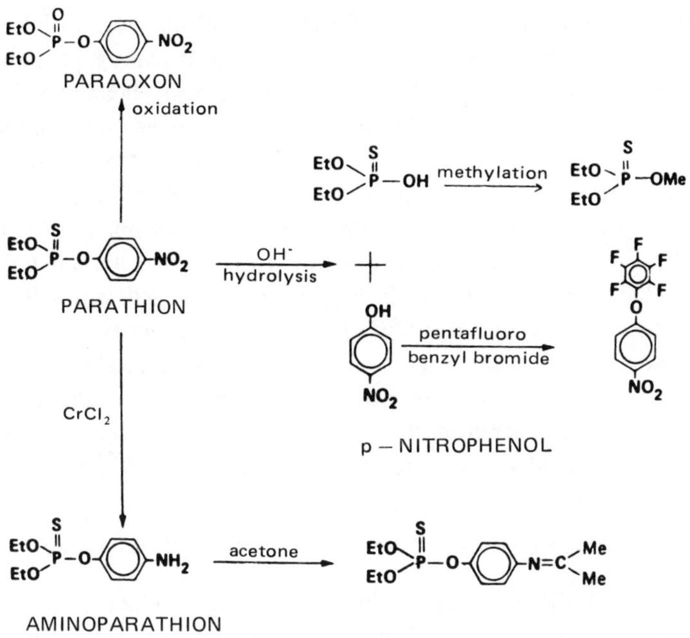

PARAOXON

oxidation

EtO, S, P—OH methylation EtO, S, P—OMe

PARATHION

OH⁻ hydrolysis

methylation

pentafluoro benzyl bromide

p — NITROPHENOL

CrCl₂

acetone

AMINOPARATHION

Fig. 15. Various reactions used in the confirmation of organophosphorus insecticides.

142°C permitted satisfactory GC separation of all the trialkyl phosphates of interest in metabolic work. This coupled with optimized NaOH hydrolysis conditions resulted in a method of confirmation of 32 OP compounds at the residue level.[166] Further improvements in methodology involved preparation of the less volatile, more FPD-sensitive amyl derivatives together with GC separation on a 5% OV-210 at 173°C.[167] In this instance further confirmation was achieved by formation of the *n*-hexyl derivatives which were separated at a GC column temperature of 190°C or by reinjection on a 4% SE-30/6% OV-210 column. It was noted that using diazopentane, O,O-dimethyl phosphorothionate (DMTP), and O,O-diethyl phosphorothionate (DETP) were isomerized producing a mixture of their respective thionate and thiolate esters. This was in contrast to the previous method[168] where the thiolate isomer of DMTP was not evident. Quantitation of DMTP and DETP was based on the thionate isomers O,O-dimethyl-O-amyl phosphorothionate (DMATP) and DEATP, respectively, because of greater interference with the thiolate peak. Still a further refinement included the quantitative extraction of OP metabolites or hydrolysis products in urine using an Amberlite CG 400 ion exchange resin microcolumn, followed by derivatization with diazopentane on the resin.[169] Other workers[170,171] have also modified the basic method[167] by including a heating procedure for amyl derivative formation and a change in the SiO$_2$ cleanup column to eliminate interfering materials[170] or using diazoethane, no cleanup, and quantitation of both S- and O-isomeric products.[171] Even with optimization of methodology this confirmatory test still suffers from two major drawbacks. The unfavorable partitioning of the highly polar dialkyl phosphates into the extraction solvent, also, if two or more parent OPs yield the same dialkyl phosphate as hydrolysis product there will be no distinction between them in the final analysis. For example, dichlorvos, trichlorfon, ronnel oxon, fenthion oxon, etc. each yield DMP (dimethyl phosphate), while temephos, fenthion, fenitrothion, and methyl parathion produce DMTP as product. Therefore, this approach which only characterizes the P moiety is more suited as a very good general screening method.

3.2. Derivatization of the Alkyl or Aryl Moiety

As shown in Figure 15, parathion on hydrolysis produces p-nitrophenol, which is further derivatized with pentafluorobenzyl bromide (PFBB) to its PFB ether, which is very EC sensitive.[172,173] The further derivatization of hydrolysis-generated phenols to various ethers and esters has been previously reviewed[8,9] and a large volume of literature also exists in connection with the quantitation analysis of phenolic herbicides or metabolites.[9,174,175] The most commonly used derivatives are the trifluoroacetyl, heptafluorobutyryl, pentafluorobenzyl,[172,173,176,179] 2,4-dinitrophenyl,[177] and 2,6-dinitro-4-trifluoromethyl phenyl[176] ethers. Seiber et al.[176] found that the relative GC retention times for a number of phenols varied in the order PFB < DNT < DNP, with PFB derivatives being the most EC sensitive. Bowman et al.[180] used the PFB ether to quantitatively analysis for the hydrolysis product of etrimfos in corn and alfalfa at levels of about 10 ppb or less. Conversion was accomplished by reaction of the dry cleaned-up extract with 1 ml pentafluorobenzyl chloride for 2 hr in a sealed tube at 60°C. This procedure differs slightly from other workers who have used PFBB in acetone in the presence of K$_2$CO$_3$.[179] Using the latter procedure levels of 0.01–0.1 ppb of ronnel, crufomate, methyl parathion, fenitrothion, and parathion in 1-l water samples were confirmed after hydrolysis/PFB ether formation.[181] The procedure was subsequently modified and extended to dyfonate, dichlorofenthion, and cyanox.[179] Also removal of excess PFBB reagent and separation of the PFB ether derivatives on a silica gel microcolumn is mandatory to remove interferences and fractionate the PFB ethers to avoid peak overlap on EC–GC.[172,180,181] As stated previously[8] the general approach of initially identifying a suspect residue using a P-, S-, or N-selective detector and subsequently confirming using the nonspecific ECD after hydrolysis, derivatization and cleanup may be viewed as a retrograde step. Especially, since one of the advantages of selective detectors is to reduce initial sample cleanup. This, of course, could be circumvented if the OP compound can be detected and confirmed, after hydrolysis–derivatization, using the same GC system. Therefore, the phosphorylation of seven aliphatic alcohols and 12 phenols with diethyl chlorophosphate in the presence of triethylamine at 60–70°C in benzene has distinct possibilities[182]:

$$\text{ArOH} + \text{ClOP(OEt)}_2 + \text{Et}_3\text{N} \xrightarrow{\text{60–70°C}} \text{ArOP(O)(OEt)}_2 + \text{Et}_3\text{N HCl}$$
$$\text{aryl diethyl phosphate}$$

Quantities as low as 10–25 ng of the alcohols/phenols were detected by P-mode FPD after separation on a 3% OV-101 column at temperatures in the 150–200°C range. The phenols studied included phenol, cresols, and xylenols, but the technique has not, so far, been applied to hydrolysis-generated OP phenols. Similarly, Jacob et al.[183] used dimethyl thiophosphinic chloride in the presence of excess triethylamine to convert primary aliphatic, aromatic, and heterocyclic amines to the corresponding N-dimethyl thiophosphinic amides:

$$\text{RNH}_2 + \underset{\overset{\|}{\text{S}}}{\text{ClP}}\text{—(OCH}_3)_2 + \text{Et}_3\text{N} \xrightarrow{-20 \text{ to } +20°C} \text{R—NH—}\underset{\overset{\|}{\text{S}}}{\text{P}}\text{—(OCH}_3)_2 + \text{Et}_3\text{N HCl}$$

Excess reagent was removed with NaHCO$_3$/MeOH and derivatives separated on SE-30 or OV-17 capillary columns. A detection limit of 0.5 pg N-dimethylthiophosphinylaniline was obtained with a rubidium-sulfate-tipped AFID. Finally it should be pointed out that the derivatives cited above are only a few of the many that can be used in quantitating phenols. For example, Dishburger et al.[184] determined the 3,5,6-trichloro-2-pyridinol moiety from chlorpyrifos as the trimethylsilyl (TMS) derivative, at the 0.05-ppm level in tissues of cattle, using a 5% DC-200 column at 135–40°C and EC detection.

Since the alkyl or aryl moiety produced on OP hydrolysis is more indicative of the parent OP than is the P moiety the above techniques constitute a more meaningful approach to confirmation of identity and especially if both the "before" and "after" chromatograms can be obtained the same column–detector system.

3.3. On-Column Transesterification

On-column ("in-block") transesterification is the technique whereby a volatile methyl derivative of an acidic compound is formed in the injector port of a GC. The methyl derivatives, and any other volatile products, are swept onto the GC column by the carrier gas and separated–detected in the normal manner. Pyrolytic methylation is another term used in this connection, the theory, technique, and applications of which have been recently reviewed.[185] Moye[186] transesterified a number of OPs by dissolving them in methanolic sodium hydroxide then injecting onto a precolumn of alkali-coated glass beads followed by a Porapak P column and AFID detector. A typical reaction would be

$$
\underset{\text{R}\overset{\text{O(S)}}{\overset{\|}{\text{OP}}}(\text{OC}_2\text{H}_5)_2 + \text{CH}_3\text{OH}}{} \xrightarrow{\text{NaOH}} \text{ROH} + \text{CH}_3\overset{\text{O(S)}}{\overset{\|}{\text{OP}}}(\text{OC}_2\text{H}_5)_2
$$

The procedure was evaluated for five OPs, namely, azodrin, ronnel, diazinon, parathion plus one other OP, representing four types of substituted phosphoric acid structures. Reproducible conversions were obtained and distinguishing GC peaks were also observed when the OPs were injected in ethanol, 1-propanol, and 1-butanol. Previous studies on the methylation of hydrolysis products of OP by Shafik et al.[165–167] had shown that a variety of base concentrations and other reactions are necessary for the solution hydrolysis of these compounds. If $0.2\ M$ sodium methoxide solution is used the on-column technique can be improved such that malathion, azinphos-ethyl, dichlorvos, fenitrothion dimethoate, ethion, ruelene, and leptophos can be included and the technique used as a general screening method for OPs in crops, etc.

It was originally demonstrated by Robb and Westbrook[187] that the methyl esters of carboxylic acids could be prepared, in 80–100% yields, by on-column thermal decomposition of a quaternary N-methyl ammonium salt of the acid at injector temperatures of 330–365°C. This procedure has been extended to the analysis of chlorphoxim residues in water and fish by on-column methylation with $0.01\ M$ trimethylanilinium hydroxide (TMAH) in methanol at an injector temperature of 280°C.[188] Limits of detection were 10 ppb for 5-g fish samples and 0.1 ppb for 300-ml samples of water. The moiety detected and quantitated by P-mode FPD was diethyl methyl thiophosphate (DEMTP):

TABLE 10. Efficiency of the In-Block Reaction of Various Pesticides in 0.01 M TMAH[a]

Compound	Derivative	Efficiency, %
Phoxim	DEMTP	61
Temephos	TMTP	64
Parathion	DEMTP	65
Chlorphoxim	DEMTP	66
Dichlorvos	TMP	68
Malathion	TMDTP	76
Methyl parathion	TMTP	76
Mevinphos	TMP	94
Chlorphyrifos-methyl	TMTP	95
Azinphosmethyl	TMDTP	100

[a] Reference 189.

The use of TMAH as a derivatizing reagent was investigated for nine other pesticides and it was found that the efficiency of the on-column reaction in 0.1 M TMAH varied from 61% for phoxim to 100% for azinphosmethyl (Table 10).[189] In addition, it has been demonstrated that trimethylphenyl-ammonium hydroxide (TMPAH) could be used for on-column transester-ification of not only the parent OP compounds but also their dialkyl phos-phorothioate hydrolysis products. Parathion is transesterified to DEMTP, while malathion and azinphosmethyl yield trimethyl dithiophosphate (TMDTP). The minimum detectable limit for malathion (Figure 16) was 400 pg using the P-mode FPD. It was suggested that TMPAH quantitation of dialkyl phosphorothioates and phosphorodithioates should provide a safe, convenient alternative to the use of diazomethane in a final methylation step in the analysis of these pesticide metabolites.

Although the on-column transesterification technique has certain time and ease of handling advantages over the stepwise hydrolysis–derivatization procedure of Shafik et al.,[165–168] to date it has only had limited application to field samples. Finally, on-column acetylation is another approach as used by Holland[190] for the quantitation of 0.1–10 ppm trichlorophon in soil and ryegrass. Here acetic anhydride was co-injected with trichlorophon residues and GC separation was carried on 3% SE-30 at 170°C with detection using a Rb_2SO_4-tipped AFID.

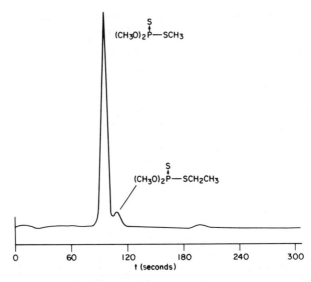

Fig. 16. GLC trace for the derivatization of malathion in 0.01 M TMPAH in ethanol.
Conditions: OV-225 column at 166°C; inlet temperature, 260°C.

3.4. Derivatization of the Intact Insecticide

Confirmatory tests usually involve specific groups, e.g., Cl, NO_2, P, S, etc. in the molecule; therefore a wide range of reactions have been used in the derivatization of intact OP compounds. Specific tests have included the 30-min low-intensity uv irradiation of crufomate in bovine blood.[191] The C–Cl bond is cleaved and the resulting radical extracts a proton from the solvent (hexane) to give the deschloro derivative. The reaction is facile and gives a 89% yield at the ng level. Parathion can readily be reduced to the amino derivative using $CrCl_2$.[192] While other NO_2-containing OPs such as fenitrothion, EPN, and methylparathion can also be confirmed, $CrCl_2$ did not reduce the CN moiety of surecide, and compounds not having a NO_2 group are unaffected, e.g., malathion. As little as 50 ng/liter in water or 50 μg/kg in fish or sediment were confirmed. The only drawback with this reaction is that a FPD or AFID should be used since there is a difference is sensitivity between the NO_2 and NH_2 group of about 400-fold on EC. Also a Zn/HCl reagent was found to have wider application since it also reduced surecide, paroxon, and fenitroxon. Oxidation of P=S to P=O is readily achieved by a variety of oxidizing agents, m-chloroperbenzoic acid has been extensively used,[193] and more recently, sodium hypochlorite.[194,195] The choice of oxidation reagent is very important. For example, oxidation of dithiophosphates and amidothiophosphates with m-chloroperbenzoic acid results in multiple products,[196] which detracts from the usefulness of oxidation as a general confirmatory test for P=S compounds. Similarly, oxidation proceeds well with thiophosphates having no other oxidizable group, but if such a group is present, then further reaction occurs such as the formation of the oxon sulfone from phorate[197] when oxidants such as $KMnO_4$ are used. This is in contrast to the hydrogen peroxide or sodium hypochlorite reactions, which both yield phorate oxon sulfoxide.[195] The sodium hypochlorite reaction has been used in the confirmation of ten OP compounds including chlorpyrifos, ethion, phorate, DMPA (Zytron), leptophos, diazinon, malathion, etc. in celery, potatoes, lettuce, tomatoes, and apples at levels of 0.25–0.5 ppm.[195]

Another reaction which has a broader application is the alkylation of OP compounds containing an amino or alkyl amino groups.[198,199] Hydrolysis of a phosphorothioate produces an intermediate ion which gives the choice of two sites for alkylation. With an alkylating agent such as dimethyl sulfate the reaction proceeds exclusively on the S atom,

$$
\begin{array}{ccc}
\overset{\displaystyle S}{\underset{\displaystyle OR}{RO-P-OR}} & \xrightarrow{\ \ SH^-\ \ } & \left[\overset{\displaystyle S}{\underset{\displaystyle OR}{RO-P\!\!=\!\!O}}\right]^{-} \xrightarrow{\ \ DMS\ \ } \overset{\displaystyle SMe}{\underset{\displaystyle OR}{RO-P\!\!=\!\!O}}
\end{array}
$$

This reaction is used to prepare the S-methyl isomers of insecticides in 91% yield. With a phosphorothioamidate, alkylating agents also react with the S atom to again give the S isomer, even though there is an NH group present:

$$
\overset{\displaystyle S}{\underset{\displaystyle NHR}{RO-P-OR}} \xrightarrow{\ \ R'I\ \ } \overset{\displaystyle SR'}{\underset{\displaystyle NHR}{RO-P\!\!=\!\!O}} + RI
$$

Miller and O'Leary[198] showed, however, that in the presence of strong base, the anion formed reacts with various alkylating reagents, e.g., methyl iodide at room temperature to give the N-methyl derivative exclusively in 80–90% yield:

$$
\overset{\displaystyle S}{\underset{\displaystyle OC_2H_5}{C_2H_5O-P-NHR}} \xrightarrow{\ \ base\ \ } \left[\overset{\displaystyle S}{\underset{\displaystyle OC_2H_5}{C_2H_5O-P-\bar{N}R}}\right] \xrightarrow{\ \ CH_3I\ \ } \overset{\displaystyle S}{\underset{\displaystyle OC_2H_5}{C_2H_5O-P-N}}\!\overset{\displaystyle CH_3}{\underset{\displaystyle R}{\diagdown}}
$$

In comparing two alkylating agents[199] it was found that reaction with a NaH–CH_3I–DMSO mixture at 50°C for 10 min was preferred for all practical purposes. Confirmation of 16 insecticides and herbicides, including

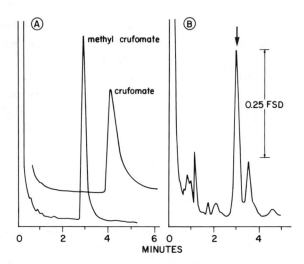

Fig. 17. Gas chromatograms of crufomate, before and after alkylation (2-μg level), and alkylated crude chloroform extract of bovine blood (0.42 ppm of crufomate): (A) 15 ng of crufomate, before and after alkylation. (B) alkylated extract of bovine blood. Column: 4% SE-30/6% QF-1 at 228°C.

Monitor, crufomate, BAY 93820, and dimethoate, was carried out at the residue level using a RbCl-tipped AFID. This reaction was shown to have general application to triazines and carbamates possessing reactive NH moieties. As shown in Figure 17 this alkylation procedure resulted in greatly improved peak shape for crufomate. Alkylation has also been used to characterize the hydroxydiazinon metabolites G 27550 and G 31144 at the 1-ppm level in canine urine.[200]

Another very useful confirmatory technique for NH-containing insecticides is trifluoroacetylation using trifluoroacetic anhydride (TFAA) as reagent. Azodrin was quantitatively converted to TFA-azodrin in 10–15 min at room temperature:

$$(CH_3O)_2{-}\overset{\overset{\displaystyle O}{\|}}{P}{-}O{-}\overset{\overset{\displaystyle CH_3}{|}}{C}{=}CH{-}CO{-}\overset{\overset{\displaystyle H}{|}}{N}{-}CH_3$$

$$\xrightarrow{\text{TFAA}} (CH_3O)_2{-}\overset{\overset{\displaystyle O}{\|}}{P}{-}O{-}\overset{\overset{\displaystyle CH_3}{|}}{C}{=}CH{-}CO\overset{\overset{\displaystyle COCF_3}{|}}{N}{-}CH_3$$

The detection limit, using a 3% OV-210 column at 170°C and a FPD, was about 2 ppb in fresh, frozen, or pureed strawberries.[201] It has also been found[202] that sulfoxides are reduced on trifluoracetylation and a trifluoroacetoxy group formed. Dasanit and dasanit oxon both reacted with TFAA at room temperature for 15 min to give trifluoroacetoxymethyl sulfide derivatives as single products, respectively, in $>90\%$ yield. Dasanit was confirmed at the 1.1 ppm in extracts of sandy loam soil. The rearrangement of sulfoxides to give α-substituted derivatives of the corresponding sulfides is well known as the Pummer reaction[203]:

$$CH_3{-}\overset{\overset{\textstyle O}{\|}}{S}{-}R \xrightarrow[-H^+]{TFAA} \overset{\overset{\textstyle OCO{-}CF_3}{|}}{\bar{C}H_2{-}S^+{-}R} \xrightarrow{Rearrangement} \overset{\overset{\textstyle O{-}CO{-}CF_3}{|}}{CH_2{-}SR}$$

This type of substitution is significant in that it determines the nature of the products—the S-acyloxy group normally migrates to the least-substituted carbon atom. The trifluoroacetylation of nemacur sulfoxide is complicated by the presence of an NH group in the molecule. Therefore, both mono-TFA and di-TFA derivatives can be formed in high yields under different reaction conditions. Oxydemeton-methyl, phorate, and Counter sulfoxides readily reacted with TFAA to give derivatives that underwent further rearrangement:

$$(C_2H_5O)_2{-}\overset{\overset{\textstyle S}{\|}}{P}{-}S{-}CH_2{-}\overset{\overset{\textstyle O}{\|}}{S}{-}C_2H_5 \xrightarrow[RT, 15 \text{ min}]{TFAA} (C_2H_5O)_2{-}\overset{\overset{\textstyle O}{\|}}{P}{-}S{-}CH_2S{-}C_2H_5$$

phorate sulfoxide   phorate oxon

The quantitative $KMnO_4$ oxidation of oxydemeton methyl (also called Metasystox-R) to its corresponding sulfone has also been used for its determination in a variety of plant and animal tissues down to a lower limit of 0.01 ppm[204] using an AFID.

As discussed in Section 3.1 methods have been developed for the residue analysis of the urinary OP metabolites, i.e., the dialkyl phosphates and thiophosphates and phenyl phosphonates, involving esterification with diazoalkanes. However, another approach is via the formation of the benzyl[205] or p-nitrobenzyl esters[206] using triazene reagents. Prior to benzylation the phosphoric acids were fully protonated by passing through a strong cation exchange resin in the H^+ form. The dry acids were then refluxed for 20 min or more with 3-benzyl-1-p-tolyltriazene in sealed vials, partitioned into cyclohexane, and determined by FPD after separation on a 5% OV-210 column, temperature programmed from $170°{-}225°C$ at $16°C/min$:

$$(RO)_2\overset{\overset{\textstyle S(O)}{\|}}{P}{-}OH + CH_3{-}C_6H_4{-}NH{-}N{=}N{-}CH_2{-}C_6H_5$$

$$\longrightarrow (RO)_2{-}\overset{\overset{\textstyle S(O)}{\|}}{P}{-}O{-}CH_2{-}C_6H_5 + N_2 + CH_3{-}C_6H_4{-}NH_2$$

Although inorganic phosphate did not interfere with urine analysis, it could also be removed by $Ca(OH)_2$ precipitation. Gentle agitation at 40°C for 2 hr in sealed vials was found suitable for the derivatization of dimethyl and diethyl derivatives of phosphoric, phosphorothioc, and phosphorodithioc acids in human urine when p-nitrobenzyltolyltriazene was used as reagent.[205] These six urinary OP metabolites could be determined at the 0.01-ppm level using a 5% OV-101 column and dual FPD. As with both these reagents, and the diazoalkane procedure, a mixture of the O- and S-isomeric derivatives are obtained. The amounts of the O and S isomer produced on methylation of dimethylthiophosphoric acid was found to be in a 1:10 ratio.[207] Shafik and co-workers[165–168] quantitated using the O isomer. Daughton et al.[204] indicated a mixed product could be avoided by oxidation of the sample with sodium hypochlorite, subsequent benzylation yielding only the corresponding phosphate or phosphonate. Takade et al.[205] separated the P=O and P=S derivatives on a SiO_2 column and determined each separately. This step was inserted into the procedure after incomplete resolution of some derivatives was observed on the OV-101 column. It would appear that a quantitation procedure based on the sum of the O and S esters could be devised if the per cent conversion to each was known and constant for a given set of conditions and reagent.

Where the formation of a mixed reaction product is not a problem, diazomethane has been the reagent of choice. For example, ethephon (2-chloroethyl phosphoric acid) has been quantitatively analyzed at the 0.05-ppm level in apples.[208] Reagents such as BF_3/MeOH, HCl/MeOH, or methyl fluorosulfonate proved ineffective. Ethephon determination in tomatoes, cherries, and apples has also been achieved via degradation to ethylene.[209]

4. CARBAMATES (INSECTICIDES AND HERBICIDES)

The carbamate class of pesticides is conveniently subdivided into the N-methylcarbamates, e.g., carbaryl, which are used as insecticides, and the N-aryl carbamates, e.g., chlorpropham (CIPC), used as herbicides. In the past the thermal instability of N-methylcarbamates has hampered the application of GC techniques to the direct determination of their residues. Therefore in the 1960s and early 1970s many indirect methods were devised based on derivatization to increase thermal stability and at the same time the addition of a group which would enhance the detectability of the carbamate to ECD, FPD, or a microcoulometric detection. For example, the use of a variety of bromo-, chloro-, fluoro-, and nitro-containing reagents was extensively studied in combination with EC detection. These procedures have been reviewed by Williams,[210] Cochrane,[8,9,211] Dorough and Thorstenson,[212] and Kuhr and Dorough.[213] Similarly, the various attempts to modify GC columns to minimize on-column decomposition, primarily the utility of Carbowax 20M surface-modified supports, has been summarized by Hall and Harris.[214]

Like the OPs there are four basic approaches to the analysis of carbamates which involve derivatization. These are (1) hydrolysis and derivatization of the N moiety, (2) hydrolysis and derivatization of the alkyl or aryl moiety, (3) on-column transesterification, and (4) derivatization of the intact carbamate.

Figure 18 outlines the above four approaches as applied to a typical *N*-methyl carbamate insecticide.

4.1. Hydrolysis and Derivatization of the N Moiety

The formation of dinitrophenyl (DNP) derivatives of $R-NH_2$ and $RR'-NH$ compounds that were amenable to GC was reported by Day *et al.*[215] They investigated eleven C_1-C_4 primary and secondary aliphatic amines after conversion, using 1-fluoro-2,4-dinitrobenzene as reagent. Levels in the range 3.6–4.5 ng could easily be quantitated by ECD. Dinitrophenylation was applied to the methyl- and dimethylcarbamate insecticides by Holden *et al.*[216] (Figure 18) and to the substituted carbamate and urea herbicides by Cohen and Wheals[217] (Table 11). Holden[216] found that background interferences varied with different crops but derivatization was essentially quantitative and a lower level of 0.05 ppm of ten carbamates could be determined. Since this nonspecific procedure tended to be time consuming and was found to give low yields, Sumida *et al.*[218,219] modified the method to avoid loss of volatile methylamine by combining the carbamate hydrolysis and 2,4-DNP derivatization reactions into a single step. The method of Sumida *et al.*[218] was further modified by Mendoza and Shields[220] for the analysis of 0.01–1-ppm methomyl in rapeseed oil. Hydrolysis of methomyl was achieved in 10 min with 0.5 ml NaOH containing 5% borax at a temperature of 82°C. Subsequently, the DNFB reagent was added and the mixture allowed to stand 10 min at 82°C when glycine was added and reaction continued for another 10 min. Quantitative recoveries of 2,4-dinitrophenylmethylamine were obtained for methomyl and carbaryl standards.

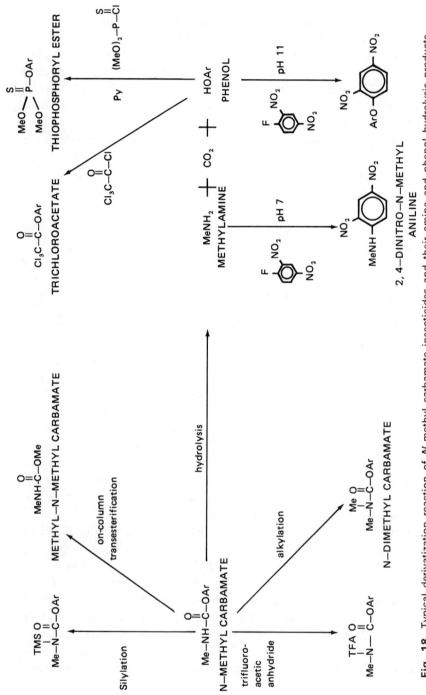

Fig. 18. Typical derivatization reaction of *N*-methyl carbamate insecticides and their amine and phenol hydrolysis products.

TABLE 11. Amines and Anilines Arising from Various Pesticides

Amine or aniline	Pesticide class	Examples
Methylamine	Carbamates	Carbaryl, carbofuran, mobam, propoxur, methomyl, oxamyl, methiocarb, matacil, Zectran, Banol
Dimethylamine	Carbamates	Dimetilan, pyrolan
	Ureas	Fenuron, diuron, monuron, fluometuron, chloroxuron
	Amides	Bidrin
Diethylamine	Dithiocarbamates	CDEC
Aniline	Carbamates	Propham (IPC), carbetamide
	Urea	Fenuron, siduron
3-Chloroaniline	Carbamates	Chlorpropham (CIPC), barban
4-Chloroaniline	Ureas	Monuron, mondenuron, Urox
4-Bromoaniline	Urea	Metabromuron
3,4-Dichloroaniline	Carbamate	Swep
	Ureas	Diuron, linuron, neburon
	Anilide	Propanil
3-Chloro-4-bromoaniline	Urea	Chlorobromuron
3-Chloro-4-methylaniline	Urea	Chlortoluron

Cohen and Wheals[217] derivatized ten herbicidal carbamates and ureas *in situ* on silica gel TLC plates and after removal, quantitated via GC–ECD. This on-plate hydrolysis to produce anilines and chloroanilines and their DNP derivatization was used successfully, down to 0.001 ppm, with surface water samples. The DNP derivatization procedure gave limited information as carbamates and ureas yielding the same aniline (or amine) (Table 11) could not be distinguished from one another or from their respective aniline (amine) metabolites. Nevertheless, levels down to 0.02–0.05 ppm were determined in 50-g plant and soil samples using this DNP derivatization technique. Similarly, Crosby and Bowers[221] utilized 4-chloro-α, α, α-trifluoro-3, 4-dinitrotoluene and α,α,α-4-tetrafluoro-3-nitrotoluene in alkaline solution to prepare substituted aniline derivatives. These di- and mono-nitrotoluene derivatives provided much greater EC detector response, better resolution, and shorter retention times than the series of DNP derivatives. For example, amounts down to 50 pg could be quantitatively detected by ECD, and the procedure was applied to molinate in water, EPTC in sugar beets, and also propham and CDEC. Other bromo- and/or nitrochloro-substituted reagents have been used with varying degrees of success[222,223]:

$$CH_3NH_2 + Cl—CO—C_6H_4—Br \xrightarrow[OH^-]{catalyst} CH_3NH—CO—C_6H_4—Br$$

a benzamide

With the 2,4-DNP derivative being preferred for the analysis of carbofuran and propoxur residues in rice, straw, and soil down to a level of 0.005 ppm.

Since both primary and secondary amines (Table 11) can result from the hydrolysis of carbamate pesticides it is not surprising that the applicability of the TFA, PFP (pentafluoropropionyl), HFB (heptafluorobutyryl), PFB (pentafluorobenzyl), and TMS (trimethyl silyl) derivatives have been investigated on combination with EC detection. Clark *et al.*[224] compared the EC properties and GC column separations of the acetyl, *N*-trifluoro- and monochloroacetyl, PFP and HFB derivatives of various amines and phenolic amines. A study of the PFB, PFP, DNP, HFB, and other halo-derivatives to the analysis of aniline metabolites in urine showed that the only derivatization reagents to give clean chromatograms at the low level were PFPA (pentafluoropropionic anhydride) and HFBA, with PFPA being the better of the two.[225] The recoveries of anilines from spiked water and urine samples at the 1.0- and 0.1-ppm levels were between 85% and 90%.

A large amount of literature exists covering the various derivatization techniques and in the analysis of primary, secondary, and tertiary amines,[226,227,228] primarily in connection with pharmaceuticals. Specificity of reaction towards a primary amine group can be achieved using 2,4-dinitrobenzenesulfonic[229] or 2,6-dinitro-4-trifluoromethylbenzene sulfonic acids[230] as reagents, with the latter compound completely derivatizing primary amines in 10 min at room temperature:

$$RNH_2 + CF_3\!-\!\!\!\underset{NO_2}{\overset{NO_2}{\bigodot}}\!\!\!-\!SO_3H \xrightarrow[\text{sodium borate}]{RT} CF_3\!-\!\!\!\underset{NO_2}{\overset{NO_2}{\bigodot}}\!\!\!-\!NHR$$

Use of the S-mode FPD can be utilized to advantage if benzene sulfonyl-chloride (BSC) is used for the identification of primary and secondary amines.[231] Since the BSC derivatives of primary amines are soluble in alkali and not extracted with organic solvent, while those of secondary amines can be extracted, this sulfonylation procedure was quantitatively applied to dimethylamine, diethylamine, n-diproplyamine, and n-dibutyl-amine:

$$R_2NH + C_6H_5SO_2Cl \xrightarrow[\text{pH 10.8}]{\text{RT}} C_6H_5SO_2NR_2$$

Finally, pentafluorobenzyl chloroformate has been treated with tertiary amines to form the corresponding carbamates.[232] The pentafluorobenzyl carbamates have high stability, are formed easily at 105°C, and show good GC separation properties. Picogram levels of the amine can be quantitated by ECD of the acylated substrate. Although applicable to tertiary methyl-amines,

$$R_2N-CH_3 + C_6F_5CH_2OCOCl \xrightarrow[\text{Na}_2\text{CO}_3]{105°C, 1/2-1 \text{ hr}} C_6F_5CH_2OCONR_2$$

this reagent failed to react with dimethylaminoalkanes, which also contain a pyridine nucleus.[233] Since a carbamate is formed use of a NP detector would have distinct advantages.

4.2. Hydrolysis and Derivatization of the Alkyl or Aryl Moiety

The ease of hydrolysis of the phenol-generating carbamates has offered a simple method for extracting the pesticide from the crop or environmental extract. Normally, the organic extract is treated with aqueous or alcoholic alkali and the phenol isolated by acidification and organic solvent extraction. As shown in Table 12, there have been at least six derivatization techniques used in the analysis of the resulting phenols. Generally, the derivative is sensitive to detection by either ECD or the FPD. Many investigations have used carbaryl and its hydrolysis product, 1-naphthol, as model compounds. For example, treatment of 1-naphthol with Br_2 at 100°C gave 2,4-dibromo-1-naphthol, which exhibited a 30-sec retention time on a DC high-vac silicone grease column at 180°C.[238] Similarly, the chloromethyl-dimethylsilyl derivative of 1-naphthol appeared at 2.70 min on 3% SE-30 at 190°C.[239] Carbamates such as methomyl and oxamyl give their respective oximes on hydrolysis, and the TMS ethers of their oximes have been used in the analysis of crops[240]:

$$\underset{\text{methomyl}}{\overset{\overset{\displaystyle SCH_3}{\mid}}{CH_3-C=NCONHCH_3}} \xrightarrow[\text{pH 10-11}]{\text{NaOH}} \underset{\text{methomyl oxime}}{\overset{\overset{\displaystyle SCH_3}{\mid}}{CH_3-C=NOH}} \xrightarrow{\text{BSA}} \underset{\text{oxime TMS ether}}{\overset{\overset{\displaystyle SCH_3}{\mid}}{CH_3C=NOSi(CH_3)_3}}$$

TABLE 12. Derivatization Procedures Used in the Analysis of Phenol- or Oxime-generating Carbamates

Parent compound	Derivatization technique	Detection limit using ECD	References
Carbaryl	Bromination (Br_2)	0.05 ppm	234–237
Carbaryl, mesurol, Zectran	Silylation	1–60 ng	238, 239
Methomyl, oxamyl	Trimethylsilylation	0.05 ppm[a]	240
Banol, carbaryl, propoxur and eight others	Monochloroacetylation	0.04 ppm	241, 246, 247
Propoxur and metabolites	Trichloroacetylation	0.02 ppm	242, 243
Carbaryl		0.005 ppm	237, 244
Carbofuran and metabolites		0.01 ppm	245
Carbofuran and phenol metabolite plus other carbamates	Thiophosphorylation	0.04 ppm[a]	248
17 Carbamates	DNP/DNT formation	0.05 ppm	249–255
	Sulfonylation	0.4 ng	256
Carbaryl, propoxur, Mobam, carbofuran and metabolites	Pentafluorobenzylation	0.01 ppm (100 ppt for water)	257–260

[a] Lower limits using a FPD.

If the initial extract contains methomyl and its oxime, separate derivatizations are done before and after hydrolysis and the amount of original methomyl calculated by difference. Since no electron-capturing groups have been added the TMS–oximes were separated on a 5% OV-1 column at 75°C (methomyl) or 130°C (oxamyl) using the S-mode FPD. The formation of trichloroacetates has been used in the carbamate field (Table 12) as well as phenols derived from the phenoxy acid herbicides (see Sections 5.3 and 6). Although the 2,4-dibromo-1-naphthyl acetate derivative of carbaryl could be determined under the same conditions as the trichloroacetate derivative, it was found to be four times less sensitive to ECD.[244] Bowman and Beroza[248] determined carbofuran and its phenol degradation product in corn silage and milk via alkaline hydrolysis and derivatization with dimethyl chlorothiophosphate to produce a compound detectable by the

P-mode FPD:

$$R{-}OH + Cl{-}\overset{\overset{\displaystyle S}{\|}}{P}{-}(OCH_3)_2 \xrightarrow[\text{reflux}]{\text{pyridine}} R{-}O{-}\overset{\overset{\displaystyle S}{\|}}{P}{-}(OCH_3)_2{-} + HCl$$

Since these derivatives contain both P and S they were also found to respond to the ECD.

Dinitrophenylation of phenols with 1-fluoro-2,4-dinitrobenzene in the presence of base (pH 11) is another well-studied procedure (Table 12 and Section 2.3, 5.3, and 6). Various carbamate residues in crops were initially determined at the 0.05-ppm level,[250] and subsequently the procedure was collaboratively studied by eight laboratories for propoxur, carbofuran, carbanolate, and carbaryl in apples, corn, green beans, and leafy vegetables at levels of 0.1, 0.2, and 0.5 ppm.[251] A standard 6-ft 10% DC-200 column at 232°C was used in this study. This derivatization procedure has been used for 0.1-ppm carbofuran in soil.[253] It has been extended to the phenolic metabolites of carbofuran in plant and animal tissues at lower levels ranging from 0.02 ppm (milk) to 0.1 ppm (potatoes) using either the N-selective microcoulometric or electrolytic conductivity detectors.[254] Satisfactory derivatization of the 3-hydroxy-7-phenol metabolite to a monosubstituted DNP derivative could only be achieved by initially forming the 3-ethoxy-7-phenol by selective reaction with ethanol. The formation of the 2,4-DNP ether derivatives of carbamate phenols was found superior to N-trifluoro-acetylation in the recovery of 12 carbamates added to 7 substrates.[255] By utilization of column and thin-layer chromatographic cleanup steps the simultaneous residue analysis of ten N-methylcarbamates at levels as low as 0.003 ppm in the presence of carbamate phenols, OP, and OC pesticides was proved feasible. A 5% OV-17 column was chosen for multiresidue analysis and the EC responses of the DNP derivatives were 2–4 times greater than those of the corresponding TFA derivatives. Attempts to analyze aldicarb and methomyl by this procedure proved unsuccessful.

Kawahara[257] demonstrated that ethers and thioethers derived from phenols and mercaptans, respectively, with α-bromo-2,3,4,5,6-pentafluoro-toluene were stable in water, amenable to GC analysis, and exhibited excellent EC response in the picogram range. The major drawback with this procedure (and with other procedures employing halogenated reagents) is the removal of excess pentafluorobenzyl bromide reagent. Johnson[258] simultaneously removed excess reagent and separated the PFB ethers on a silica gel microcolumn. This procedure was applied to the analysis of carbofuran, 3-ketocarbofuran, carbaryl, propoxur, and Mobam in water samples.[176,259] The determination of mexacarbate (Zectran) and aminocarb

(Matacil) was not possible since their phenolic hydrolysis products were not extracted from the acidified hydrolysis solution. Archer et al.[260] used a combination of PFB and TFA derivatives to determine the fate of carbofuran and its metabolites on strawberries. After initial separation on a SiO_2 column, carbofuran, 3-hydroxycarbofuran, 3-ketocarbofuran, and the 3,7-diol were quantitated as their TFA derivatives, while the 7-phenol and 3-keto-7-phenol were converted to the PFB ethers since derivatization with TFAA did not give reproducible results. The method was sensitive to 0.05 ppm for carbofuran and 0.1 ppm for the other compounds.

As mentioned previously (Section 3.2) Seiber et al.[176] compared and evaluated the PFB, DNP, and DNT derivatives of a series of phenol-generating carbamates, OPs, and herbicides. The three types were found to be similar in their ease of preparation, stability, and EC response characteristics. Relative retention times varied in the order PFB < DNT < DNP. Either potassium carbonate or KOH could be used as base, with K_2CO_3 giving higher yields in the case of 1-naphthol. On the other hand KOH was preferred with the OP compounds, since K_2CO_3 gave inefficient hydrolysis. The relative EC response did not vary greatly among the three series of derivatives, although the PFB responses were consistently higher than the others, which may be due in part to the PFB derivatives having the shorter retention times on the 5% SE-30 column used.

4.3. On-Column Methylation

Moye[261] described the on-column transesterification of eight N-methylcarbamate insecticides to methyl-N-methylcarbamate by coinjection of the carbamate with methanol containing catalytic amounts of sodium hydroxide (Figure 18) at an injector block temperature of 215°C. Reproducible high yields were obtained with efficient separation being performed on Poropak P and detection by a Rb_2SO_4 AFID at the nanogram level. This procedure has been applied to crop and soil extracts,[261] to carbofuran residues in field-treated lettuce,[262] tobacco,[212] and soil,[263] and to methomyl in tobacco.[264] The applicability of this procedure was extended to the N-arylcarbamates by Wien and Tanaka,[265] who used trimethylanilinum hydroxide (Methelute®) as catalytic base in a 2:1 ratio with the carbamate. The product derived from N-arylcarbamates (e.g., Propham) was the expected N-methyl-N-arylcarbamate, but the N-methylcarbamates (e.g., carbaryl, carbofuran) yielded a methyl substitution product (arylmethyl ether) rather than the anticipated N,N-dimethyl carbamate:

$$Ar{-}O{-}CO{-}NH{-}CH_3 \xrightarrow[220°C]{TMAH} Ar{-}O{-}CH_3$$

A similar reaction was observed by Bromilow and Lord[266] when oxime carbamates yielded methoximes and substituted phenyl-N-methyl carbamates yielded anisoles (arylmethyl ethers) when coinjected with trimethylphenylammonium hydroxide (TMPAH) at injection port temperatures of 180–300°C. The reaction of carbamates with TMPAH appears to proceed in two steps: initial loss of the carbamate group to give the hydroxyl moiety (oxime or phenol), which is then methylated:

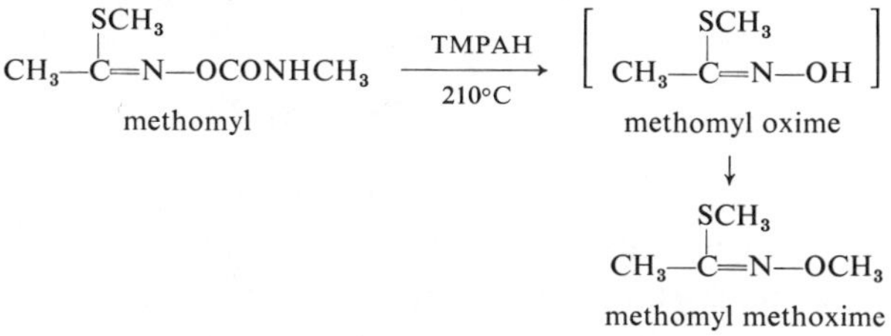

$$
\underset{\text{methomyl}}{CH_3\overset{\overset{\displaystyle SCH_3}{|}}{-}C=N-OCONHCH_3} \quad \xrightarrow[210°C]{TMPAH} \quad \left[\underset{\text{methomyl oxime}}{CH_3\overset{\overset{\displaystyle SCH_3}{|}}{-}C=N-OH}\right]
$$

$$
\downarrow
$$

$$
\underset{\text{methomyl methoxime}}{CH_3\overset{\overset{\displaystyle SCH_3}{|}}{-}C=N-OCH_3}
$$

Aldoxime carbamates, such as aldicarb and its sulfoxide, and sulfone gave varying yields owing to competing nitrile formation or the instability of the resulting methoxime. This TMPAH reaction was applied to the residue analysis of oxamyl in a wide range of soil and crop samples at the 0.01-ppm level using the S-mode FPD.[267]

Although this on-column derivatization technique has had only limited application it has distinct advantages as a rapid screening procedure for a wider range of carbamates and OP insecticides.

4.4. Derivatization of the Intact Carbamate

The derivatization of intact carbamates can be separated into two distinct areas: reactions involving the NH group and those associated with the S moiety in those carbamates possessing a thioether group, e.g., methiocarb. Unless special precautions are taken,[214] most N-methyl carbamates are either retained on the GC column or are decomposed to methyl isocyanate and the corresponding phenol:

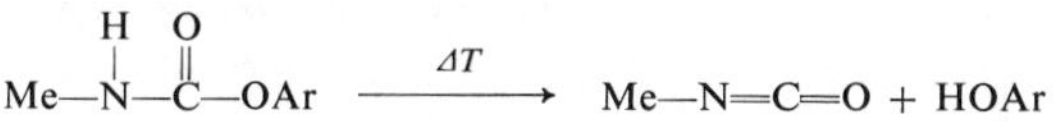

$$
\underset{}{Me-\overset{\overset{\displaystyle H}{|}}{N}-\overset{\overset{\displaystyle O}{\|}}{C}-OAr} \quad \xrightarrow{\varDelta T} \quad Me-N=C=O + HOAr
$$

Similarly, the thermal stability of the N-trichloroacetyl derivatives of halogenated alkyl and aryl carbamates was investigated by Fishbein and Zielinski.[268] Only the 2,2,2-trifluoroethyl derivative appeared as a single peak on 4% QF-1 at 110°C. The tribromo- and trichloroethyl derivatives underwent degradation with release of the amide and the corresponding trihaloethanol:

$$Cl_3C-\overset{\overset{O}{\|}}{C}-NH-\overset{\overset{O}{\|}}{C}-O-CH_2-CBr_3 \quad \xrightarrow{\Delta} \quad Cl_3C-\overset{\overset{O}{\|}}{C}-NH_2 + CO_2 + HOCH_2-CBr_3$$

Of the 13 chlorophenyl derivatives studied, all cleaved to trichloro-acetamide and the respective chlorophenol. This thermal instability is related to the presence of a reactive NH (see also Section 5.2) which when substituted may lead to derivatives amenable to GC analysis.

4.4.1. Reactions Involving the NH Group

Early attempts at obtaining thermally stable derivatives included N-silylation[238,269] and N-acetylation.[270,271] Although attempted silylation of carbaryl, methiocarb, Zectran, and CIPC with bromo- and chloro-methyldimethylchlorosilane proved unsuccessful,[238] success has been reported using a hexamethyldisilazane–trichlorosilane mixture in pyridine[268] (Figure 18). The TMS derivatives of Zectran, 1-naphthol, propham, chloropropham, and various herbicidal ureas were separated on 4% QF-1 or 4% SE-30 at 130°C or 3% Carbowax 20M at 170°C using a FID. On all the columns Zectran produced two peaks and it was suggested the major peak was Zectran or TMS–Zectran and the secondary peak as 4-dimethyl-amino-3,5-xylenol. Methylcarbamates also react with acetic anhydride to give N-acetyl derivatives[270,271] which are GC stable. The thermal stability of N-acetyl carbaryl over the N-trichloroacetyl, N-dichloroacetyl, and N-monochloroacetyl derivatives, which extensively decomposed on various GC columns, was reported by Magallona and Gunther.[272] For these three chloroacetyl derivatives, thermal stability appears to be related to the degree of chlorination. The response of N-acetyl carbaryl was adequate for residue when the electrolytic conductivity or EC detectors were used.

Following this early work, a number of workers have carried out detailed studies into the direct derivatization of carbamates via N-alkylation or N-perhaloacetylation and the EC comparison of the various derivatives formed. Lau and Marxmiller[273] quantitated Landrin at concentrations as low as 0.02 ppm in corn after N-acetylating overnight with trifluoroacetic anhydride in ethyl acetate. The general applicability of trifluoroacetylation to eight methylcarbamates and the metabolites of carbofuran was subsequently demonstrated.[274] The derivatives were obtained in high yield using excess TFAA in benzene at 100°C for 2 hr. It was concluded that two factors, solvent polarity and reaction temperature, governed the rate of reaction. The EC chromatogram of seven TFA derivatives is shown in

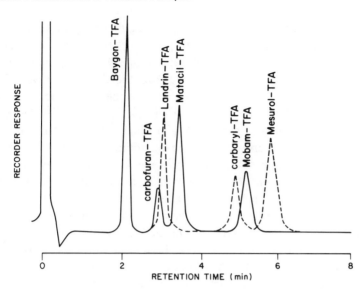

Fig. 19. Composite gas chromatogram of seven methyl carbamate TFA derivatives; 2 ng of each injected using a 7-ft 6% SE-30 column at 230°C with electron capture detection.

Figure 19. A 7-ft 6% SE-30 column at 230°C was used and this high temperature indicates the thermal stability of these derivatives. The pentafluoropropionyl (PFP) and heptafluorobutyryl (HFB) derivatives were investigated and found almost identical to the TFA derivatives in thermal stability, GC retention time, and EC response. For example, the relative EC response increased only twofold for carbaryl and fivefold for carbofuran in going from the TFA to the HFB derivatives. Subsequently, Bose[275] described the influence of trifluoroacetic acid on the extent of reaction of carbofuran with trifluoroacetic anhydride and the use of pyridine to accelerate the reaction and increase product yield. Khalifa and Mumma[276] also evaluated the TFA, HFB, and chloroacetyl derivatives of carbaryl and various aglycone metabolites. Carbaryl readily formed TFA and HFB derivatives in 15 min at room temperature using either the respective anhydride or imidazole reagents in benzene with pyridine as catalyst. Interestingly chloroacetic anhydride did not acylate the NH group. Preference was given to the TFA derivatives owing to their ease of purification, short GC retention times on 3% SE-30 at 190°C, and high sensitivity to ECD. Conversely, the TFA and HFB derivatives of the corresponding carbamate phenols gave poor separations at all temperatures and for phenols the chloromethyl dimethylsilyl ether derivatives were preferred.

TFA derivative formation has been applied to the quantitation of carbofuran and metabolites in strawberries at the 0.1-ppm level,[260] and in animal tissue in the range 0.07–0.5 ppm.[277] These latter workers also showed that the TFA derivstives of carbofuran, 3-hydroxycarbofuran, and 3-ketocarbofuran could be distinguished from the TFA derivatives of aminocarb, propoxur, carbaryl, and methiocarb on the basis of retention time difference on 3% DC-200 at 165°C. Ueji and Kanazawa[278] derivatized eight N-methylcarbamates with TFAA in ethyl acetate at 50°C for 2 hr. They then applied this procedure to the quantitation of MTMC (3-methylphenyl-N-methylcarbamate) residues in rice grain and rice straw down to a lower limit of 0.005 ppm. Also, eight carbamates were determined in water samples, at the 1–10-ppb level, after conversion to their TFA derivatives (2 hr at 100°C in toluene). Separation was accomplished on a 6% SE-30 column at 180°C. Similarly, residues of methiocarb and its metabolites at the 0.02-ppm level in blueberries have been determined using TFA derivatization but S-mode FPD quantitation.[280,281] Seven carbamates were derivatized, using PFPA, as part of a multiresidue procedure for pesticide residues in air.[282] Reaction was carried out at room temperature in isooctane containing pyridine as catalyst. Because of background interferences on the 5% SE-30 column the analyses of aminocarb, propoxur, and Zectran were unsuccessful. Johansson[283] determined six carbamates in apples, as their TFA derivatives, as part of a multiresidue method for the simultaneous extraction of OC, OP, dinitrophenyl, and carbamate pesticides. Trifluoroacetylation was carried out in toluene, containing pyridine, at 63–65°C for 45 min and the TFA derivatives were found stable for 3 hr. Detection limits using a 10% DC-200/15% QF-1 column at 130–170°C and EC quantitation were 3–300 ppb for barban, carbaryl, methiocarb, promecarb, propham, and propoxur. Recoveries were in the range 73–100%.

Lawrence and co-workers[284–286] investigated the use of the TFA, PFP, HFB, and PFO (pentadecafluoro-octanyl) derivatives of carbofuran and linuron and their sensitivities were found to be 10–50 times greater by EC detection than by electrolytic conductivity detection (halogen mode). For EC detection the HFB or PFO derivatives were preferred, while the PFO derivative was superior via electrolytic conductivity detection.[284] Derivatization was carried in trimethylamine–benzene at 70°C for 1.5 hr using the respective anhydride. The TFA derivatives were the least stable, the HFB derivatives decomposed to the extent of 30% after three days, while the PFO derivatives decomposed only 10% after one week in the final benzene solution. Although the HFB derivatives of propoxur and carbofuran were more sensitive to EC than electrolytic conductivity detection when applied to the analysis of foods, detection limits were similar owing to the selectivity of the conductivity detector allowing a greater quantity of sample to be injected.[285] As low as 30–50 ppb of both carbofuran and 3-ketocarbofuran in turnips could be detected by either detector. It was also demonstrated that as little as 200 pg N-HFB carbofuran, 400 pg N-HFB 3-ketocarbofuran, and 100 pg of the di-HFB derivative of 3-hydroxycarbofuran could be detected by the electrolytic conductivity detector[286] with the ECD being 20 times more sensitive. The minimum detectable level for carbofuran was 0.005–0.02 ppm and for 3-ketocarbofuran was 0.01–0.05 ppm depending upon the background interference from the crops studied. With HPLC at 254 nm, 3-ketocarbofuran could be detected at the 0.02-ppm level in turnip.

The alkylation of NH-containing pesticides was applied to carbamate and urea herbicides in foods.[287,288] Using a methyl iodide–sodium hydride–dimethyl sulfoxide mixture for 15 min at room temperature, propham, chloropropham, and Swep could be detected at the 1–2-ng level in foods using the electrolytic conductivity detector in the N-mode:

$$C_6H_5-NH-CO-O-iPr- \xrightarrow[\text{RT}]{\text{NaH–CH}_3\text{I–DMSO}} \quad C_6H_5-\overset{\overset{\textstyle CH_3}{|}}{N}-CO-O-iPr-$$

propham N-methyl propham

No column cleanup was required and the minimum detectable levels were in the range 0.005–0.01 ppm in the foods studied. Alkylation of carbamates such as methiocarb, propham, and terbutol requires very mild reaction conditions (5 min at room temperature) to give single derivatives in 83–97% yields. A lower level of 0.55-ppm propham in potatoes was easily confirmed without interference from the alkylation of impurities also present in the extract. Monoethylation occurs at the NH group in all cases and the order of reactivity appears to be

carbamates > phosphoroamidates–amides > phosphoroamidothioates
> triazines > ureas

In addition to alkylation, trifluoroacetylation, etc. the N-thiomethyl derivative of carbaryl was used for the S-mode FPD analysis of vegetables fortified at the 5–10-ppm level.[289] These workers used methylsulfonyl chloride after observing no reaction between the intact carbamate and methane or trifluoromethane sulfonyl chlorides:

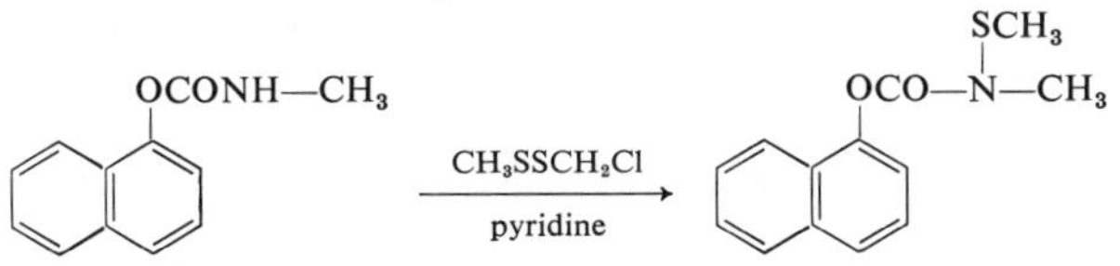

carbaryl

Originally, Moye[290] developed a residue procedure for carbaryl and carbofuran in vegetables by reaction of their corresponding phenols with 2,5-dichlorobenzene sulfonyl chloride at 80°C for 15 min at pH 8:

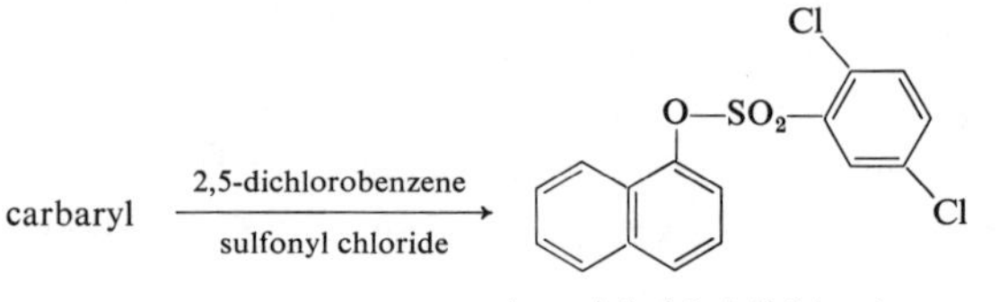

$$
\text{carbaryl} \xrightarrow[\text{sulfonyl chloride}]{\text{2,5-dichlorobenzene}}
$$

1-naphthyl-2,5-dichlorobenzene
sulfonate

It was observed that 1 pg of 1-naphthyl-2,5-dichlorobenzene sulfonate could easily be detected by EC and only 10 ng responded well to the S-mode FPD. However, the procedure based on the FPD was not demonstrated on crops. The S-mode FPD was also used to quantitate the mesylate derivatives of propoxur, Landrin, carbaryl, carbofuran and metabolites, and methiocarb and its metabolites.[291] The carbamates were hydrolyzed with methanolic KOH and the resultant phenols reacted for 15 min at room temperature with methane sulfonyl chloride in benzene containing pyridine:

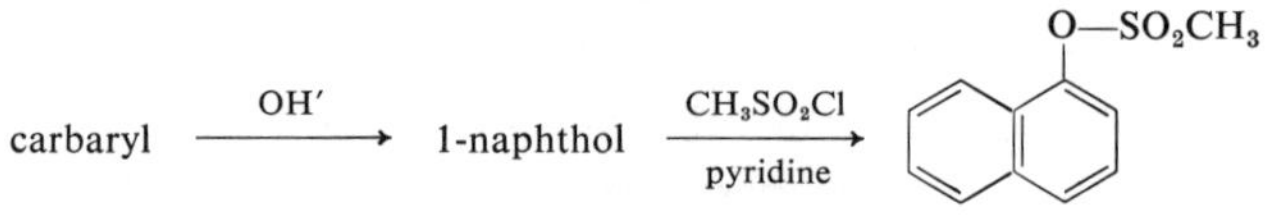

$$
\text{carbaryl} \xrightarrow{\text{OH}'} \text{1-naphthol} \xrightarrow[\text{pyridine}]{\text{CH}_3\text{SO}_2\text{Cl}}
$$

Recoveries of carbaryl from lentil straw fortified at the 0.1-ppm level averaged 103%. This method was found comparable to the trichloroacetyl chloride method of Butler and McDonough.[237]

4.4.2. Reactions Involving the S Moiety

Carbamates containing a thioether (sulfide) group readily undergo oxidation to their sulfoxide and sulfone analogs. Generally, sulfoxides do not GC well due to their polar nature and are analyzed by conversion to their sulfones:

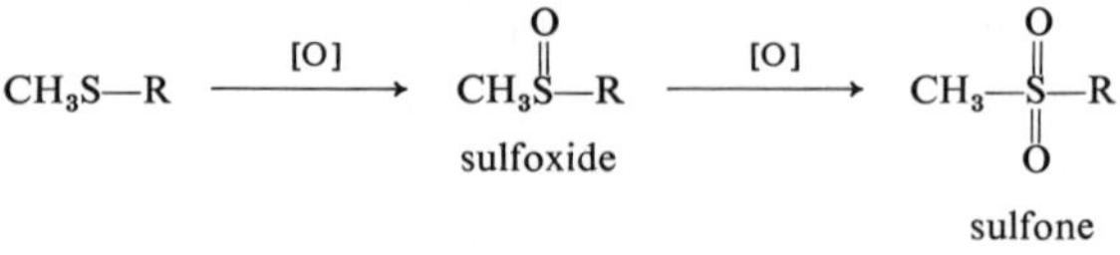

$$
\text{CH}_3\text{S}-\text{R} \xrightarrow{\text{[O]}} \underset{\text{sulfoxide}}{\text{CH}_3\overset{\text{O}}{\underset{}{\text{S}}}-\text{R}} \xrightarrow{\text{[O]}} \underset{\text{sulfone}}{\text{CH}_3-\overset{\text{O}}{\underset{\text{O}}{\text{S}}}-\text{R}}
$$

Thornton and Drager[292] oxidized methiocarb to the sulfone using potassium permanganate followed by overnight silylation to give a product detectable by S-mode FPD. A lower limit of 0.03 ppm in plant and animal tissues was obtained. However, in this reaction silylation of the carbamate had yielded the TMS phenol with loss of methyl isocyanate. Similarly, aldicarb can be oxidized using peracetic acid.[293] Woodham et al.[294] determined aldicarb residues in soil, cotton seed, and cotton lint by GC–FPD after oxidation of the extracts with 40% peracetic acid at room temperature for 30 min. A lower limit of 0.01 ppm could be obtained for aldicarb and its metabolites which were measured as aldicarb sulfone to give "total residue." The levels of aldicarb and its oxidation products in potato tubers was determined after reaction with m-chloroperbenzoic acid in acetone for 1 hr at 50–55°C.[295] Oxidation of aldicarb and its sulfoxide to the sulfone was achieved in 96–100% yields, and the reaction also performed a cleanup function on otherwise interfering GC peaks. Aldicarb sulfone was separated on a 4% Carbowax column and had a detection limit of 2.5 ng with a Pye FPD which resulted in a lower limit of 0.005 ppm for the complete procedure. This procedure was subsequently used to study the conversion rates of aldicarb and its oxidation products in soil.[296] The individual metabolites were separated on SiO_2 columns prior to oxidation. Although methomyl and oxamyl contain sulfide groups, they have not been utilized in their analysis. Instead, methomyl is normally base hydrolyzed to 1-(methylthio)-acetaldoxime followed by analysis of the oxime by GC–FPD.[297,298] This technique has been applied to methomyl residues in hay and oil at lower levels of 0.02–0.04 ppm,[299] 0.01 ppm in soil,[300] and 0.001 ppm in water.[300] Similar results have been obtained for oxamyl in vegetable crops.[297]

Also mentioned previously (Section 3.4) the rearrangement of acylated sulfoxides to give α-substituted derivatives of the corresponding sulfides is known as the Pummer reaction.[203] Reaction of methiocarb sulfoxide (and its phenol analog) with TFAA yields a di-TFA derivative due to reaction with both the NH (or OH) and S=O moieties[280,281]:

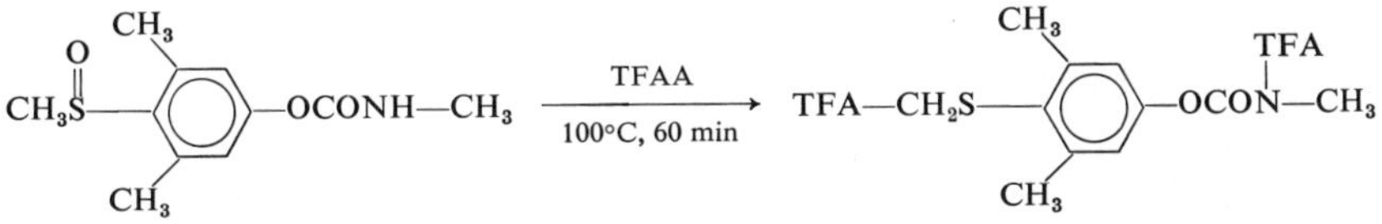

This reaction was applied to fensulfothion and its oxon, oxydemeton–methyl, and the sulfoxides of aldicarb, carboxin, Nemacur, Counter, and phorate.[301] Reaction conditions are important. Methiocarb sulfoxide and aldicarb sulfoxide gave predominantly mono-TFA derivatives at 100°C

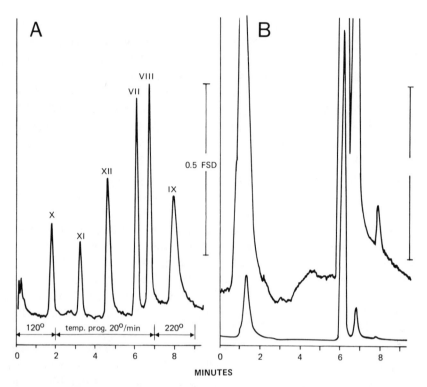

Fig. 20. Gas chromatograms of the TFA derivatives of Mesurol and its metabolites and a derivatized extract of field-treated blueberries. (*A*) Mesurol phenol TFA (X) 3 ng, Mesurol sulfoxide phenol TFA (XI) 8 ng, Mesurol sulfone phenol TFA (XII) 6 ng, Mesurol TFA (VII) 5 ng, Mesurol sulfoxide di-TFA (VIII) 10 ng, and Mesurol sulfone TFA (IX) 12 ng. (*B*) Acetone extract of field-treated blueberries after partition and derivatization. Column: 5% DC-200 at 170°C (*B*).

for 30 min, initial reaction occurring at the S=O group. Application to methiocarb and its metabolites in blueberries is shown in Figure 20.[280] It indicates the presence of methiocarb (Mesurol), the sulfoxide, and a small amount of sulfone at estimated levels of 5, 0.9, and 0.03 ppm, respectively.

5. HERBICIDES

Herbicides constitute a large and diverse group of organic and inorganic compounds. Unlike the insecticides, which can be conveniently classed into three groups (i.e., OCs, OPs, and carbamates), herbicides can only be roughly categorized into about ten major and minor groups. For convenience, the herbicidal carbamates have been considered with the insecticidal carbamates in Section 4. The use of chemical derivatization in herbicidal analysis has been extensively reviewed by Cochrane and Purkayastha,[211] Khan,[174] and updated by Cochrane.[8,9] The major classes of herbicides considered here are the triazines, ureas, chlorophenoxy acids (including other organic acids, e.g., picloram), and the carbamates (Section 4), with the amides, 2,6-dinitroanilines, uracils, etc. being covered under miscellaneous.

5.1. Triazines

The majority of the triazines can be analyzed intact by GC with good response being obtained via flame ionization (e.g., formulations), ECD (e.g., atrazine residues), or selective nitrogen detection. Therefore, chemical derivatization has been used in the confirmation of identity of the parent triazine or to render the polar dehalogenated or dealkylated metabolites more amenable to GC analysis. Most derivatizations involve reaction with the NH or OH moieties followed by selective nitrogen detection. Using atrazine and hydroxyatrazine as typical examples, various alkylation, silylation, chlorination, methoxylation, and DNP derivatization of cleaved amino side chains are illustrated in Figure 21.

Silylation has been used extensively to derivatize the S-triazines and the hydroxy and dealkylated metabolites.[302-304] Hydroxysimazine and its metabolites were treated with HMDS and trichlorosilane for identification purposes,[302] while nine rat urinary metabolites of prometone were characterized by mass spectrometry after silylation.[302] The reaction of the hydroxy derivatives of simazine, atrazine, and propazine with BSFTA in closed vials at 150°C for 15 min led to a mixture of products, the ratio of which varied with time and the temperature of reaction.[303] Soil and corn samples fortified with hydroxyatrazine could be analyzed down to a lower level of 1 ppm. This silylation reaction was reexamined,[304] the reaction time being fixed at 20 min and the reaction temperature varied from 90 to 210°C. Three products were obtained. Initially, the mono- and disilyl derivatives were obtained with the yield of the monoderivative decreasing above 130°C. The disilyl compound was the major product at 160°C, at higher temperatures the trisilyl derivative was formed. Silylation of hydroxyatrazine at 150°C for 30 min gave 43.5%, 41.3%, and 2% yields, respectively, of the mono, di-, and trisilyl derivatives. Atrazine could be derivatized at temperatures above 180°C to give a single derivative in about 90% yield. Monosilylation having occurred at the less sterically hindered N-ethyl group.

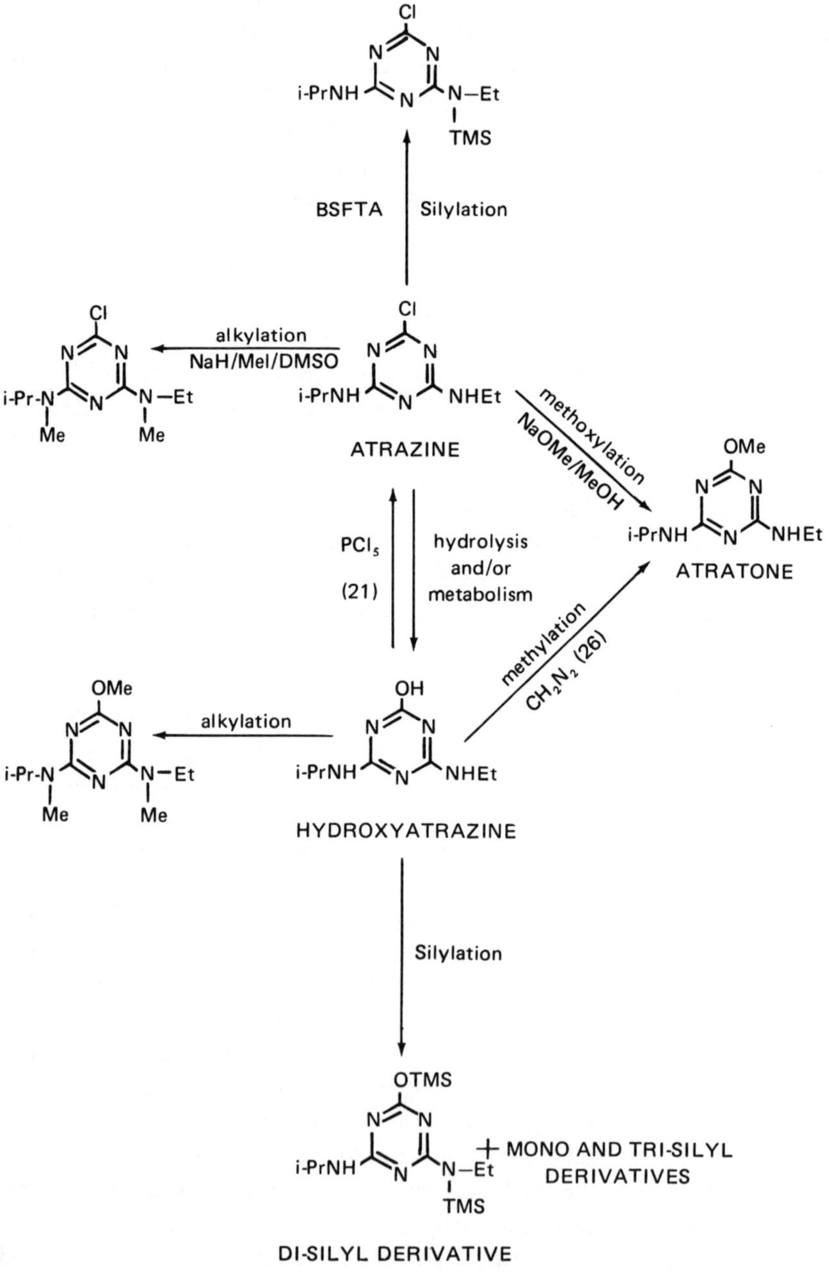

Fig. 21. Confirmatory tests for s-triazine herbicides using atrazine and hydroxy atrazine as typical examples.

The lower limit of sensitivity for both atrazine and hydroxyatrazine in soil samples was approximately 0.05 ppm. Using BSTFA at 150°C for 30 min the TMS derivatives of six suspected metabolites of simazine were analyzed by GC–MS.[306] Three of the suspected metabolites (cyanuric acid, ammelide, and ammeline) yielded trisilyl derivatives, while N-ethyl-ammeline and N-ethyl-ammelide gave predominantly disilyl products with hydroxysimazine giving a monosilyl derivative as major product plus a lesser amount of the disilyl derivative. Similarly, the TMS derivatives of cyanuric acid, ammelide, ammeline, and melamine were prepared using trifluoroacetic acid, as solvent, and MSTFA (N-methyl-N-trimethylsilyl-trifluoro acetamide) as silylation reagent at 150°C for 5 min.[307] Each metabolite formed a trisilyl derivative, and no evidence was found for N,N-bis-silyl derivatization. The minimum detectable amounts with a 3% OV-17 column at 165–195°C using a FID was 0.5–0.7 ng at a signal-to-noise ratio of 5. In an attempt to increase detector sensitivity, the HFB derivatives of 10 triazines were prepared by reacting the herbicides with HFBA in benzene in the presence of trimethylamine or pyridine as catalyst.[308] The major derivative was the di-HFB product with the mono-HFB derivative appearing with some triazines. The derivatives were 300-fold more sensitive to ECD and 5–10-fold more sensitive to electrolytic conductivity detection than the underivatized triazines. This HFB method was successfully applied to the analysis of potatoes, peas, and tomatoes spiked with various triazines at levels of 0.13–0.86 ppm.

Prometone, atrazine, atratone, and prometryne can be alkylated with NaH/CH$_3$I/DMSO at 50°C for 10 min in 95–98% yields at the residue level.[199,309] Alkylation occurs at both NH groups (Figure 21), unlike silylation where it is restricted to the NH-ethyl group. Hydroxyatrazine can be alkylated to a trimethyl derivative identical to that obtained from atratone. Hydroxyatrazine can also be converted to atratone in 50–75% yield using a large excess of diazomethane.[305] These three methods, silylation, alkylation, and methylation, were examined and compared for the confirmation of hydroxyatrazine in soil using AFID detection. Methylation and alkylation were the reactions of choice.[305] The residue levels of three hydroxy-S-triazines in soil were monitored using the semiquantitative diazomethane procedure[310,311] as well as determining atrazine and its metabolites in chicken tissues.[312] Similarly, triazuril, an as-triazine, was N-methylated with diazomethane prior to EC–GC analysis of its residues and metabolites in mammals.[313]

Replacement of the labile chlorine atom in chloro-S-triazines by a methoxyl group is the basis for another direct confirmatory test.[314] Alkylation and methoxylation were compared with an indirect method, involving hydrolysis to the dealkylated triazine and derivatization with 2,4-dinitrofluorobenzene for the confirmation of metribazin, simazine, atrazine, promazine, and prometone in plants (Figure 21). The direct confirmatory tests were preferred, methoxylation being the more facile reaction (NaOMe for 5 min) but restricted to chloro-S-triazines, whereas alkylation is a more general reaction.[314] The alkylation technique was successfully applied to the analysis of atrazine in turnips, peas, beets, and parsnips at the 0.05 ppm level. In formulation analysis, metribuzin, which has a free NH_2 group, was found to react with ketones, such as acetone, to form Schiffs bases.[315] Therefore, analysis was carried out using CH_2Cl_2 as solvent.

Two other confirmatory tests have been used for hydroxytriazines. Hydroxycyprazine was chlorinated with PCl_5 to give the parent compound (Figure 21)[316] after the hydroxy metabolite had been separated from the parent compound by column chromatography. Also the water-soluble metabolites of cyanazine were quantitatively determined by the simultaneous cleavage of the S-triazine ring and cyclization of the characteristic side chain to give 5,5-dimethyl hydantoin, which is further derivatized with trichloromethane sulfenylchloride to facilitate EC–GC analysis.[317] Recoveries at 0.2–1.0 ppm averaged 90% for soil and water and averaged 80% for crops.

5.2. Ureas

The GC analysis of intact urea herbicides has paralleled that of the carbamates in that on-column decomposition has been a major problem. This can be traced from Reiser,[318] who reported that only alkyl-substituted ureas could be chromatographed undecomposed. Many of the ureas with herbicidal activity have the structure aryl—NH—CO—N—$(CH_3)_2$ or aryl—NH—CO—N—$CH_3(OCH_3)$, and Spengler and Hamroll[319] were of the opinion that most N-phenyl ureas analyzed by the method of McKone and Hance[320] were pyrolyzed and eluted as phenylisocyanates and aliphatic amines:

$$p\text{-Cl}—C_6H_4—NH—CO—CH_3(OCH_3) \longrightarrow p\text{-Cl}—C_6H_4—NCO + NH—CH_3(OCH_3)$$
monolinuron

Using FID detection, a neutral GC column and an injector temperature of 350°C, fenuron, linuron, and monolinuron were found to thermally decompose, while monuron and diuron remained intact. Thermogravimetric analysis has also shown that partial decomposition of linuron occurs at 165°C in nitrogen and is complete at 220°C.[321] Using a short XE-60 column, on-column injection, and injector port temperature <165°C made it possible to detect intact linuron in soil samples down to a lower limit of 0.1 ppm. This confirmed the original work of Katz and Strusz,[322] who also used short XE-60 columns to reduce decomposition. While the N-phenyl-N,N-dimethyl ureas such as monuron and diuron are much more thermally stable than the N-phenyl-N-methoxy-N-methyl ureas (e.g., linuron),[323] Buchert and Lokke[324] showed that the thermal stability of the mono-, di-, and trisubstituted ureas was related to the presence of an amide–H group. Since tetrasubstituted phenyl ureas are thermally stable, most derivatization reactions have involved replacement of the amide–H.

These derivatization reactions, like those discussed in detail for carbamates (Section 4.4), have included silylation,[268,269] trifluoroacetylation, and similar reactions[284,285,325] and alkylation.[199,287,288,309,323,325,326]

Efforts to prepare halomethyldimethylsilyl derivatives of monuron and diuron were unsuccessful.[238] However, the TMS derivatives of monuron chromatographed well on Carbowax 20M at 170°C, while diuron and its TMS derivative were not observed and it was postulated that this was due to thermal decomposition.[269] Chromatography of diuron was successful at the lower temperatures, 130°C, used for QF-1 and SE-30 columns.

Saunders and Vanetta[325] showed that both the trifluoroacetyl and -alkyl derivatives of fenuron, monuron, and 1-(5-*tert*-butyl-1,3,4-thiadiazol-2-yl)-1,3-dimethyl urea were thermally stable and eluted intact from a 2% OV-17 column as determined by GC–MS. Diuron did not give a TFA derivative; also yields were poor, possibly owing to incorrect conditions. Lawrence[284,285] prepared the TFA, PFP, HFB, and PFO derivatives (see Section 4.4) of linuron and compared their sensitivities on ECD and by electrolytic conductivity detection. The PFO and HFB products were preferred because of their sensitivity and greater resistance to decomposition upon storage. In common with some other ureas, diflubenzuron cannot be analyzed by GC owing to thermal decomposition to a phenylisocyanate and a benzamide.[327–329] Although it can be N,N-dimethylated using NaH/CH$_3$I/ DMSO at room temperature for >15 min,[328] a derivative with increased EC response was obtained via trifluoroacetylation using TFAA in ethyl acetate at 50°C for 30 min.[327] At first the reaction product was thought to be the N,N-bis-TFA derivative, but in fact the compound being quantitatively determined was the N-TFA derivative of 4-chloroaniline:

diflubenzuron

N-TFA-4-chloroaniline

This derivative probably arises from the decomposition of diflubenzuron during the reaction and *in situ* trifluoroacetylation of the intermediate 4-chlorophenyl isocyanate.[329] This trifluoroacetylation procedure could measure down to 0.02 ppm in pond water,[327,329] while the alkylation technique was capable of 0.2 ppm in spruce foliage.

Alkylation of the NH moiety has been achieved at the macro level using *t*-BuOK/CH$_3$I/THF [325] at 52°C for 2–16 hr and at the residue level with NaH/CH$_3$I/DMSO [199] in a sealed tube at 50°C for 10 min. The latter reaction could be carried out at the 0.5-µg urea/ml level with linuron, monuron, and fenuron and was applied at the 2-ppm level for the confirmation of fenuron in soil.[309] Monomethylation occurs at the NH group in all cases. Lawrence and Laver[287,288] extended this to the analysis of urea herbicides in nine foods. Using the electrolytic conductivity detector in the N mode, no column cleanup was required for any samples down to the 0.01-ppm level using a 4% SE-30/6% QF-1 column at 199°C. In this method, alkylation was achieved at room temperature for 15 min. The identities of the methylated urea herbicides were further confirmed by cleaving with NaOMe to the methyl aniline moiety which was chromatographed on the same GC column as the parent compound[288]:

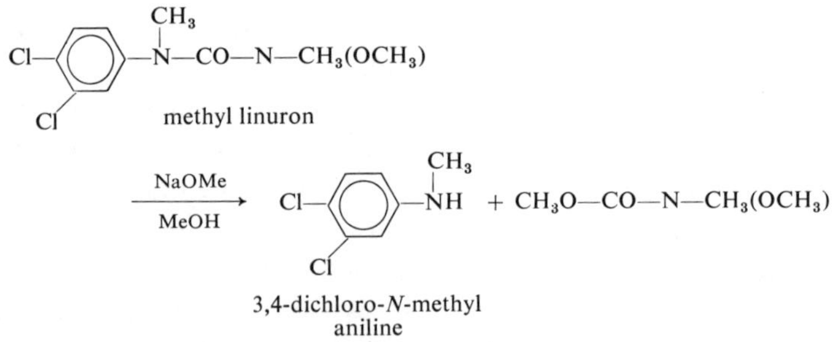

methyl linuron

3,4-dichloro-*N*-methyl aniline

For example, using a 3% OV-1 column at 195°C the *N*-methyl aniline derivative of methyl linuron displayed a retention time of 0.39 relative to the parent compound.

Since *N*-methylation of the parent urea and certain of its metabolites can result in the same derivative this limits the applicability of the CH$_3$I alkylation procedure. Glad *et al.*[326] utilized NaH/C$_2$H$_5$I/DMSO at 65°C for 20 min, while the use of trideutero methyl iodide followed by GC–MS has been suggested by others.[323]

Finally, Tanaka and Wein[330] achieved on-column *N*-methylation of seven phenyl ureas using trimethyl anilinium hydroxide (Methelute®). Two methods were used to introduce the sample into the GC: (a) premixing

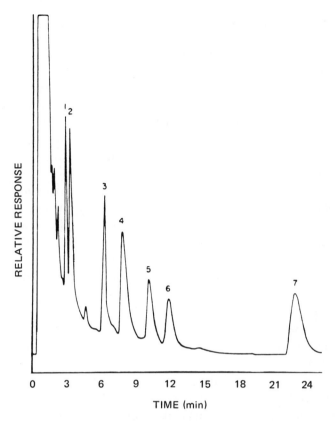

Fig. 22. The separation of a mixture of substituted phenylureas by on-column (flash-heater) methylation on a 2% DC-LSX-3-0295 and 8% SE-30 column. (1) Fluometuron; (2) fenuron; (3) monuron; (4) metobromuron; (5) linuron; (6) diuron; and (7) neburon.

sample plus TMAH and (b) drawing TMAH followed by the sample into the injecting syringe. If the TMAH and sample were reversed in (b) methylation did not occur. Method (a) was preferred for quantitative work and maximum methylation was obtained at a TMAH: urea ratio of 2.5:1. A total of six different GC columns were evaluated in this study; an example of the separations achieved is shown in Figure 22. All results were obtained with an FID since use of the ECD was unsatisfactory. Lok[331] also utilized an on-column technique for the determination of tebuthiuron and its metabolites in grass and sugar cane at the 0.01–0.03-ppm level. The total residue of the two metabolites was obtained after prior separation from tebuthiuron on an alumina column since parent and metabolites each gave the same decomposition product.

5.3. Chlorophenoxy Acids

Originally esterification procedures were devised and developed for fatty acids and some were successfully applied to herbicides. The methyl esters have been preferred because of their ease of preparation and their reasonably short GLC retention times. Basically four different methods of esterification are currently available–diazoalkanes (RN_2), dimethyl sulfate (Me_2SO_4), pyrolysis of tetra-alkylammonium salts, and acid-catalyzed alcohol reagents. It is in the latter category that most variation is possible and encountered. Here the alcohol can be in combination with a mineral acid, most commonly sulfuric acid, or a Lewis acid, primarily BF_3 or BCl_3. Other catalysts that have been tried and found of passing interest are perchloric acid and thionyl chloride. Obviously the type of alcohol used can also be varied, and commercial preparations of BF_3 or BCl_3 with methanol, propanol, butanol, and 2-chloroethanol are available.

Since these derivatization techniques are well established and have been extensively used and reported[8,9,174,211] for the quantitative analysis of the chlorophenoxy acids in a vast array of substrates, this aspect will not be completely covered in this review. The use of new and more EC-sensitive esterification reagents for general use and novel techniques–reactions will be examined.

In an effort to increase the EC sensitivity of MCPA, Gutenmann and Lisk[332] esterified with BF_3/2-chloroethanol at 138°C for 10 min. Normally, the methyl esters of MCP are about 100 times less sensitive to EC than the 2,4-D methyl esters. They applied this method to MCPA residues in soil down to 0.5 ppm. Subsequently, Yip[333] tested 2-chloroethanol and 2,2,2-trifluoroethanol in search of a suitable transesterification derivative for use in the confirmation of MCPA, 2,4-D, 2,4,5-T, fenoprop, and 2,4-DB residues. The background pattern of the 2-chloroethanol derivative gave too many peaks and was discarded. Trifluoroethanol increased the response of 2,4,5-T and fenoprop but not MCPA or 2,4-D. A TFAA/2,2,2-trifluoroethanol reagent has been used in a quantitative esterification technique for ten model benzenoid acids with EC detection in the picogram range.[334] Although the 2-chloroethyl ester of 2,4-D was found to have increased sensitivity and gave a longer retention time than the methyl ester, it was still subject to interferences from certain soil types.[335] In applying the 10% BCl_3/2-chloroethanol method to other acid herbicides it was found that fenac and 2,4,5-T were very sensitive, with amiben and picloram being somewhat less sensitive when compared to the 2,4-D 2-chloroethyl ester. The lower limit of sensitivity for this method in soils was approximately

0.01 ppm for 2,4-D. The BCl_3/2-chloroethanol reagent produced little or no product for 2,3,6-TBA or dicamba owing to the expected steric hindrance of the o-chlorine atoms.[336] Therefore a new esterification procedure was developed to form the 2-chloroethyl esters of 10 of the more common herbicidal acids. This involved reaction with dicyclohexyl carbodi-imide/2-chloroethanol overnight at room temperature. Milder conditions, i.e., 1–2-hr reaction, were all that was required for 2,4-D, fenoprop, MCPA, and MCPB. Except for the 2-chloroethyl esters of 2,3,6-TBA which had a retention time close to MCPA, all nine other esters could be separated on OV-17/QF-1, OV-101/OV-210, or OV-225 columns. The practical limits of detection in natural waters ranged from a lower limit of 0.01 μg/l fenoprop to a high of 2–5 μg/l MCPB.

Since the 2-chloroethyl ester showed promise it naturally followed that the 2,2,2-trihaloethanols would be investigated. Alley et al.[337] showed that of the two catalysts, pyridine and HFBA, it was the latter that promoted the greatest amount of esterification with 2,2,2-trichloroethanol (TCE). The next best catalyst was a mixture of TFAA/trifluoroacetic acid (3:1) followed by TFAA itself, for the TCE esterification of aliphatic acids (C_2–C_8) which permitted their EC detection in the low nanogram range. Applied to 2,4-D and MCPA, Mierzwa and Witek[338] showed that while the 2,2,2-trifluoroethyl derivative gave a marginal increase in sensitivity over the 2-chloroethyl derivative it was the TCE ester that gave the greater increase in sensitivity. Esterification was carried out with 20% (V/V) TCE in TFAA in the presence of sulfuric acid at 100°C for 15 min or room temperature for at least 2 hr. By using ^{63}Ni-ECD the detection limit for MCPA (150 pg) was lower than that for 2,4-D (240 pg) when using a 15% QF-1/10% DC 200 column at 195°C. A lower limit of 0.096 ppb 2,4-D and 0.06 ppb MCPA in 1-l water samples was achieved.

In an effort to increase the sensitivity of the chlorophenoxy acids even more, Chau and Terry[339] compared the pentafluorobenzyl (PFB) esters of ten herbicidal acids and eventually compared the three esterification techniques pentafluorobenzylation, BCl_3/2-chloroethanol, and DCC/2-chloroethanol.[340] The optimum condition for PFB ester formation was reaction at room temperature for at least 5 hr in the presence of potassium carbonate. Three GC columns were investigated and various peak overlaps occurred on all, including the 3% OV-225 column as shown in Figure 23, MCPA 2,4-DP, and 2,4-D and fenoprop (Silvex) coeluting, respectively. However, nine of the ten derivatives could be separated on 3–6% OV-101/5–5% OV-210. The PFB ester of MCPA has a 2–5 times greater EC response than the 2-chloroethyl ester and 1000 times greater than the methyl ester. Using the

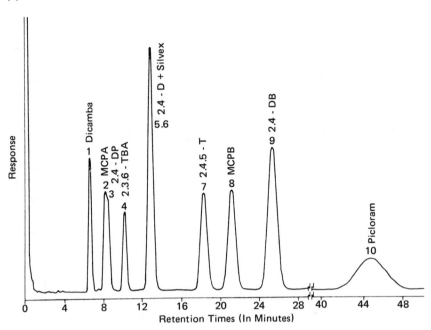

Fig. 23. Separation characteristics of ten herbicidal acids as their PFB derivatives on 3% OV-225 at 200°C.

PFB ester derivatives, concentrations as low as 0.1 μg/l MCPA and 0.2 μg/l MCPB could be determined[341] in natural waters. DeBeer *et al.*[342] also studied the PFB esterification of ten chlorophenoxy acids and compared retention indices on nine frequently used GC columns. The separations achieved on the DC-200 and OV-17 columns are illustrated in Figure 24, with the DC-200 giving complete separation of the ten PFB esters. As in most derivatization reactions, removal of excess reagent is advisable prior to EC analysis. In the determination of free fatty acids, Glyllenhaal *et al.*[343] removed excess PFBB by coupling it with a phenolalkylamine and extraction of the product into an acidic aqueous phase. A similar procedure had been used by Sumida *et al.*[218] for the removal of excess 2,4-dinitrobenzene. PFB ester formation has been applied to the analysis of MCPB, mecoprop, and MCPA residues in soils down to 0.1 ppm.[344,345] It would appear from the above studies that the sensitivities of the PFB and TCE esters are about equivalent for, at least, MCPA and 2,4-D.

Originally Gutenmann and Lisk[346] rendered MCPA and MCPB more EC sensitive by bromination followed by BF_3/MeOH esterification at 100°C for 2 min and obtained two peaks for each acid. More recently Kahn[347] increased the EC response of mecoprop by first esterifying with diazomethane then brominating at 50°C for 30 min. Here two monobromo and two dibromo derivatives were observed. The major monobrominated derivative was used for quantitative purposes making the method sensitive to about 0.5 ppm mecoprop in 20-g soil samples.

It was shown by Robb and Westbrook[187] that the methyl esters of carboxylic acids can be prepared in the injector block of a gas chromatograph by pyrolysis of their tetramethylammonium (TMA) derivatives. Yields in the range 80–100% were obtained by thermal decomposition of these TMA derivatives at injector temperatures of 330–365°C. Henkel[348] applied this technique to the determination of chlorophenoxyalkanoic acid residues in soil extracts. When the acids were dissolved directly in 10% aqueous TMA hydroxide and pyrolyzed at 380–400°C, ester yields varied substantially from 20% for MCPP to 85% for MCPA. Nevertheless, reproducible on-column esterification was achieved using 20% aqueous tetraethylammonium (TEA) hydroxide in alcoholic solution and free acid concentrations up to 1%. Quantitative determinations and identifications were carried out without difficulties down to the 0.05-ppm level. Hyman[349] investigated the use of TMAH for the analysis of MCPA and 2,3,6-TBA formulations. Compared with diazomethane esterification the MCPA methyl ester peaks obtained by TMAH on-column methylation exhibited appreciable tailing. This was attributed to insufficient pyrolysis, and by attention

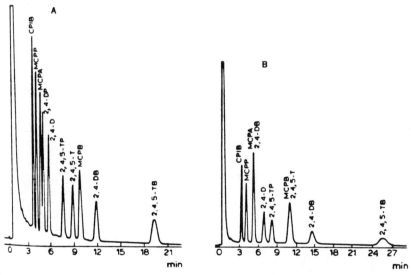

Fig. 24. Gas chromatographic separation of the PFB esters of various herbicidal acids on (A) 5% DC-200 on Varaport 30 and (B) 3% OV-17 on Chromosorb W AW DMCS both at 210°C.

to temperature and length of heating zone a maximum yield of 94% could be obtained at 240°C. This procedure was used by Middleditch and Desiderio[350] for the formation of C_{14}–C_{24} fatty acid methyl esters using trimethylanilinium hydroxide at 265°C. Greely[351] described a very interesting general reaction, capable of a rapid esterification procedure, using nonacidic reagents. The technique is very reminiscent of the t-BuOK/CH_3I/THF or NaH/CH_3I/DMSO alkylation reagents mentioned earlier (Sections 3.4 and 4.4). Greely[351] used a TMAH/1-iodobutane/N,N-dimethylacetamide mixture to form the butyl ester of stearic acid in 99.4% yield. The base can vary as can the alkyl halide. There were two limitations to this procedure: (a) the solvent should be anhydrous and (b) the order of addition must be acid–solvent–base mixture added to the alkyl halide. This technique could be applied to acids, barbiturates, and NH compounds, and preparation of deuterated esters for GC–MS use. A modification of this procedure was applied to the esterification of dicamba, 2,4-D, 2,4,5-T, and mecoprop by Thio *et al.*[352] Using cesium carbonate/CH_3I/acetone at room temperature for 10–100 min (depends on acid used) the methyl esters were obtained in >90% yields. Butyl esters could be prepared in an analogous manner. In contrast to acid-catalyzed esterification, this base-catalyzed esterification is facilitated by the presence of steric bulk around the carboxylate moiety.

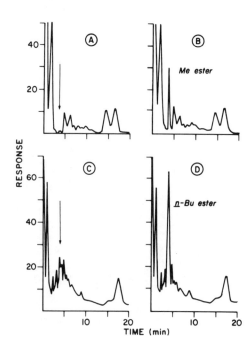

Fig. 25. Gas chromatograms derived from 5-μl injections of (A) methylated control urine extract, arrow indicates retention time of 2,4-D methyl ester; (B) methylated extract containing 0.1 ppm 2,4-D; (C) *trans*-butylated extract (A), arrow indicates retention time of 2,4-D *n*-butyl ester and (D) trans-butylated extract (B). Column: Ultra-Bond 20M at 150°C.

This procedure would therefore have distinct advantages in multiresidue herbicide analysis which includes the chlorophenoxy acid as well as 2,3,6-TBA, dicamba, etc.

Yip[333] transesterified the methyl esters of 2,4-D, 2,4,5-T, fenoprop, MCPA, and 2,4-DB to their respective propyl esters as a means of confirming identity. McKone and Hance[353] used the butyl esters because of their longer GC retention times. The use of the BF_3/n-BuOH reagent has subsequently been used for 2,4,5-T in Vietnamese soils[354] and 2,4-D in air,[355] formulations,[356] and urine.[357] In the latter case transbutylation was carried out by adding 2 ml n-BuOH to 20 ml hexane extract, removing hexane by evaporation, adding four drops H_2SO_4, and heating at 100°C for 1 hr. On cooling, 25 ml water and 10 ml n-hexane were added. Typical chromatograms obtained are shown in Figure 25. Normally, 85–90% yields of the n-butyl esters of 2,4-D were obtained by this procedure. A slight drawback was that the transbutylated extracts derived from the methylated control urine samples showed some interference with the same retention time as the n-butyl ester of 2,4-D (Figure 25, chromatograms C and D).

5.3.1. Other Organic Acids

Originally, trichloroacetic acid (TCA) in crops, at the 0.01-ppm level, was determined by decomposition to $CHCl_3$ which was distilled, collected, and quantitated by ECD.[358,360] A much simpler procedure is used for water analysis, that is, via its methyl ester.[359,360] The minimum level of detection being 1 ppb TCA in 500 ml water, Cotterill[361] utilized the longer retention time of the butyl ester of dalapon to eliminate GC interferences and thus eliminate elaborate "cleanup" procedures. With an overall recovery of 96%, dalapon residues could be determined down to 0.05 ppm in soil. Similarly, 3,6-dichloropicolinic acid residues in soil were also determined using the 1-butyl ester to avoid volatilization losses.[362] Previously, Pik and Hodgson,[363] using the methyl ester, obtained recoveries from 84% to 94% at the 6–1000-ppb level in soils. Ether-containing extracts of the mono- and diacid metabolites were converted to their propyl esters overnight at room temperature using 1-n-propyl-β-tolyl-triazine as reagent.[364] The method was sensitive to 0.02, 0.08, and 0.2 ppm of Dacthal and its mono- and diacid metabolites, respectively, in carrots on a fresh weight basis. Methylation of flamprop-methyl,[365] a wild oat herbicide, was used in characterization studies, while the aquatic herbicide endothall was converted to endothall N-chloroethylimide by reaction with 2-chloroethylamine at 120°C for 1 hr.[366] The amide could be determined at the 0.1–5.0 ppm level (in

water and hydrosoil extracts) using an electrolytic conductivity detector in
the N mode:

endothall   endothall N-chloroethylimide

5.4. Miscellaneous

A number of approaches have been tried in an effort to analyze paraquat
and diquat by GC. Paraquat was determined in water by its catalytic hydrogenation and analysis using a FID.[367] Attempts to apply this procedure to
crops were unsuccessful as no recovery of paraquat was obtained at 1.0 ppm.
Khan[368,369] was able, using a Pt_2O catalyst, to hydrogenate the acid extracts
of paraquat and diquat residues and determine them in soil[368] and subsequently paraquat in lettuce, carrots, and onions[369] down to lower levels
of 0.05 ppm. This method was extended to cyperquat residues in soil.[370]
Recoveries of soil fortified at the 0.5- and 1.0-ppm levels were 77.1% and
85.2%, respectively. Normally, two GC peaks are obtained via catalytic
hydrogenation of diquat and cyperquat.

Another approach has been the reduction of paraquat and diquat with
sodium borohydride. van Dijk et al.[371] determined paraquat in plasma
after $NaBH_4$ reduction at pH > 10 at 60°C for 12.5 min in a stoppered
tube. The major product was N,N- dimethyl-2,2′,3,3′,6,6′-hexahydro-4,4′-
dipyridyl as previously identified by Ukai et al.[372] A minor monoene product was also observed. The $NaBH_4$ reduction of diquat, in contrast to catalytic hydrogenation, furnishes only one major product, N,N-ethylene-
1,1′,2,2′,3,3′,6,6′-octahydro-6,6′-bipyridyl.[372,373] Recoveries averaged 87.4%
of diquat added to potatoes in the 0.05–1.0-ppm range, while a 64% recovery
was obtained from soil in the 0.1–5-ppm range.[373] In this procedure, $NaBH_4$
reduction was carried out in 95% ethanol for 30 min. In contrast, Pryde[374]
was able to determine paraquat in urine by HPLC analysis, at 258 nm, on a
chemically bonded 20-μm spherical alumina by direct injection of the urine.

Using Regisil-TMCS, terbacil and three metabolites were determined in
plants and animal tissues with a sensitivity of 0.04 ppm for all four compounds.[375] Separation was achieved on a 5% XE-60/0.2% Epon 1001
column by programming from 100 to 230°C at 2°C/min. A method capable
of determining 0.5 ppb fluridone, a pre-emergent herbicide, in water and
10–30 ppb in other substrates includes a novel bromination step using a
PBr_3/pyridine reagent.[376] Essentially, a stable ketone moiety is deriva-

tized with PBr$_3$ to form an EC-responsive derivative capable of low-level detection (0.04–0.20 ng). Residues of oxyfluorfen in crops were determined by reducing the NO$_2$ group with aluminum in 10% aqueous NaOH to give NH$_2$, which is further converted to the NH–HFB derivative capable of EC–GC analysis.[377]

Normally, the 2,6-dinitroaniline herbicides, e.g., trifluralin, are analyzed by EC–GC as the intact compounds. However, chromous chloride[28] is a very facile reagent for use in their confirmation. Trifluralin was confirmed in spiked potato extracts, and dinoseb in peas at levels in the 0.5–1.0-ppm range.[32] Use of the electrolytic conductivity detector in the N mode ensures no loss in sensitivity in going from NO$_2$ to NH$_4$. Oryzalin in methanol was converted overnight to its dimethyl derivative using anhydrous Na$_2$SO$_4$ and methyl iodide.[378] No cleanup by column chromatography was required, such that residual oryzalin was determined at the 1-ppb level in water–sediment systems.

6. PHENOLS

Although the chlorophenols are used in the pesticide area as fungicides–herbicides they are also extensively used in antirotting agents for nonwoolen textiles, paints, and industrial biocides. In addition, as seen in Sections 3.2, 4.2, and 5.3, they can arise as phenolic metabolites of the OP, carbamate, and chlorophenoxy acid herbicides. Although various chlorinated and non-chlorinated phenols in water have been successfully quantitated at the ppb level using FID[379] and OV-17 separation, normally, column deactivation[380] or derivatization is required in order to quantitate picogram amounts by EC or MS. Figure 26 illustrates some of the more commonly used reactions. The trimethylsilyl (TMS) protecting group was first introduced in 1958 for the GC analysis of alkyl phenols.[381] Esposito[382] subsequently devised an on-column procedure for acids and alcohols in water using Silyl-8 (a mixture of BSA, hexamethyldisilazane, and trimethylsily diethylamine). A GC procedure for the separation of 14 chlorinated aromatic acids, chlorophenols (including 2,4-dichloro- and 2,4,6-trichlorophenol), and their precursors as their TMS derivatives was reported by Larson and Rockwell.[383] A 3% SP 2401 column programmed from 100–200°C at 6°C/min and FID detection were used. Pentachlorophenol, bisphenol, and aromatic acids were identified in river water by GC–MS after ethyl acetate extraction by either silylation with N,O-bis(trimethyl silyl) acetamide (BSA) in acetonitrile (phenols) or BF$_3$/MeOH esterification

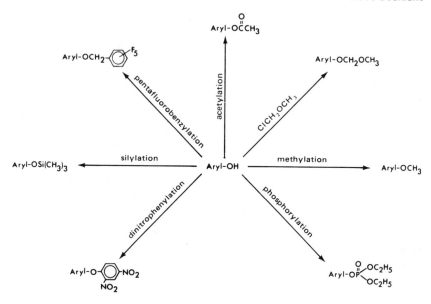

Fig. 26. Typical reactions of chloro- and nitrophenols as used in their analysis.

(acids).[384] Silylation with BSA at 50°C for 5 min has also been used in the identification of the plant growth regulators, such as abscisic acid and the gibberellins, on a 3% OV-17 at 125°C.[385]

The more commonly used derivative is the methyl ether, from reaction with diazomethane or dimethyl sulfate, after extraction and/or an acid–base partitioning cleanup step. This procedure has been applied to the combined determination of chlorinated phenol fungicides, hydroxyether bactericides, and chlorophenoxy acid herbicides in organic tissue, soil, and water after initial separation on an ion exchange column.[386] Methylation using diazomethane has also been applied to the analysis of PCP in urine,[387,388] excreta,[389] water and sediment,[386,390] and organic tissue.[386,391] Similarly, DNBP residues in feed, tissue and excreta,[392] dinoseb and DNOC in foods at the 0.1–1.0-ppm level,[393] and dinoterb in soil with a detection limit of 1 μg/kg[394] were also quantitated as their respective methyl esters. Shakif *et al.*[395] described a multiresidue procedure for halo- and nitrophenols using diazoethane to give the ethyl ethers. This method was used to measure the levels of these phenols–metabolites in urine after exposure to biodegradable pesticides yielding such metabolites. For example, 0.01-ppm PCP levels were routinely found in control urine samples. Similarly, PCP was analyzed to 0.01 ppm in marine biota and sea water[396] using the ethyl derivative, but as low as 0.002 ppb in sea water could be detected by formation of the amyl (pentyl) derivative. Salomonsson *et al.*[397] used the same reagent, $NaH/C_2H_5I/DMSO$ at 50°C for 20 min, to ethylate various phenolic acids, as utilized in the ethylation of linuron. The method was used to analyze cereal straws for eight phenolic acids such as coumaric, syringic acids, etc. The method of Shakik[395] was used to determine *para*-nitrophenol in human urine and the resulting ether was then confirmed by first reducing with CrCl$_2$, then converted to the amide by reacting with heptafluorobutyric anhydride:

Para-nitrobenzene could be detected and confirmed at concentrations of 10 ppb or more. The GC characteristics of the methyl, ethyl, propyl, etc. ethers of PCP were studied on seven different GC columns using ^{3}H-ECD.[388]

Another well-studied procedure is acetylation (Figure 26). Chau and Coburn[399] determined PCP in natural and waste waters at the 0.01-ppb/liter level by the addition of acetic anhydride to an aqueous alkaline solution after first extracting with benzene and then into a K$_2$CO$_3$ solution. The PCP acetate was then extracted with hexane and quantitated by ECD. The retention times of 17 other phenolic acetates were obtained on OV-225, OV-101/OV-210, and OV-17/QF-1 columns. The technique was subsequently extended to 3-trimethyl-4-nitrophenol in natural waters,[400] PCP in milk at the 0.02-ppm level with modifications,[401] PCP in adipose tissues,[402] and determination of chlorophenols by SE-30 capillary column GC.[403] The detection limit of PCP acetate using a ^{63}Ni-ECD is about 1–2.5 pg.[402,403] A very similar procedure was used by Coutts *et al.*[404] The acetate esters of six phenols (cresols, 2,4-dichlorophenol, etc.) were formed by the addition of 500 μl acetic anhydride to 250 ml dilute aqueous phenolic solution (concentration range 0.08–0.24 μmol/liter) containing 10 g NaHCO$_3$. The dinitrophenylation (Figure 26) of various phenols has been discussed in Sections 3.2 and 4.2. Farrington and Manday[405] reacted 6 tri-, 3 tetra-, and 1 pentachlorophenols with 2,4-DNFB in the presence of pyridine as catalyst. Using a 1.7% OV-17 column at 230°C all the ethers studied were well resolved from each other with the 2,4-DNP ether of PCP appearing at 60 min. The method was applied to trace chlorophenol levels in chicken flesh. Phosphorylation of phenols to produce compounds that can be detected by P-mode FPD has been discussed in Section 3.2 of the OPs.[182] Another interesting reaction technique is the on-column reaction of chloromethyl methyl ether (CMME) with the sodium salt of 2,4,6-trichlorophenol [406] (Figure 26). This technique was developed to determine CMME in ambient air down to a lower level of 1 ppb using ECD.

A number of reports have appeared on the increased EC sensitivity of the PFB ethers of various phenols and acids (Sections 3.2, 4.2, and 5) (Figure 26). This derivative has been employed in the determination of o-phenyl phenol in citrus fruit to a lower level of 0.005 ppm.[407] An interesting adaptation of this technique is to perform the extraction of the phenol (acid), and derivatization in one step. Here the phenol (or acid) anion is partitioned from the aqueous phase, as an ion pair by using a quaternary ammonium ion, into the organic phase which contains the perfluorobenzylating reagent.[408,409] Phenols have been extractively derivatized with PFBB in a biphasic methylene chloride–sodium hydroxide system in the absence of any phase transfer catalyst such as quaternary ammonium hydroxide.[409] It should be noted that degradation of the PFBB reagent can occur and depends upon the pH, buffer system, or quaternary ammonium ion used.[410] The technique has been applied to the extractive pentafluorobenzylation of 2,4-D and MCPA in water with a detection limit of 1-3 µg/liter.[411] It should also be generally applicable to chlorophenols in water.

Finally, Lamparski and Nestrick[412] determined concentration of ≥ 10 ng/ml substituted phenols in aquous media by EC–GC analysis of the HFB derivatives. Sample preconcentration was not required and a single benzene extraction removed the phenolics prior to derivatization with the heptafluorobutyrylimidazole reagent at 65°C for 15 min.

7. FUNGICIDES

The ethylene-bis-dithiocarbamates (EBDCs) which include nabam, mancozeb, zineb, metiram, and maneb, form an important class of fungicides. During recent years ethylene thiourea (ETU) residues have been found as surface residues on agricultural produce treated with EBDCs[413] as well as in commercial EBDC formulations.[414] Procedures for the determination of ETU residues have been summarized by IUPAC[415] and much work has been performed in this area since the FAO/WHO Meeting of Experts on Pesticide Residues drew attention to the inadequacies in ETU methodology. Although ETU can be chromatographed intact[416,417] the majority of methods using GC quantitation include a derivatization step to enhance sensitivity or improve its GC characteristics. In all cases derivatization involves alkylation of the thio carbonyl group and in some instances a double derivatization involving the NH group.

Table 13 outlines the types of derivatives used and the detection limits achieved. In nearly all methods, the derivatization step including work-up

TABLE 13. Derivatives Used in the Analysis of ETU

Derivative	GLC		Limit of detection, ppm	References
	Column	Detector		
imidazoline ring, N=C, NH; S—C₄H₉	30% DC-200	AFID	0.02	418, 419, 420, 421
	5% SE-30	FPD		
	20% SE-30	FPD	0.01	422
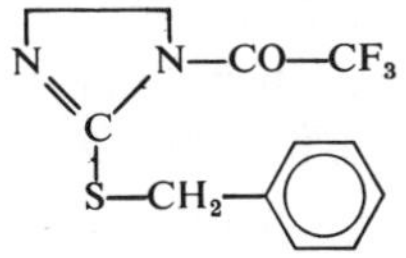 imidazoline ring, N=C, N—CO—CF₃; S—CH₂—⟨phenyl⟩	2% BDS	ECD	0.005	413, 423, 424
	4% SE-30/6% QF-1			
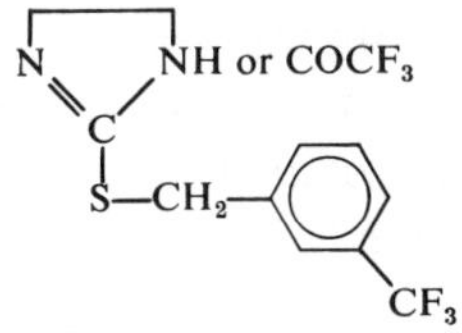 imidazoline ring, N=C, NH or COCF₃; S—CH₂—⟨phenyl⟩—CF₃	3% OV-275	ECD	0.01	425, 426, 427
imidazoline ring, N=C, N—CO—⟨phenyl⟩F₅; S—CH₂—⟨phenyl⟩ H or Cl	3% OV-17	ECD	0.005	428, 429
	3% OV-1			
	3% XE-60			

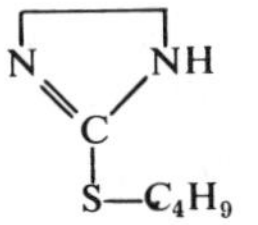

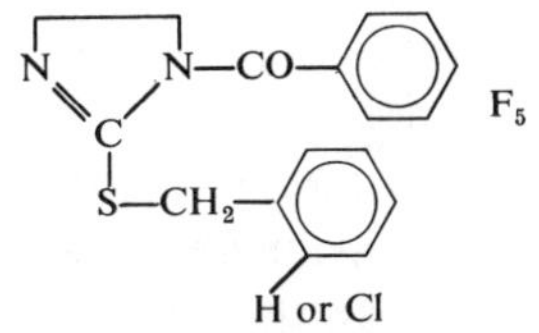

appears to be the most critical one. The method using the *S*-butyl derivative[418] required extensive cleanup, and recoveries from fortified vegetables (0.01–0.05 ppm) appear to be somewhat low, 65–78%. Owing to the poor sensitivity of ETU and its derivative large samples have to be used, 40–100 g. The *S*-benzyl was formed by refluxing the extract for 30 min with benzyl chloride with further derivatization of the NH with TFAA.[423] Prolonged reflux will convert some *S*-benzyl ETU to dibenzyl disulfide which cochromatographs with *S*-benzyl N-TFA ETU on 2% BDS. Nash[428,249] used both *S*-benzylation and *S*-*o*-chlorobenzylation as a first step and perfluorobenzoylation as a second derivatization step. These ETU derivatives displayed GC retention times of ~30 min on the 3% XE-60 column used (at 220°C). As with other reactions using pentafluorobenzylation a cleanup column is required to remove excess reagent. Newsome has also extended his earlier ETU work and developed a method for 0.01–5-ppm levels in food to be determined after conversion to a pentafluorobenzamide.[430]

Pentafluorobenzoyl chloride in acetone was used as reagent at 75°C for 1 hr, followed by silicic acid column cleanup. It was established that only the monoacetylated product was produced. Similarly, p-nitrobenzoyl chloride was used to derivatize 2-imidazoline residues in food crops at the 0.1–1-ppm level.[431] Quantitation in this instance was via HPLC and uv detection at 254 nm. The above procedures are invariably lengthy and require reflux temperatures to achieve acceptable yields of the various derivatives. Recently, the room temperature extractive N-acylation of ETU from water was described.[432] Reaction was carried using dichloroacetic anhydride in CH_2Cl_2 with acetonitrile being added (as phase transfer agent) in a 1:10 ratio with the water. Reaction is complete in 3 min to give the monodichloroacetyl derivative which eliminates HCl together with ring closure on GC injection, resulting in a sharp single peak on OV-17 or OV-330 columns. The identity of the GC derivative was 2,3,5,6-tetrahydro-6-chloroimidazo(2,1-b)thiazol-3-one. This technique was used as a rapid screening procedure for ETU in water at the 0.01–0.05-ppm level and can be extended to vegetable juices.

Other fungicides which have included a derivatization step in their analysis include thiabendazole and methyl 2-benzimidazole (MBC), which were derivatized with PFBB.[433] Derivatives could be EC detected at the 5–10-pg level. Previous workers had determined MBC residues by trifluoroacetylation[434] or determination of thiabendazole as N-methyl thiabendazole by FID and determination of PFB derivative by EC.[435] In the latter case the minimum detectable amount of thiabendazole in grapefruit was 0.01 ppm. Benomyl is also readily hydrolyzed to benzimidazol-2-yl carbamate in acid media, which upon esterification gives MBC and 2-aminobenzimidazole.[436] By using the composite value of MBC and 2-aminobenzimidazole, residues of benomyl in various fruits and vegetables could be detected to a lower level of 0.01 ppb using capillary column GC with NP and EC detection.

Originally, carboxin (Vitavax) was extracted from cereal seeds base hydrolyzed to release aniline which was distilled, collected, and quantitated via a microcoulometric nitrogen detector to give a method sensitive to less than 0.2 ppm.[437] Tafuri et al.[438] employed an AFID and quantitated carboxin directly on an OV-1/OV-210 column at 210°C. For oxycarboxin (Plantvax), the sulfone derivative of carboxin, a reduced derivative was formed using $LiAlH_4$ in THF. Carboxin could be detected in grains down to 0.005 ppm, while for oxycarboxin the level was higher, i.e., 0.5 ppm. Onley[439] described an EC–GC method for captan and its two metabolites in milk and meat down to 0.02 ppm and 0.04 ppm, respectively. After extraction captan and its metabolites are separated and captan determined directly by GC, while the metabolites are separately derivatized in acetone with PFBB. The resultant derivatives were cleaned up on alumina.

Chlorothalonil was shown to be unstable to light when dissolved in benzene or toluene,[440,441] giving rise to one and three products, respectively. The product obtained in benzene was shown by isolation[441] and GC–MS to be 2,4-dicyano-3,5,6-trichlorobiphenyl. Chlorothalonil was easily photolyzed in other aromatic solvents to give one (in o-xylene) or two (in mesitylene) products, while in acetone, ether, or hexane no photoproducts were obtained. It would therefore appear that chlorothalonil residues could easily be confirmed photolytically in benzene.

REFERENCES

1. F. A. Gunther and R. C. Blinn, *Analysis of Insecticides and Acaracides*, Interscience Publishers, New York (1955).
2. G. Zweig, *Analytical Methods for Pesticides, Plant Growth Regulators and Food Additives*, Vol. II, Academic Press, New York (1964).
3. D. M. Coulson and L. A. Cavanagh, *Anal. Chem.* **32**, 1245–1247 (1960).
4. J. E. Lovelock, *Anal. Chem.* **33**, 162 (1961).
5. B. J. Gudzinowicz, *Gas Chromatographic Analysis of Drugs and Pesticides*, Marcel Dekker, New York (1967).
6. W. P. Cochrane and A. S. Y. Chau, *Adv. Chem. Ser.* **104**, 11–26 (1971).
7. L. M. Reynolds, *Res. Rev.* **34**, 27–57 (1970).
8. W. P. Cochrane, *J. Chromatogr. Sci.* **13**, 246–253 (1975).
9. W. P. Cochrane, *J. Chromatogr. Sci.* **17**, 124–137 (1979).
10. G. Vander Velde and J. F. Ryan. *J. Chromatogr. Sci.* **13**, 322–328 (1975).
11. K. E. Elgar, *Adv. Chem. Ser.* **104**, 151–161 (1971).
12. J. P. Minyard and E. R. Jackson, *J. Agric. Food Chem.* **13**, 50–56 (1965).
13. A. K. Klein, J. O. Watts, and J. N. Damico, *J. Assoc. Off. Anal. Chem.* **46**, 165–171 (1963).
14. A. Klein and J. O. Watts, *J. Assoc. Off. Anal. Chem.* **47**, 311–316 (1964).
15. J. H. Hammence, P. S. Hall, and D. J. Caverley, *Analyst* **90**, 649–656 (1965).
16. W. W. Sans, *J. Agric. Food Chem.* **15**, 192–198 (1967).
17. H. B. Pionke, G. Chesters, and D. E. Armstrong, *Analyst* **94**, 900–903 (1969).
18. S. J. V. Young and J. A. Burke, *Bull. Environ. Contam. Toxicol.* **7**, 160–167 (1972).
19. R. T. Krause, *J. Assoc. Off. Anal. Chem.* **55**, 1042–1052 (1972).
20. C. E. Mendoza, P. J. Wales, H. A. McLeod, and W. P. McKinley, *J. Assoc. Off. Anal. Chem.* **51**, 1095–1101 (1968).
21. J. W. P. Miles, *J. Assoc. Off. Anal. Chem.* **55**, 1039 (1972).
22. G. A. Miller and C. E. Wells, *J. Assoc. Off. Anal. Chem.* **52**, 548–553 (1969).

23. A. S. Y. Chau and M. Lanouette, *J. Assoc. Off. Anal. Chem.* **55**, 1057–1066 (1972).
24. R. Gothe, *Bull. Environ. Contam. Toxicol.* **11**, 451–455 (1974).
25. A. S. Y. Chau and W. P. Cochrane, *J. Assoc. Off. Anal. Chem.* **52**, 1220–1226 (1969).
26. W. P. Cochrane, and A. S. Y. Chau, *Bull. Environ. Contam. Toxicol.* **5**, 251–254 (1970).
27. A. S. Y. Chau and W. P. Cochrane, *Bull. Environ. Contam. Toxicol.* **5**, 133–138 (1970).
28. W. P. Cochrane and M. A. Forbes, in *Methods in Residue Analysis*, A. S. Tahori, ed., Vol. IV of *Pesticide Chemistry*, Gordon and Breach, London (1971), pp. 385–402.
29. A. S. Y. Chau, H. Won, and W. P. Cochrane, *Bull. Environ. Contam. Toxicol.* **6**, 481–484 (1971).
30. A. S. Y. Chau and W. P. Cochrane, *J. Assoc. Off. Anal. Chem.* **54**, 1124–1131 (1971) and references therein.
31. M. A. Forbes, B. P. Wilson, R. Greenhalgh, and W. P. Cochrane, *Bull. Environ. Contam. Toxicol.* **13**, 141–148 (1975).
32. J. F. Lawrence, D. Lewis, and H. A. McLeod, *J. Agric. Food Chem.* **25**, 1359–61 (1977).
33. K. A. Banks and D. D. Bills, *J. Chromatogr.* **33**, 450–455 (1968).
34. W. M. Kaufman, D. D. Bills, and E. J. Hannan, *J. Agric. Food Chem.* **20**, 628–631 (1972).
35. D. E. Glotfelty, *Anal. Chem.* **44**, 1250–1254.
36. E. J. Hannen, D. D. Bills, and J. L. Herring, *J. Agric. Food Chem.* **21**, 87–90 (1973).
37. R. A. Leavitt, G. C. C. Su, and M. J. Zabik, *Anal. Chem.* **45**, 2130–2131 (1973).
38. P. M. Ward, *J. Assoc. Off. Anal. Chem.* **60**, 673–678 (1977).
39. D. R. Erney, Anal. Lett. **12**, 501–22 (1979).
40. M. F. Grammer and M. F. Copeland, *Bull. Environ. Contam. Toxicol.* **9**, 186–192 (1973).
41. M. B. Abou-Donia, *J. Assoc. Off. Anal. Chem.* **51**, 1247–1260 (1968).
42. *Pesticide Analytical Manual, Vol. 1*, Food and Drug Administration, Washington D. C., revised 1978, Sections 33–331c.
43. S. B. Soloway, U. S. Patent 2, 676, 131; *Ca* **48**, 8473c (1954).
44. A. S. Y. Chau and W. P. Cochrane, *J. Assoc. Off. Anal. Chem.* **52**, 1092–1100 (1969).
45. K. Noren, *Analyst* **93**, 39–41 (1968).
46. M. Osadchuk and E. B. Wanless, *J. Assoc. Off. Anal. Chem.* **51**, 1264–1267 (1968).
47. W. W. Wiencke and J. A. Burke, *J. Assoc. Off. Anal. Chem.* **52**, 1277–1280.
48. D. W. Woodham, C. D. Loftis, and C. Collier, *J. Agri. Food Chem.* **20**, 163–165 (1972).
49. R. B. Maybury and W. P. Cochrane, *J. Assoc. Off. Anal. Chem.* **56**, 36–40 (1973).
50. A. S. Y. Chau and W. P. Cochrane, *Bull. Environ. Contam. Toxicol.* **5**, 515–520 (1970).
51. E. J. Skerrett and E. A. Baker, *Analyst* **84**, 376–380 (1959).
52. E. A. Baker and E. J. Skerrett, *Analyst* **85**, 184–187 (1960).
53. A. S. Y. Chau and W. P. Cochrane, *Chem. Ind.* 1568–1569 (1970).
54. J. ApSimon, J. A. Buccini and A. S. Y. Chau, *Tetrahedron Lett.*, 539–542 (1974).
55. W. R. Benson, *J. Agric. Food Chem.* **19**, 66–72 (1971).
56. C. L. Henderson and D. C. Crosby, *J. Agr. Food Chem.* **15**, 888–893 (1968).
57. P. Lombardo, K. H. Pomerantz, and I. G. Ergy, *J. Agric. Food Chem.* **20**, 1288–1289 (1972).
58. A. S. Y. Chau and W. P. Cochrane, *Bull. Environ. Contam. Toxicol.* **5**, 435–439 (1970).

59. A. S. Y. Chau and R. J. Wilkinson, *Bull. Environ. Contam. Toxicol.* **8**, 105–18 (1972).
60. A. S. Y. Chau, *Bull. Environ. Contam. Toxicol.* **8**, 169–176 (1972).
61. A. S. Y. Chau, *J. Assoc. Off. Anal. Chem.* **55**, 519–525 (1972).
62. A. S. Y. Chau, *J. Assoc. Off. Anal. Chem.* **52**, 1240–1248 (1969).
63. P. A. Greve and S. C. Wit, *J. Agric. Food Chem.* **19**, 372–374 (1971).
64. A. S. Y. Chau and K. Terry, *J. Assoc. Off. Anal. Chem.* **55**, 1228–1231 (1972).
65. A. S. Y. Chau, *J. Assoc. Off. Anal. Chem.* **55**, 1233–1238 (1972).
66. C. J. Musial, M. E. Peach, and D. A. Stiles, *Bull. Environ. Contam. Toxicol.* **16**, 98–100 (1976).
67. T. B. Putnam, D. D. Bills, and L. M. Libbey, *Bull. Environ. Contam. Toxicol.* **13**, 662–665 (1975).
68. W. P. Cochrane and A. S. Y. Chau, *J. Assoc. Off. Anal. Chem.* **51**, 1267–1270 (1968).
69. W. P. Cochrane, *J. Assoc. Off. Anal. Chem.* **52**, 1100–1105 (1969).
70. W. H. Dennis, Jr. and W. J. Cooper, *Bull. Environ. Contam. Toxicol.* **16**, 425–530 (1976).
71. J. Singh, *Bull. Environ. Contam. Toxicol.* **4**, 77–79 (1969).
72. W. P. Cochrane and M. A. Forbes, *Chemosphere* **1**, 41–46 (1974).
73. W. P. Cochrane, M. Forbes, and A. S. Y. Chau, *J. Assoc. Off. Anal. Chem.* **53**, 769–774 (1970).
74. E. G. Alley, B. R. Layton, and J. P. Minyard Jr., *J. Agric. Food Chem.* **22**, 727–729 (1974).
75. R. G. Lewis, R. G. Hanisch, K. E. MacLeod, and G. W. Sovocool, *J. Agric. Food Chem.* **24**, 1030–1035 (1976).
76. R. H. Lane, R. M. Grodner, and J. L. Graves, *J. Agric. Food Chem.* **24**, 192–193 (1976).
77. J. Mes, D. J. Davies, and W. Miles, *Bull. Environ. Contam. Toxicol.* **19**, 564–570 (1978).
78. A. S. Y. Chau, J. M. Carron, and H. Tse, *J. Assoc. Off. Anal. Chem.* **61**, 1475–1480 (1978).
79. W. R. Lusby and K. R. Hill, *Bull. Environ. Contam. Toxicol.* **22**, 567–569 (1979).
80. M. V. H. Holdsrinat, *Bull. Environ. Contam. Toxicol.* **21**, 46–52 (1979).
81. R. J. Norstrom, H. T. Won, M. V. H. Holdrinet, P. G. Calway, and C. D. Naftel, *J. Assoc. Off. Anal. Chem.* **63**, 37–42 (1980).
82. K. V. Scherer, Jr., R. S. Lunt III, and G. A. Ungefug, *Tetrahedron Lett.*, 1199–1205 (1965).
83. W. L. Dilling, H. P. Broendling, and E. T. McBee, *Tetrahedron* **23**, 1211–1224 (1967).
84. R. F. Moseman, M. K. Ward, H. L. Crist, and R. D. Zehr, *J. Agric. Food Chem.* **26**, 965–968 (1978).
85. H. Parlar, W. K. Lein, and F. Korte, *Chemosphere* **3**, 129–132 (1972).
86. B. Proszynska, *Rocz. Panstw. Zakl. Hig.* **28**, 201–207 (1977).
87. J. R. Pearson, F. D. Aldrich, and A. W. Stone, *J. Agric. Food Chem.* **15**, 938–939 (1967).
88. D. F. Goerlitz and L. M. Law, *Bull. Environ. Contam. Toxicol.* **6**, 9–10 (1972).
89. R. E. Johnson and L. Y. Mansell, *Bull. Environ. Contam. Toxicol.* **17**, 573–576 (1977).
90. J. F. Lester and J. W. Smiley, *Bull. Environ. Contam. Toxicol.* **7**, 43–44 (1971).
91. D. L. Strauble, *Bull. Environ. Contam. Toxicol.* **11**, 231–237 (1974).
92. B. Schwemmer, W. P. Cochrane, and P. B. Polon, *Science* **169**, 1087 (1970).

93. J. H. Lawrence, R. P. Barron, J. Y. T. Chen, P. Lombardo, W. R. Benson, *J. Assoc. Off. Anal. Chem.* **53**, 261 (1970).
94. P. B. Polen, M. Hester, and J. Benziger, *Bull. Environ. Contam. Toxicol.* **5**, 521–528 (1971).
95. D. W. Cinder, P. C. Oloffs, and Y. S. Szelo, *Bull. Environ. Contam. Toxicol.* **7**, 33–35 (1972).
96. T. Miyazaki, K. Akiyama, S. Kaneko, S. Horii, and T. Yamagishi, *Bull. Environ. Contam. Toxicol.* **23**, 631–635 (1979).
97. T. Miyazaki, K. Akiyama, S. Kaneko, S. Horii, and T. Yamagishi, *Bull. Environ. Contam. Toxicol.* **24**, 1–8 (1980).
98. W. P. Cochrane and M. A. Forbes, *Can. J. Chem.* **49**, 3569–3571 (1971).
99. W. P. Cochrane and R. Greenhalgh, *J. Assoc. Off. Anal. Chem.* **59**, 698–702 (1976).
100. G. Wayne Sovocool, R. G. Lewis, R. L. Harless, N. K. Wilson, and R. D. Zehr, *Anal. Chem.* **49**, 734–740 (1977).
101. W. P. Cochrane, H. Parlar, S. Gäb, and F. Korte, *J. Agric. Food Chem.* **23**, 882–886 (1975).
102. S. Gäb, L. Bom, H. Parlar, and F. Korte, *J. Agric. Food Chem.* **25**, 1365 (1977a).
103. S. Gäb, H. Parlar and F. Korte, *J. Agric. Food Chem.* **25**, 1224 (1977).
104. G. T. Brooks, *Residue Rev.* **27**, 81–138 (1969).
105. G. T. Brooks, in Pesticide Terminal Residues, suppl. to *Pure Appl. Chem.*, 111–136 (1971).
106. H. Parlar and F. Koste, *Chemosphere* **6**, 665–705 (1977).
107. F. I. Onuska and M. E. Comba, *J. Assoc. Off. Chem.* **58**, 6–9 (1975).
108. F. I. Onuska and M. E. Comba, *J. Chromatogr.* **119**, 385–399 (1976).
109. D. A. Carlson, K. D. Konyha, W. B. Wheeler, G. P. Marshall, and R. G. Zaylskie, *Science* **194**, 939 (1976).
110. G. W. Ivie, H. W. Dorough, and E. G. Alley, *J. Agric. Food Chem.* **22**, 933–935 (1974).
111. B. Zimmerli, H. Sasler, and B. Marek, *Mitt. Geb. Lebensmittelunters. Hyg.* **62**, 60–71 (1971).
112. W. P. Cochrane and R. B. Maybury, *J. Assoc. Off. Anal. Chem.* **56**, 1324–1329 (1973).
113. W. H. Dennis, Jr. and W. J. Cooper, *Bull. Environ. Contam. Toxicol.* **18**, 57–59 (1977).
114. G. B. Collins, D. C. Holmes, and M. Wallen, *J. Chromatogr.* **69**, 198–200 (1972).
115. Mr. V. H. Holdrinet, *J. Assoc. Off. Anal. Chem.* **57**, 580–583 (1974).
116. B. E. Baker, *Bull. Environ. Contam. Toxicol.* **10**, 279–284 (1973).
117. K. T. Rosewell and B. E. Baker, *Bull. Environ. Contam. Toxicol.* **21**, 470–477 (1979).
118. H. L. Crist, R. F. Moseman, and J. W. Noneman, *Bull. Environ. Contam. Toxicol.* **14**, 273–280 (1975).
119. E. M. Brevik, *Bull. Environ. Contam. Toxicol.* **19**, 281–286 (1978).
120. R. L. Holmstead, S. Khalefa, and J. E. Casida, *J. Agric. Food Chem.* **22**, 939–944 (1974).
121. W. P. Cahill, B. J. Estevan, and G. W. Ware, *Bull. Environ. Contam. Toxicol.* **5**, 260–262 (1970).
122. E. D. Gomes, *Bull. Environ. Contam. Toxicol.* **17**, 456–462 (1977).
123. R. D. Enery, *Bull. Environ. Contam. Toxicol.* **12**, 717–720 (1974).
124. J. A. Armour and J. A. Burke, *J. Assoc. Off. Anal. Chem.* **53**, 761–768 (1970).
125. D. R. Erney, *Bull. Environ. Contam. Toxicol.* **12**, 710–716 (1974).

126. N. J. Kveseth and E. M. Brevik, *Bull. Environ. Contam. Toxicol.* **21**, 213-
127. M. V. H. Holdrinet, *J. Assoc. Off. Anal. Chem.* **57**, 580–584 (1974).
128. D. L. Stalling, A. Smith, and J. N. Hackins, *J. Assoc. Off. Anal. Chem.* **61**, 32–28 (1978).
129. M. Doguchi, S. Fukano, and F. Ushio, *Bull. Environ. Contam. Toxicol.* **11**, 157–158 (1974).
130. M. Ahnoff and B. Josefson, *Bull. Environ. Contam. Toxicol.* **13**, 159–166 (1975).
131. J. Jan and S. Malnersic, *Bull. Environ. Contam. Toxicol.* **19**, 772–780 (1978).
132. A. S. Y. Chau, *J. Assoc. Off. Anal. Chem.* **57**, 585–591 (1974).
133. G. G. Sims, J. R. Campbell, F. Zemlyak, and J. M. Graham, *Bull. Environ. Contam. Toxicol.* **18**, 697–705 (1977).
134. G. Keck and J. Raffenot, *Bull. Environ. Contam. Toxicol.* **21**, 689–696 (1979).
135. D. Stainken and J. Rollwagen, *Bull. Environ. Contam. Toxicol.* **23**, 690–697 (1979).
136. W. J. Trotter, *J. Assoc. Off. Anal. Chem.* **58**, 461–465 (1975).
137. J. C. Underwood, *Bull. Environ. Contam. Toxicol.* **21**, 787–790 (1979).
138. J. R. W. Miles, *J. Assoc. Off. Anal. Chem.* **55**, 1039 (1972).
139. Y. A. Greichus, J. J. Woman, M. A. Pearson, and D. J. Call, *Bull. Environ. Contam. Toxicol.* **11**, 113–120 (1974).
140. B. Luckas, H. Pscheidl, and D. Haberland, *J. Chromatogr.* **147**, 41–46 (1978).
141. V. Zitko, *J. Assoc. Off. Anal. Chem.* **57**, 1253–1259 (1974).
142. W. H. Dennis, Jr., Y. H. Chang, and W. J. Cooper, *Bull. Environ. Contam. Toxicol.* **22**, 750–753 (1979).
143. J. H. Carey, J. Lawrence, and H. M. Tosine, *Bull. Environ. Contam. Toxicol.* **16**, 697–701 (1976).
144. L. L. Ruzo and M. J. Zabik, *Bull. Environ. Contam. Toxicol.* **13**, 181–182 (1975).
145. M. Mansour and H. Parlar, *J. Agric. Food Chem.* **26**, 483–485 (1978).
146. D. Feiedman and P. Lombardo, *J. Assoc. Off. Anal. Chem.* **58**, 703–706 (1975).
147. P. Lombardo, J. L. Dennison, and W. W. Johnson, *J. Assoc. Off. Anal. Chem.* **58**, 707–710 (1975).
148. D. R. Erney, *J. Assoc. Off. Anal. Chem.* **58**, 1202–1205 (1975).
149. W. J. Trotter, *Bull. Environ. Contam. Toxicol.* **18**, 726–733 (1977).
150. M. Gulan, D. D. Bills, and T. B. Putnam, *Bull. Environ. Contam. Toxicol.* **11**, 438–441 (1974).
151. O. W. Berg, P. L. Diosady, and G. A. V. Rees, *Bull. Environ. Contam. Toxicol.* **7**, 338–347 (1972).
152. J. A. Armour, *J. Assoc. Off. Anal. Chem.* **56**, 987–993 (1973).
153. W. J. Trotter and S. J. V. Young, *J. Assoc. Off. Anal. Chem.* **58**, 466–468 (1975).
154. O. W. Berg, G. A. V. Rees, and M. S. Ali, paper No. 579 presented at the 29th Pittsburgh Conference, *The State of the Art in Analytical Chemistry and Applied Spectroscopy*, Cleveland, March 1978.
155. C. L. Stratton, J. M. Allen and S. A. Whitlock, *Bull. Environ. Contam. Toxicol.* **21**, 230–237 (1979).
156. J. Mes and D. J. Davies, *Bull. Environ. Contam. Toxicol.* **21**, 381–387 (1979).
157. A. L. Robbins and C. R. Willhite, *Bull. Environ. Contam. Toxicol.* **21**, 428–431 (1979).
158. D. L. Stalling and J. W. Hogan, *Bull. Environ. Contam. Toxicol.* **20**, 35–43 (1978).
159. J. R. Haas, L. T. Jao, N. K. Wilson, and H. B. Matthews, *J. Agric. Food Chem.* **25**, 1330–1333 (1977).

160. V. Zitko, O. Hutzinger, and P. M. K. Choi, *Bull. Environ. Contam. Toxicol.* 12, 649–653 (1974).

161. R. Greenhalgh and W. P. Cochrane, *Int. J. Environ. Anal. Chem.* 3, 213–228 (1974).

162. W. P. Cochrane and R. Greenhalgh, *Chromatographia* 9, 255–265 (1976).

163. L. E. St. John, Jr. and D. J. Lisk, *J. Agric. Food Chem.* 16, 408–410 (1968).

164. J. Askew, J. H. Ruzicka, and B. B. Wheals, *J. Chromatogr.* 41, 180–187 (1969).

165. M. T. Shafik and H. F. Enos, *J. Agric. Food Chem.* 17, 1186–1189 (1969).

166. M. T. Shafik, D. Bradway, and H. F. Enos, *Bull. Environ. Contam. Toxicol.* 6, 55–66 (1971).

167. M. T. Shafik, D. E. Bradway, H. F. Enos, and A. R. Yobs, *J. Agric. Food Chem.* 21, 625–628 (1973).

168. M. T. Shafik, D. E. Bradway, F. J. Biros, and H. F. Enos, *J. Agric. Food Chem.* 18, 1174–1175 (1970).

169. E. M. Lores and D. E. Bradway, *J. Agric. Food Chem.* 25, 75–79 (1977).

170. J. B. Knaak, K. T. Maddy, and S. Khulifa, *Bull. Environ. Contam. Toxicol.* 21, 375–380 (1979).

171. R. Greenhalgh, personal communication, 1980.

172. L. G. Johnson, *J. Assoc. Off. Anal. Chem.* 56, 1503 (1973).

173. J. A. Coburn and A. S. Y. Chau, *J. Assoc. Off. Anal. Chem.* 57, 1272–1278 (1974).

174. S. U. Khan, *Residue Rev.* 59, 21–50 (1975).

175. W. P. Cochrane, *ACS Symposium Series*, Vol. 136, J. Harvey, Jr. and G. Zweig eds., ACS, Washington, DC (1980), pp. 231–249.

176. J. N. Seiber, D. G. Crosby, H. Fonda, and C. J. Sanderquist, *J. Chromatogr.* 73, 89–97 (1972).

177. I. C. Cohen, J. Norcup, J. H. A. Ruzicka, and B. B. Wheals, *J. Chromatogr.* 44, 251–255 (1969); 49, 215–221 (1970).

178. F. H. Kawahara, *Environ. Sci. Technol.* 5, 235–239 (1971).

179. J. A. Coburn and A. S. Y. Chau, *Environ. Lett.* 10, 225–236 (1975).

180. M. C. Bowman, C. L. Holden, and L. G. Rushing, *J. Agric. Food Chem.* 26, 35–42 (1978).

181. J. A. Coburn and A. S. Y. Chau, *J. Assoc. Off. Anal. Chem.* 57, 1272–1278 (1974).

182. P. G. Deo and P. H. Howard, *J. Assoc. Off. Anal. Chem.* 61, 210–213 (1978).

183. K. Jacob, C. Falkner, and W. Vogt, *J. Chromatogr.* 167, 67–75 (1978).

184. H. J. Dishburger, R. L. McKellar, J. Y. Pennington, and J. R. Rice, *J. Agric. Food Chem.* 25, 1325–1329 (1977).

185. W. C. Kossa, J. MacGee, S. Ramachandren, and A. J. Webber, *J. Chromatogr. Sci.* 17, 177–187 (1979).

186. H. A. Moye, *J. Agric. Food Chem.* 21, 621–625 (1973).

187. E. W. Robb and J. J. Westbrook III, *Anal. Chem.* 35, 1644–1647 (1963).

188. W. E. Dale, J. W. Miles, and F. C. Churchill II, *J. Assoc. Off. Anal. Chem.* 59, 1088–1093 (1976).

189. J. W. Miles and W. E. Dale, *J. Agric. Food Chem.* 26, 480–482 (1978).

190. P. T. Holland, *Pestic. Sci.* 8, 354–358 (1977).

191. R. Greenhalgh, J. Dokladalova, and W. O. Haufe, *Bull. Environ. Contam. Toxicol.* 7, 237–242 (1972).

192. M. A. Forbes, B. P. Wilson, R. Greenhalgh, and W. P. Cochrane, *Bull. Environ. Contam. Toxicol.* 13, 141–148 (1975).

193. R. L. Schutzmann and W. F. Barthel, *J. Assoc. Off. Anal. Chem.* 52, 151–156 (1969).

194. J. Singh and M. R. Lapointe, *J. Assoc. Off. Anal. Chem.* **57**, 1285–1287 (1974).
195. J. Singh and W. P. Cochrane, *J. Assoc. Off. Anal. Chem.* **62**, 751–756 (1979).
196. E. M. Bellet and J. E. Casida, *J. Agric. Food Chem.* **22**, 207–211 (1974).
197. R. Blinn, *J. Assoc. Off. Anal. Chem.* **47**, 641–645 (1964).
198. B. Miller and T. P. O'Leary, *J. Org. Chem.* **27**, 3382 (1962).
199. R. Greenhalgh and J. Kovacicova, *J. Agric. Food Chem.* **23**, 325–329 (1975).
200. J. F. Lawrence and F. Iverson, *J. Chromatogr.* **103**, 341–347 (1975).
201. J. F. Lawrence and H. A. McLeod, *J. Assoc. Off. Anal. Chem.* **59**, 637–640 (1976).
202. R. Greenhalgh, R. R. King, and W. D. Marshall, *J. Agric. Food Chem.* **26**, 475–480 (1978).
203. R. R. King, R. Greenhalgh, and W. D. Marshall, *J. Org. Chem.* **43**, 1262–1263 (1978).
204. J. S. Thornton, T. J. Olson, and K. Wagner, *J. Agric. Food Chem.* **25**, 573–576 (1977).
205. C. G. Daughton, A. M. Cook, and M. Alexander, *Anal. Chem.* **51**, 1949–1953 (1979).
206. D. Y. Takade, J. M. Reynolds, and J. H. Nelson, *J. Agric. Food Chem.* **27**, 746–753 (1979).
207. I. P. Hectepoba, *Zh. Anal. Khim.* **32**, 1790–1796 (1977).
208. W. P. Cochrane, R. Greenhalgh, and N. E. Looney, *J. Assoc. Off. Anal. Chem.* **59**, 617–621 (1976).
209. J. Hurter, M. Manser, and B. Zimmerli, *J. Agric. Food Chem.* **26**, 472–475 (1978).
210. I. H. Williams, *Residue Rev.* **38**, 1–20 (1971).
211. W. P. Cochrane and R. Purkayastha, *Toxicol. Environ. Chem. Rev.* **1**, 137–268 (1973).
212. H. W. Dorough and J. H. Thorstenson, *J. Chromatogr. Sci.* **13**, 212–224 (1975).
213. R. J. Kuhr and H. W. Dorough, *Carbamate Insecticides: Chemistry, Biochemistry and Toxicology*, CRC Press, Cleveland, Ohio (1976), Chap. 9.
214. R. C. Hall and D. E. Harris, *J. Chromatogr.* **169**, 245–259 (1979).
215. E. W. Day Jr., T. Golab, and J. R. Koons, *Anal. Chem.* **38**, 1053–1057 (1966).
216. E. R. Holden, W. M. Jones, and M. Beroza, *J. Agric. Food Chem.* **17**, 56–59 (1969).
217. I. C. Cohen and B. B. Wheals, *J. Chromatogr.* **43**, 233–240 (1969).
218. S. Sumida, M. Takaki, and J. Miyamoto, *J. Agr. Biol. Chem.* **34**, 1576–1580 (1970).
219. S. Sumida, M. Takaki, and J. Miyamoto, *Bochu-Kagaku* **35**, 72–76 (1970).
220. C. E. Mendoza and J. B. Shields, *J. Agric. Food Chem.* **22**, 255–258 (1974).
221. D. G. Crosby and J. B. Bowers, *J. Agric. Food Chem.* **16**, 839–843 (1969).
222. R. L. Tilden and C. H. Van Middelem, *J. Agric. Food Chem.* **18**, 154–157 (1970).
223. T. Aizawa, I. Kamiyama, and S. Sawafaji, *Noyaku Kagaku* **2**, 60–63 (1974).
224. D. D. Clark, S. Wilk, and S. E. Gitlow, *J. Gas Chromatogr.* **4**, 310–313 (1966).
225. D. E. Bradway and T. Shafik, *J. Chromatogr. Sci.* **15**, 322–328 (1977).
226. J. Drozd. *J. Chromatogr. (Chromatogr. Rev.)* **113**, 303–356 (1975).
227. S. Ahuja, *J. Pharm. Sci.* **65**, 163–182 (1976).
228. K. Blau and G. S. King, *Handbook of Derivatives for Chromatography*, Heyden & Sons, London (1977).
229. D. J. Edwards and K. Blau, *Anal. Biochem.* **45**, 387–391 (1972).
230. P. S. Doshi and D. J. Edwards, *J. Chromatogr.* **176**, 359–366 (1979).
231. T. Hamano, A. Hasegawa, K. Tanaka, and Y. Matsuki, *J. Chromatogr.* **179**, 346–350 (1979).
232. P. Hartvig, W. Handl, J. Vessman, and C. M. Svahn, *Anal. Chem.* **48**, 390–393 (1976).
233. L. A. Stemson and A. D. Cooper, *J. Chromatogr.* **150**, 257–258 (1978).
234. J. W. Ralls and A. Cortes, *J. Gas Chromatogr.* **2**, 132–134 (1964).

235. W. H. Gutenmann and D. J. Lisk, *J. Agric. Food Chem.* **13**, 48–50 (1965).
236. C. H. Van Middelem, T. L. Norwood, and R. E. Waites, *J. Gas Chromatogr.* **3**, 310–313 (1965).
237. L. I. Butler and L. M. McDonough, *J. Agric. Food Chem.* **16**, 403–407 (1968).
238. C. A. Bache and D. L. Lisk, *Anal. Chem.* **40**, 1241–1242 (1968).
239. S. Khalifa and R. O. Mumma, *J. Agric. Food Chem.* **20**, 632–634 (1972).
240. R. A. Chapman and C. R. Harris, *J. Chromatogr.* **171**, 249–262 (1979).
241. R. J. Argauer, *J. Agric. Food Chem.* **17**, 888–892 (1969).
242. C. W. Stanley, J. S. Thornton, and D. B. Katague, *J. Agric. Food Chem.* **20**, 1265–1268 (1972).
243. C. W. Stanley and J. S. Thornton, *J. Agric. Food Chem.* **20**, 1269–1273 (1972).
244. L. I. Butler and L. M. McDonough, *J. Assoc. Off. Anal. Chem.* **53**, 495–498 (1970).
245. L. I. Butler and L. M. McDonough, *J. Assoc. Off. Anal. Chem.* **54**, 1357–1360 (1971).
246. C. W. Miller, M. T. Shalif, and F. J. Biros, *Bull. Environ. Contam. Toxicol.* **8**, 339–344 (1972).
247. Y. Ishii and Y. Yamashita, *Bull. Agr. Chem. Insp. Sta. Jpn.* **12**, 63–66 (1972).
248. M. C. Bowman and M. Beroza, *J. Assoc. Off. Anal. Chem.* **50**, 926–933 (1967).
249. I. C. Cohen, J. Norcup, J. H. A. Ruzicka, and B. B. Wheals, *J. Chromatogr.* **44**, 251–255 (1969); **49**, 215–221 (1970).
250. E. R. Holden, *J. Assoc. Off. Anal. Chem.* **56**, 713–717 (1973).
251. E. R. Holden, *J. Assoc. Off. Anal. Chem.* **58**, 562–565 (1975).
252. B. C. Turner and J. H. Caro, *J. Environ. Qual.* **2**, 245–247 (1973).
253. J. H. Caro, D. E. Glotfelty, H. P. Freeman, and A. W. Taylor, *J. Assoc. Off. Anal. Chem.* **56**, 1319–1323 (1973).
254. R. F. Cook, J. E. Jackson, J. M. Shuttleworth, O. H. Fullmer, and G. H. Fujie, *J. Agric. Food Chem.* **25**, 1013–1017 (1977).
255. J. Miyamoto, Y. Takimoto, and J. Ohnishi, *J. Pestic. Sci.* **3**, 119–127 (1978).
256. A. H. Blagg and J. L. Rawls, *Am. Lab.* **4**, 17–21 (1972).
257. F. K. Kawahara, *Anal. Chem.* **40**, 1009–1010 (1968); **40**, 2073–2075 (1968).
258. L. G. Johnson, *J. Assoc. Off. Anal. Chem.* **56**, 1503–1505 (1973).
259. J. A. Coburn, B. D. Ripley, and A. S. Y. Chau, *J. Assoc. Off. Anal. Chem.* **59**, 188–196 (1976).
260. T. E. Archer, J. D. Stokes, and R. S. Bringhurst, *J. Agric. Food Chem.* **25**, 536–541 (1977).
261. H. A. Moye, *J. Agric. Food Chem.* **19**, 452–455 (1971).
262. C. H. Van Middelem, H. A. Moye, and M. J. Hanes, *J. Agric. Food Chem.* **19**, 459–462 (1971).
263. D. H. Hubbell, D. F. Rothwell, W. B. Wheeler, W. B. Tappan, and F. M. Rhoads, *J. Environ. Qual.* **2**, 93–96 (1973).
264. W. B. Tuppan, W. B. Wheeler, and H. W. Lundy, *J. Econ. Entomol.* **66**, 197–201 (1973).
265. R. G. Wien and F. S. Tanaka, *J. Chromatogr.* **130**, 55–63 (1977).
266. R. H. Bromilow and K. A. Lord, *J. Chromatogr.* **125**, 495–502 (1976).
267. R. H. Bromilow, *Analyst* **101**, 982–985 (1976).
268. L. Fishbein and W. L. Zielinski Jr., *J. Chromatogr.* **23**, 298–301 (1966).
269. L. Fishbein and W. L. Zielinski Jr., *J. Chromatogr.* **20**, 9–14 (1965).
270. L. J. Sullivan, J. M. Elridge, and J. B. Knaak, *J. Agric. Food Chem.* **15**, 927–931 (1967).

271. G. D. Paulson, R. Zayskie, M. V. Zehr, C. E. Portnoy, and V. J. Feil, *J. Agric. Food Chem.* **18**, 110–115 (1970).

272. E. D. Magallona and F. A. Gunther, *Arch. Environ. Contam. Toxicol.* **5**, 185–190 (1977).

273. S. C. Lau and R. L. Marxmiller, *J. Agric. Food Chem.* **18**, 413–415 (1970).

274. J. N. Seiber, *J. Agric. Food Chem.* **20**, 443–446 (1972).

275. R. J. Bose, *J. Agric. Food Chem.* **25**, 1209–1210 (1977).

276. S. Khalifa and R. O. Mumma, *J. Agric. Food Chem.* **20**, 632–634 (1972).

277. L. Wong and F. M. Fisher, *J. Agric. Food Chem.* **23**, 315–318 (1975).

278. M. Ueji and J. Kanazawa, *Bun. Kagaku.* **22**, 16–20 (1973).

279. H. G. Lobering, L. Weil, and K. E. Quentin, *Vom Wasser* **51**, 267–271 (1978).

280. R. Greenhalgh, W. D. Marshall, and R. R. King, *J. Agric. Food Chem.* **24**, 266–270 (1976).

281. R. Greenhalgh, G. W. Wood, and P. A. Pearce, *J. Environ. Sci. Health* **1312**, 229–244 (1977).

282. J. Sherma and T. M. Shafik, *Arch. Environ. Contam. Toxicol.* **3**, 55–61 (1975).

283. C. E. Johansson, *Pestic. Sci.* **9**, 313–322 (1978).

284. J. J. Ryan and J. F. Lawrence, *J. Chromatogr.* **135**, 117–122 (1977).

285. J. F. Lawrence and J. J. Ryan, *J. Chromatogr.* **130**, 97–102 (1977).

286. J. F. Lawrence, D. A. Lewis, and H. A. McLeod, *J. Chromatogr.* **138**, 143–150 (1977).

287. J. F. Lawrence and G. A. Laver, *J. Agric. Food Chem.* **23**, 1106–1109 (1975).

288. J. F. Lawrence, *J. Assoc. Off. Anal. Chem.* **59**, 1061–1065 (1976).

289. S. B. Mathur, Y. Iwata, and F. A. Gunther, *J. Agric. Food Chem.* **26**, 769–770 (1978).

290. H. A. Moye, *J. Agric. Food Chem.* **23**, 415–418 (1975).

291. J. C. Maitlen and L. M. McDonough, *J. Agric. Food Chem.* **28**, 78–82 (1980).

292. J. S. Thornton and G. Drager, *Int. J. Environ. Anal. Chem.* **2**, 229–234 (1973).

293. J. A. Durden Jr., W. J. Bartley, and J. F. Stephen, *J. Agric. Food Chem.* **18**, 454–458 (1970).

294. D. W. Woodham, R. R. Edwards, R. G. Reeves, and R. L. Schutzmann, *J. Agric. Food Chem.* **21**, 303–307 (1973).

295. J. H. Smelt, N. W. H. Houx, T. M. Lexmond, and H. M. Nollen, *Agric. Envir.* **3**, 337–347 (1977).

296. J. H. Smelt, M. Leistra, N. W. H. Houx, and A. Dekker, *Pestic. Sci.* **9**, 279–285, 286–292, and 293–300 (1978).

297. R. H. Holt and H. L. Pearse, *J. Agric. Food Chem.* **24**, 263–266 (1976).

298. R. G. Reeves and D. W. Woodham, *J. Agric. Food Chem.* **22**, 76–80 (1974).

299. U. Kiigemagi, D. Wellman, E. J. Cooley, and L. C. Terriere, *Pestic. Sci.* **4**, 89–99 (1973).

300. K. K. H. Fang, *Pestic. Sci.* **7**, 571–574 (1976).

301. R. Greenhalgh, R. R. King, and W. D. Marshall, *J. Agric. Food Chem.* **26**, 475–480 (1978).

302. M. L. Montgomery, D. L. Botsford, and V. H. Freed, *J. Agric. Food Chem.* **17**, 1241–1243 (1969).

303. J. E. Bakke and C. E. Price, *J. Agric. Food Chem.* **21**, 640–644 (1973).

304. G. T. Flint and W. A. Aue, *J. Chromatogr.* **52**, 487–490 (1970).

305. S. U. Khan, R. Greenhalgh, and W. P. Cochrane, *J. Agric. Food Chem.* **23**, 430–434 (1975).

306. W. R. Lusby and P. C. Kearney, *J. Agric. Food Chem.* **26**, 635–638 (1978).

307. P. G. Stoks and A. W. Schwartz, *J. Chromatogr.* **168**, 455–460 (1979).

308. R. Bailey, G. LeBel, and J. F. Lawrence, *J. Chromatogr.* **161**, 251–257 (1978).

309. R. Greenhalgh and J. Kovacicova, *Bull. Environ. Contam. Toxicol.* **14**, 47–48 (1975).

310. D. C. G. Muir and B. E. Baker, *J. Agric. Food Chem.* **26**, 420–424 (1978).

311. D. C. G. Muir and B. E. Baker, *Weed Res.* **18**, 111–120 (1978).

312. S. U. Khan and T. S. Foster, *J. Agric. Food Chem.* **24**, 768–771 (1976).

313. M. J. Lynch and S. K. Figdor, *J. Agric. Food Chem.* **25**, 1344–1353 (1977).

314. J. F. Lawrence, *J. Agric. Food Chem.* **22**, 936–938 (1974).

315. W. R. Betker, C. F. Smead, and R. T. Evans, *J. Assoc. Off. Anal. Chem.* **59**, 278–283 (1976).

316. P. S. Schroeder, N. R. Patel, L. W. Hedrick, W. C. Doyle, J. R. Riden, and L. V. Phillips, *J. Agric. Food Chem.* **20**, 1286–1290 (1972).

317. S. C. Lau, D. B. Katague, and D. J. Stoutamire, *J. Agric. Food Chem.* **21**, 1091–1095 (1973).

318. R. W. Reiser, *Anal. Chem.* **36**, 96–98 (1964).

319. D. Spengler and B. Hamroll, *J. Chromatogr.* **49**, 205–214 (1970).

320. C. E. McKone and R. J. Hance, *J. Chromatogr.* **36**, 234–247 (1968).

321. S. U. Khan, R. Greenhalgh, and W. P. Cochrane, *Bull. Environ. Contam. Toxicol.* **13**, 602–610 (1975).

322. S. E. Katz and R. F. Strusz, *J. Agric. Food Chem.* **17**, 1409–1412 (1969).

323. H. Buser and K. Grolimund, *J. Assoc. Off. Anal. Chem.* **57**, 1294–1299 (1974).

324. A. Buchert and H. Lokke, *J. Chromatogr.* **115**, 682–686 (1975).

325. D. G. Saunders and L. E. Vanetta, *Anal. Chem.* **46**, 1319–1321 (1974).

326. G. Clad, T. Popoff, and O. Theander, *J. Chromatogr. Sci.* **16**, 118–122 (1978).

327. B. L. Worobey and G. R. B. Webster, *J. Assoc. Off. Anal. Chem.* **60**, 213–217 (1977); 986 (1977) (correction).

328. J. F. Lawrence and K. M. S. Sundaram, *J. Assoc. Off. Anal. Chem.* **59**, 938–941 (1976).

329. A. B. DeMilo, P. H. Terry, and D. M. Rains, *J. Assoc. Off. Anal. Chem.* **61**, 629–635 (1978).

330. F. S. Tanaka and R. G. Wein, *J. Chromatogr.* **87**, 85–93 (1973).

331. A. Lok, S. D. West, and T. D. Macy, *J. Agric. Food Chem.* **26**, 410–413 (1978).

332. W. H. Gutenmann and D. J. Lisk, *J. Assoc. Off. Anal. Chem.* **47**, 353–354 (1964).

333. G. Yip, *J. Assoc. Off. Anal. Chem.* **54**, 343–345 (1971).

334. R. V. Smith and S. L. Tsai, *J. Chromatogr.* **61**, 29–34 (1971).

335. D. W. Woodham, W. G. Mitchell, C. D. Loftis, and C. W. Collier, *J. Agric. Food Chem.* **19**, 186–188 (1971).

336. A. S. Y. Chau and K. Terry, *J. Assoc. Off. Anal. Chem.* **58**, 1294–1301 (1975).

337. C. C. Alley, J. B. Brooks, and G. Choudhary, *Anal. Chem.* **48**, 387–390 (1976).

338. S. Mierzwa and S. Witek, *J. Chromatogr.* **136**, 105–111 (1977).

339. A. S. Y. Chau and K. Terry, *J. Assoc. Off. Anal. Chem.* **59**, 633–636 (1976).

340. H. Agemain and A. S. Y. Chau, *J. Assoc. Off. Anal. Chem.* **60**, 1070–1076 (1977).

341. H. Agemain and A. S. Y. Chau, *Analyst* **101**, 732–737 (1976).

342. J. DeBeer, C. V. Peteghem, and A. Heyndrickx, *J. Chromatogr.* **157**, 97–110 (1978).

343. O. Glyllenhaal, H. Brotell, and P. Hartvig, *J. Chromatogr.* **129**, 295–302 (1976).

344. M. A. Saltar, M. L. Hattula, M. Laktipera, and J. Paasivirta, *Chemosphere* **11**, 747–751 (1977).

345. E. G. Cotterill, *J. Chromatogr.* **171**, 478–481 (1979).

346. W. H. Gatenmann and D. J. Lisk, *J. Assoc. Off. Anal. Chem.* **46**, 859–862 (1963).
347. S. U. Khan, *J. Assoc. Off. Anal. Chem.* **58**, 1027–1031 (1975).
348. H. G. Henkel, *J. Chromatogr.* **22**, 446–451 (1966).
349. A. S. Hyman, *J. Chromatogr.* **45**, 132–138 (1969).
350. B. S. Middleditch and D. M. Desiderio, *Anal. Lett.* **5**, 605–609 (1972).
351. R. H. Greely, *J. Chromatogr.* **88**, 229–233 (1974).
352. A. P. Thio, M. J. Kornet, H. S. I. Tan, and D. H. Tompkins, *Anal. Lett.* **12**, 1009–1017 (1979).
353. C. E. McKone and R. J. Hance, *J. Chromatogr.* **69**, 204–206 (1972).
354. T. H. Byast and R. J. Hance, *Bull. Environ. Contam. Toxicol.* **14**, 71–76 (1975).
355. S. S. Que Hee, R. G. Sutherland, and M. Velter, *Environ. Sci. Technol.* **9**, 62–66 (1975).
356. B. Henshaw, S. S. Que Hee, R. G. Sutherland, and C. C. Lee, *J. Chromatogr.* **106**, 33–39 (1975).
357. A. E. Smith and B. J. Hayden, *J. Chromatogr.* **171**, 482–485 (1979).
358. V. W. Kadis, W. Yarish, E. S. Molberg, and A. E. Smith, *Can. J. Plant Sci.* **52**, 674–676 (1972).
359. R. D. Comes, P. A. Frank, and R. J. Demint, *Weed Sci.* **23**, 207–210 (1975).
360. R. J. Demint, J. C. Pringle Jr., A. Hattrup, V. F. Bruns, and P. A. Frank, *J. Agric. Food Chem.* **23**, 81–84 (1975).
361. E. G. Cotterill, *J. Chromatogr.* **106**, 409–411 (1975).
362. E. G. Cotterill, *Bull. Environ. Contam. Toxicol.* **19**, 471–474 (1978).
363. A. J. Pik and G. W. Hodgson, *J. Assoc. Off. Anal. Chem.* **59**, 264–268 (1976).
364. M. Gilbert and D. J. Lisk, *Bull. Environ. Contam. Toxicol.* **20**, 180–183 (1978).
365. A. J. Dutton, T. R. Roberts, and A. N. Wright, *Chemosphere* **3**, 195–200 (1976).
366. H. C. Sikka and C. P. Rice, *J. Agric. Food Chem.* **21**, 842–846 (1973).
367. C. J. Soderquist and D. G. Crosby, *Bull. Environ. Contam. Toxicol.* **8**, 363–366 (1972).
368. S. U. Khan, *J. Agric. Food Chem.* **22**, 863–867 (1974).
369. S. U. Khan, *Bull. Environ. Contam. Toxicol.* **14**, 745–749 (1975).
370. S. U. Khan and K. S. Lee, *J. Agric. Food Chem.* **24**, 684–686 (1976).
371. A. van Dijk, R. Ebberink, G. de Groot, R. A. A. Maes, J. M. C. Douze, and A. N. P. van Heyst, *J. Anal. Toxicol.* **1**, 151–154 (1977).
372. S. Vkai, K. Hirose, and S. Kawase, *J. Hyg. Chem.* **19**, 281–286 (1973).
373. R. R. King, *J. Agric. Food Chem.* **26**, 1460–1463 (1978).
374. A. Pryde, *Biochem. Soc. Trans.* **3**, 864–867 (1975).
375. R. F. Holt and H. L. Pease, *J. Agric. Food Chem.* **25**, 373–377 (1977).
376. S. D. West, *J. Agric. Food Chem.* **26**, 644–646 (1978).
377. I. L. Adler, L. D. Haines, and B. M. Jones, *J. Assoc. Off. Anal. Chem.* **61**, 636–639 (1978).
378. S. Smith and G. H. Willis, *J. Agric. Food Chem.* **26**, 1473–1474 (1978).
379. C. D. Chriswell, R. C. Chang, and J. S. Fritz, *Anal. Chem.* **47**, 1325–1329 (1975).
380. M. A. White and K. R. Parsley, *Biomed. Mass Spectrom.* **6**, 570–572 (1979).
381. S. H. Langer, P. Pantages, and I. Wender, *Chem. Ind.* 1664–1665 (1958).
382. G. C. Esposito, *Anal. Chem.* **40**, 1902–1904 (1968).
383. R. A. Larson and A. L. Rockwell, *J. Chromatogr.* **139**, 186–190 (1977).
384. G. Matsumute, R. Ishi Watari, and T. Hanya, *Weed Res.* **11**, 693–698 (1977).
385. V. A. Jolliffe, C. W. Coggins Jr., and W. W. Jones, *J. Chromatogr.* **179**, 333–336 (1979).

386. L. Renberg, *Anal. Chem.* **46**, 459–461 (1974).
387. T. R. Edgerton and R. F. Moseman, *J. Agric. Food Chem.* **26**, 425–428 (1978).
388. M. Crammer and J. Freal, *Life Sci.* **9**, 121–128 (1970).
389. R. S. H. Yang, F. Coulston, and L. Goldberg, *J. Assoc. Off. Anal. Chem.* **58**, 1197–1201 (1975).
390. R. H. Pierce Jr., C. R. Bent, H. P. Williams, and S. G. Reeves, *Bull. Environ. Contam. Toxicol.* **18**, 251–258 (1977).
391. H. J. Hoben, S. A. Ching, L. J. Casarett, and R. A. Young, *Bull. Environ. Contam. Toxicol.* **15**, 78–85 (1976).
392. T. R. Edgerton and R. F. Moseman, *J. Agric. Food Chem.* **26**, 425–428 (1978).
393. J. F. Lawrence, *Int. J. Environ. Anal. Chem.* **5**, 95–101 (1978).
394. W. H. Dekker and H. A. Selling, *J. Agric. Food Chem.* **23**, 1013 (1975).
395. T. M. Shafik, H. C. Sullivan, and H. R. Enos, *J. Agric. Food Chem.* **21**, 295–298 (1973).
396. Linda F. Faas and J. C. Moore, *J. Agric. Food Chem.* **27**, 554–557 (1979).
397. A. C. Salomonsson, O. Theander, and P. Aman, *J. Agric. Food Chem.* **26**, 830–835 (1978).
398. K. W. Kirby, J. E. Keiser, J. Greene, and E. F. Slack, *J. Agric. Food Chem.* **27**, 757–759 (1979).
399. A. S. Y. Chau and J. A. Coburn, *J. Assoc. Off. Anal. Chem.* **57**, 389–393 (1974).
400. J. A. Coburn and A. S. Y. Chau, *J. Assoc. Off. Anal. Chem.* **59**, 862–865 (1976).
401. D. R. Erney, *J. Assoc. Off. Anal. Chem.* **61**, 214–216 (1978).
402. T. Ohe, *Bull. Environ. Contam. Toxicol.* **22**, 287–292 (1979).
403. W. Krijgsman and C. G. Van de Kamp, *J. Chromatogr.* **131**, 412–416 (1977).
404. R. T. Couts, E. E. Hargesheimer, and F. M. Pasutto, *J. Chromatogr.* **179**, 291–299 (1979).
405. D. S. Farrington and J. W. Manday, *Analyst* **101**, 639–643 (1976).
406. G. J. Kallos, W. R. Albe, and R. A. Soloman, *Anal. Chem.* **49**, 1817–1820 (1977).
407. N. Nose, S. Kobayashi, A. Hirose, and A. Watanabe, *J. Chromatogr.* **125**, 439–443 (1976).
408. B. Bruce, *Anal. Chem.* **49**, 832–834 (1977).
409. J. M. Rosenfeld and J. L. Grocco, *Anal. Chem.* **50**, 701–704 (1978).
410. O. Gyllenhaal, *J. Chromatogr.* **153**, 517–520 (1978).
411. M. Akerblom, *Fourth International Congress of Pesticide Chemistry (IUPAC)*, Abstract Volume, Zurich 1978, paper No. VI-702.
412. L. L. Lamparski and T. J. Nestrick, *J. Chromatogr.* **156**, 143–151 (1978).
413. Z. Pecka, P. Baulu, and H. Newsombe, *Pestic. Monit. J.* **8**, 232–234 (1975).
414. W. R. Bontoyan, J. B. Looker, T. E. Kaiser, P. Giang, and B. M. Olive, *J. Assoc. Off. Anal. Chem.* **55**, 923–925 (1972).
415. *Ethylenethiourea*, IUPAC Special Report, *Pure Appl. Chem.* **49**, 675–689 (1977).
416. S. Otto, W. Keller, and N. Drescher, *J. Environ. Sci. Health* **B12**, 179–191 (1977).
417. T. Hirvi, H. Pyysalo, and K. Savolainen, *J. Agric. Food Chem.* **27**, 194–195 (1979).
418. J. H. Onley and G. Yip, *J. Assoc. Off. Anal. Chem.* **54**, 165–169 (1971).
419. J. H. Onley, *J. Assoc. Off. Anal. Chem.* **60**, 1111–1115 (1977).
420. R. R. Watts, R. W. Storherr, and J. H. Onley, *Bull. Environ. Contam. Toxicol.* **12**, 224–226 (1974).
421. J. H. Onley, L. Guiffrida, N. F. Ives, R. R. Watts, and R. W. Storherr, *J. Assoc. Off. Anal. Chem.* **60**, 1105–1110 (1977).

422. L. D. Haines and J. L. Adler, *J. Assoc. Off. Anal. Chem.* **56**, 333–337 (1973).
423. W. H. Newsome, *J. Agric. Food Chem.* **20**, 967–969 (1972).
424. W. H. Newsome, *Bull. Environ. Contam. Toxicol.* **11**, 174–176 (1974).
425. R. R. King, *J. Agric. Food Chem.* **25**, 73–75 (1977).
426. B. D. Ripley, D. F. Cox, J. Weibe, and R. Frank, *J. Agric. Food Chem.* **26**, 134–136 (1978).
427. B. D. Ripley and D. F. Cox, *J. Agric. Food Chem.* **26**, 1137–1143 (1978).
428. R. G. Nash, *J. Assoc. Off. Anal. Chem.* **57**, 1015–1021 (1974).
429. R. G. Nash, *J. Assoc. Off. Anal. Chem.* **58**, 566–571 (1975).
430. W. H. Newsome, *J. Agric. Food Chem.* **26**, 1325–1327 (1978).
431. W. H. Newsome and L. G. Panapio, *J. Agric. Food Chem.* **26**, 638–640 (1978).
432. J. Singh, W. P. Cochrane, and J. Scott, *Bull. Environ. Contam. Toxicol.* **23**, 470–474 (1979).
433. G. H. Tjan and J. T. A. Jansen, *J. Assoc. Off. Anal. Chem.* **62**, 769–773 (1979).
434. J. P. Rouchaud and J. R. Decallonne, *J. Agric. Food Chem.* **22**, 259–260 (1974).
435. N. Nose, S. Kobayashi, A. Tanaka, A. Hirose, and A. Watanabe, *J. Chromatogr.* **130**, 410–413 (1977).
436. H. Pyysale, *J. Agric. Food Chem.* **25**, 995–997 (1977).
437. H. R. Sisken and J. E. Newell, *J. Agric. Food Chem.* **19**, 738–741 (1971).
438. F. Tafuri, M. Patumi, M. Businelli, and C. Marucchini, *J. Agric. Food Chem.* **26**, 1344–1346 (1978).
439. J. H. Onley, *J. Assoc. Off. Anal. Chem.* **60**, 679–681 (1977).
440. R. S. T. Loeffler, *Pestic. Sci.* **9**, 310–312 (1978).
441. Y. Kawamura, M. Takeda, and M. Uchiyama, *J. Pestic. Sci.* **3**, 397–400 (1978).

Chapter 2

Cyclic Boronates as Derivatives for the Gas Chromatography–Mass Spectrometry of Bifunctional Organic Compounds

Walter C. Kossa

1. INTRODUCTION

Many organic compounds, particularly those of biological origin such as, for example, the amino acids, the carbohydrates, the corticosteroids, and the prostaglandins, possess two or more polar functional groups. Since many of these compounds lack the required volatility and thermal stability, they are usually analyzed by gas chromatography only after conversion to less polar, more volatile, and more thermally stable derivatives. Suitable derivatives can be produced in one of essentially three ways: by chemical or thermal degradation to structurally simpler, less polar species; by reaction of the individual polar functional groups with *monofunctional* derivatization reagents to yield less polar structures; and, in the special case where two polar functional groups are in close proximity, by simultaneous reaction of the two groups with a *bifunctional* reagent to form a volatile cyclic derivative.

Walter C. Kossa • Applied Science Division, Laboratory Group, Milton Roy Company, P.O. Box 440, State College, Pennsylvania 16801.

TABLE 1. Some Cyclic Methane- and *n*-Butaneboronates Used in Gas Chromatography–Mass Spectrometry

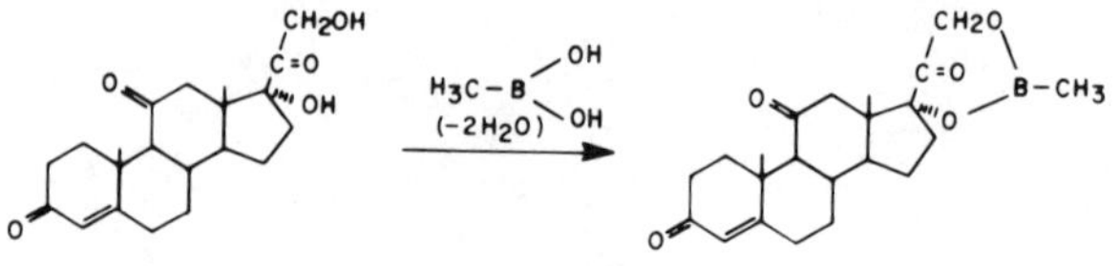

cortisone

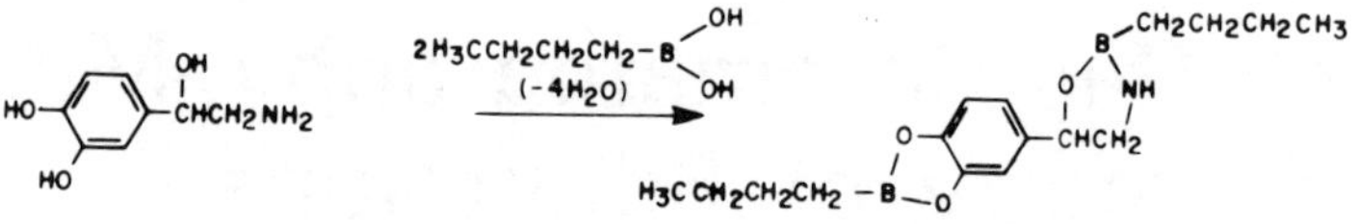

norepinephrine
(a catecholamine)

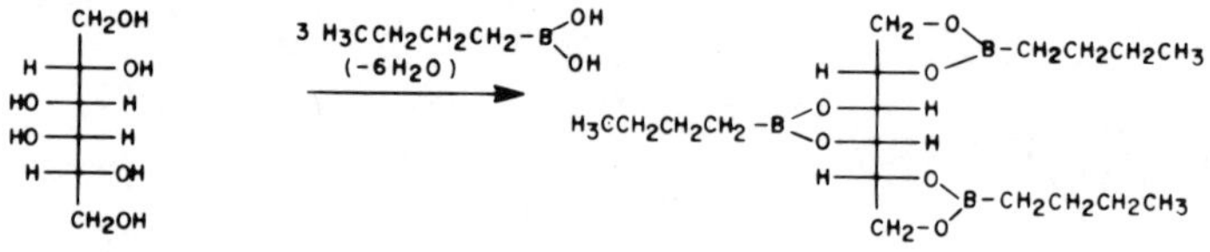

D-galactitol
(D-dulcitol)

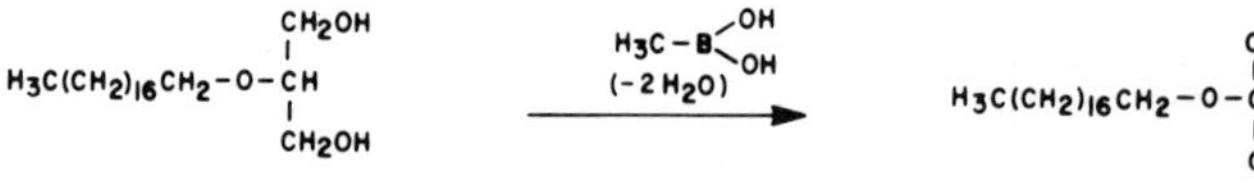

2-octadecylglycerol

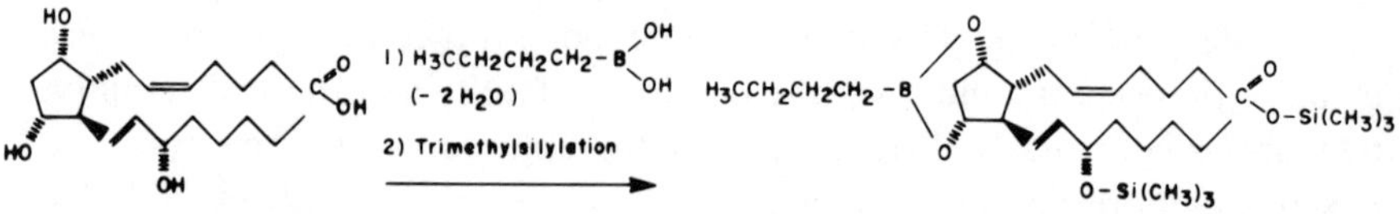

prostaglandin F$_{2\alpha}$

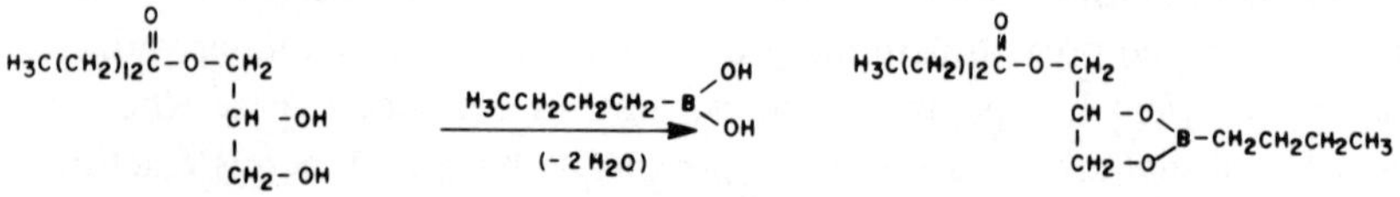

glyceryl-1-myristate

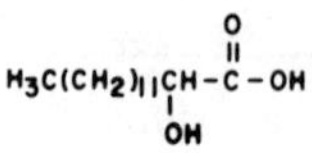

α-hydroxymyristic acid

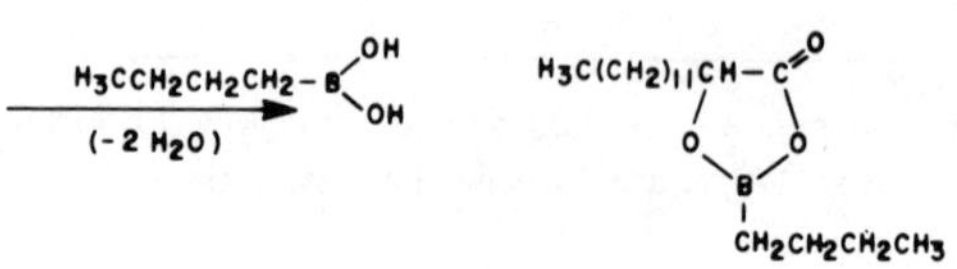

TABLE 1 (continued)

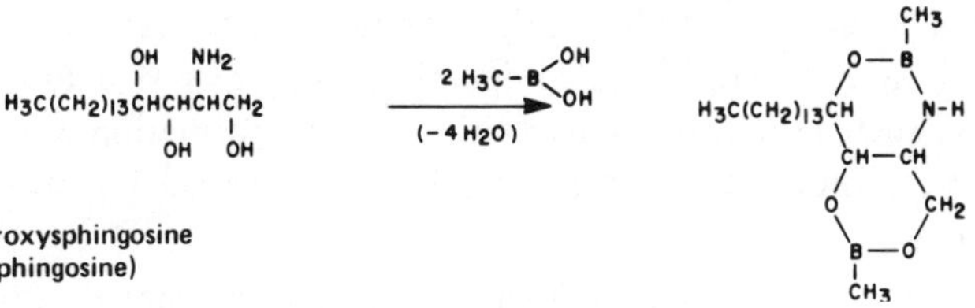

4D-hydroxysphingosine
(phytosphingosine)

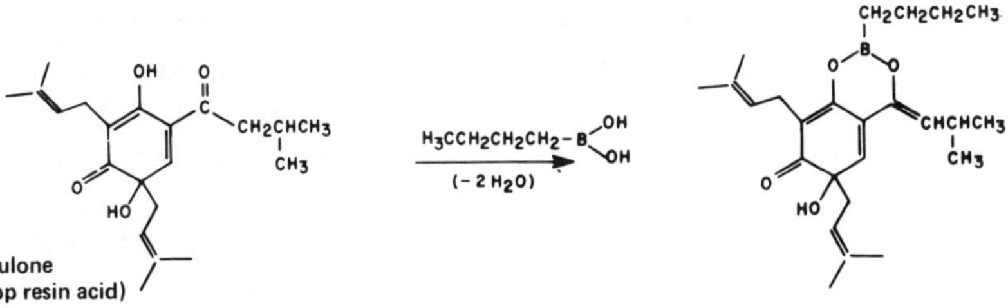

humulone
(a hop resin acid)

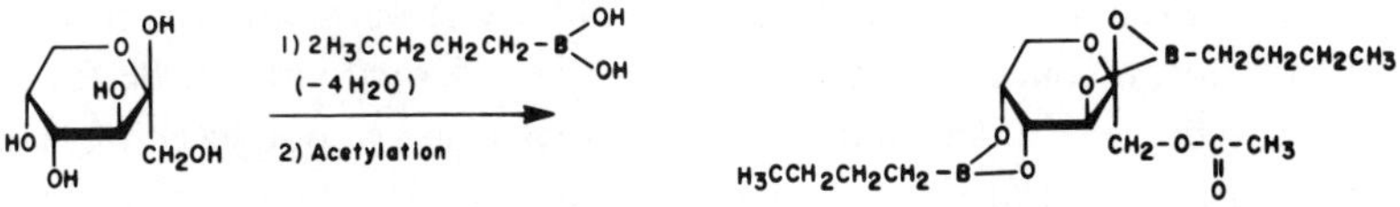

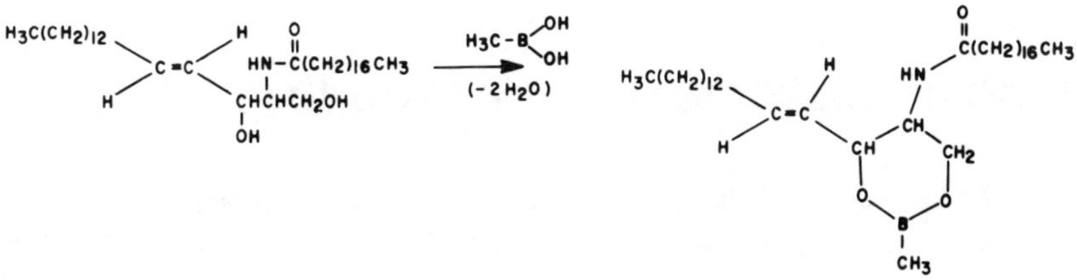

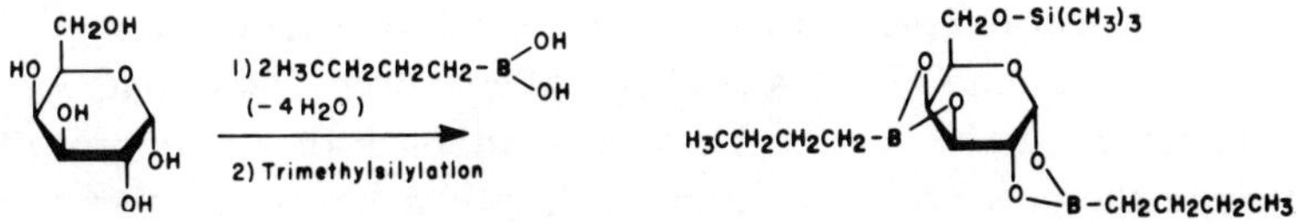

In the last instance, the derivatives that are produced often have special properties. The formation of a five- or six-membered ring usually restricts the number of conformations a molecule can assume, and, consequently, confers upon it distinctive chemical, chromatographic, and mass spectrometric properties. Indeed, with a compound of unknown structure, the mere formation of such a cyclic derivative yields useful information, since obvious geometrical requirements must be met in order for such a species to be produced at all.

A number of useful bifunctional derivatization reagents exist. Most, however, are of value only for the derivatization of specific pairs of functional groups. Examples include alkyl isothiocyanates, which react with α-amino acids to form alkylthiohydantoins[1-4]; diacetoxydimethylsilane, which reacts whith 1,2- and 1,3-diols to give cyclic siliconides[5-7]; various aldehydes and ketones, which under certain conditions, react with cis-1,2-diols to yield dioxolanes,[8-11] and with β-hydroxyamines to afford oxazolidines.[12-14] However, when applicability to a wide range of functional groups is required, undoubtedly the most versatile of such reagents are the boronic acids.

Alkane- and areneboronic acids, $RB(OH)_2$, react with a variety of 1,2- and 1,3-diols, diamines, and aminols to yield cyclic boronates, five- or six-membered boron-containing heterocycles. These derivatives can be formed rapidly in high yield under mild conditions, they display good gas chromatographic properties, and they yield useful mass spectra. Some examples of cyclic boronates that have been used in gas chromatography–mass spectrometry are shown in Table 1.

2. SOME HISTORICAL BACKGROUND

The first report of the synthesis of cyclic boronates appeared in 1954.[15] Therein, Kuivila and co-workers described the cyclic esters formed by the treatment of mannitol and other cis-diols with areneboronic acids such as benzeneboronic acid and p-tolueneboronic acid. Later work showed that boronic acids can react with a number of 1,2- and 1,3-diols,[16-18] diamines,[19-22] and aminols[21,22] to yield cyclic derivatives. These compounds proved, in general, to be fairly stable species, although they were prone to hydrolysis in the presence of moisture, and, in common with other compounds possessing carbon–boron bonds, were susceptible to autoxidation. Of particular interest was the observation that many cyclic boronates were sufficiently volatile to permit their purification by distillation or vacuum sublimation.

An early example that showed that the volatility of cyclic boronates could be of analytical utility was provided by Brown and Zweifel.[23] They treated mixtures of pairs of *cis*- and *trans*-cycloalkanediols, for example, *cis*- and *trans*-cyclohexane-1,3-diol, with *n*-butaneboronic acid, and then fractionally distilled the resulting mixture of *n*-butaneboronates. In most instances, the distillate consisted exclusively of the cyclic boronate of the *cis*-diol. The boronate of the *trans*-diol remained undistilled as a nonvolatile, presumably polymeric, species.

3. THE SCOPE OF APPLICATION OF CYCLIC BORONATES IN GAS CHROMATOGRAPHY

Brooks and co-workers[24-27] were the first to convincingly demonstrate that many classes of bifunctional compounds—1,2- and 1,3-diols, enolizable 1,2- and 1,3-ketols, α- and β-hydroxy acids, enolizable α-oxoacids, *o*-phenolic acids, β- and γ-hydroxyamines—could be readily converted to cyclic boronates and analyzed by gas chromatography. Coincidentally, and independently of Brooks, Kresze and Schäuffelhut[28] found that benzeneboronates could be applied to the gas chromatographic analysis of 1-monoglycerides. This early work and subsequent studies showed that cyclic boronates were useful derivatives not only for the gas chromatography of "simple" 1,2- and 1,3-diols and aminols, but also that they could be used in the analysis of structurally more complex compounds which possessed the required proximate functional groups. Among such compounds are cannabinoids[29,30]; carbohydrates[31-51]; catecholamines[26,27,52-54] and their metabolites[55-59]; ceramides[60-62]; corticosteroids[24-27,60,63-69]; the 1,2-diol metabolites of certain tricyclic drugs[70-72]; the 1,2-diols formed by osmium tetroxide oxidation of monoolefin mixtures[73]; the diol and aminol degradation products of heptaene antibiotics[74]; the diol metabolites of polychlorinated biphenyls[75]; duvatrienediols from tobacco leaf lipids[76]; ethyleneglycol, the simplest 1,2-diol[77]; hop resin acids[78-79]; α- and β-hydroxy fatty acids[26,27,68,80,81]; hydroxysalicylanilides[82]; menthoglycols[83]; monoglycerides[24,28,60,62,68]; monoglyceryl ethers[60,68,80,81]; polyhydroxylated fatty acids[27,68,84,85]; prostaglandins and prostanoids[85-101]; short-chain lipid diols[102]; sphingosines[60,68,103,104]; and styreneglycol obtained from the incubation of styrene and microsomal styrene monooxygenase and styrene epoxide hydratase.[105]

4. GAS CHROMATOGRAPHIC PROPERTIES OF CYCLIC BORONATES

The most popular cyclic boronates used in gas chromatography–mass spectrometry, since they proved to be the most volatile, have been those of methane- and n-butaneboronic acid. Both acids are commercially available; methaneboronic acid is, however, somewhat more expensive.

The cyclic n-butaneboronates were the first to be studied[24–26] because of the ready commercial availability of n-butaneboronic acid at the time. Later, gas chromatographic examination of the cyclic methaneboronates showed them to display, as expected, shorter retention times, on supports coated with liquid phases of low-to-moderate polarity like OV-1 and OV-17, than their n-butane homologs.[27,52,63–65] The methaneboronates also afforded simpler mass spectra.[27,52,63–65] On the other hand, they proved to be somewhat less stable to hydrolysis.[27,64] While both the methane- and n-butaneboronates appear to be of equal utility in the gas chromatography of bifunctional compounds of relatively low molecular weight, for species of high molecular weight like ceramides,[61,62] polyhydroxylated steroids[66]

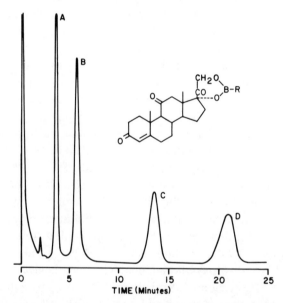

Fig. 1. Compounds: (A) cyclic 1,1-dimethylethane-; (B) cyclic n-butane-; (C) cyclic cyclohexane-; and (D) cyclic benzeneboronate of cortisone. Column: 6 ft × 4 mm i.d. glass packed with 1% OV-17 on 100/120 mesh GAS-CHROM Q. Column temperature: 250°C. Copyright: Elsevier Scientific Publishing Company (*Journal of Chromatography*).

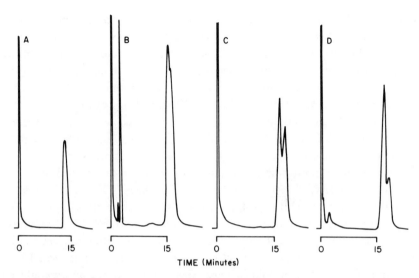

Fig. 2. Attempted separation of the diastereomers, ephedrine, and ψ-ephedrine, as (*A*) O-trimethylsilyl ethers; (*B*) cyclic *n*-butaneboronates; (*C*) cyclic cyclohexane-boronates; and (*D*) cyclic benzeneboronates. Column: 6 ft × 1/8 in. i.d. glass packed with 1% OV-17 on 100/120 mesh GAS-CHROM Q. Column temperatures: (*A*) 110°C; (*B*) 120°C; (*C*) 150°C; and (*D*) 160°C. Copyright: Preston Publications, Incorporated (*Journal of Chromatographic Science*).

and sphingosines,[103,104] the greater volatility and, in mass spectrometric applications, the simpler mass spectra of the methaneboronates make these derivatives of greater value.

Reports of the use of other boronic acids have also appeared. Benzene-boronic acid was employed in much of the early synthetic work on cyclic boronates,[15-22] and has also enjoyed some utility in the synthetic and ana-lytical chemistry of carbohydrates.[106,107] Its derivatives[24-29,34-36,38,40-42, 51-54,63-65,73,82,95,96,102-104] and those of cyclohexaneboronic acid[26,27,63-65, 95,96] display longer retention times than the cyclic boronates of the acids of lower molecular weight, as illustrated in Figure 1.[64] Consequently, they have not been as widely used as the methane- and *n*-butane boronates. They have, however, been shown to be of value in the separation of certain di-astereomers, such as ephedrine and Ψ-ephedrine (Figure 2).[26,53] Both benzene- and cyclohexaneboronic acid are commercially available.

Cyclic derivatives of 1,1-dimethylethaneboronic acid ("*tert*-butylbo-ronic acid")have been prepared, and were originally studied in the hope that the steric properties of the bulky *tert*-butyl group would make them hydrolytically more stable than the corresponding methane- or *n*-butanebo-ronates. Somewhat surprisingly, no such stabilizing effect was observed.[64]

Indeed, the boronic acid itself proved to be sensitive to air, and was best manipulated under nitrogen or in ethyl acetate solution. However, because of the "spherical" nature of the *tert*-butyl group, the retention times of the cyclic 1,1-dimethylethaneboronates proved to be similar to those of the corresponding methaneboronates, although the former were of higher molecular weight. Under the conditions of Figure 1, for example, the methane- and 1,1-dimethylethaneboronates of cortisone could not be separated.

Wiecko and Sherman,[51] in the report of a study of the cyclic boronates of the diastereomeric inositols, have described what appears to be the first use of *n*-octaneboronic acid in gas chromatography–mass spectrometry.

A comparatively recent development has been the use of halogen-containing cyclic boronates for the trace analysis of bifunctional compounds by gas chromatography with electron capture detection.[108–111] Heretofore, cyclic boronates were generally analyzed with chromatographs equipped with flame ionization detectors, although it had been demonstrated[35,67] that an approximately 50-fold increase in sensitivity to boronates could be obtained through the use of alkali flame ionization detection. An alternative method, flame photometric detection,[112] also showed increased sensitivity to boron, compared to flame ionization detection, but has been applied only to the analysis of boron hydrides.

Brooks and Watson[25] were the first to allude to the desirability of halogenated boronic acids as derivatization reagents for use in conjunction with electron capture detection. The boronic acid that they mentioned, trifluoromethaneboronic acid, proved, however, to yield derivatives that were extremely unstable. Another logical candidate, pentafluorobenzeneboronic acid, also gave cyclic boronates that displayed extreme sensitivity to attack by water and other nucleophiles.[67,108] Poole and co-workers[109–111] finally succeeded in synthesizing a series of boronic acids which afforded stable derivatives that were some 4000-fold more sensitive to electron capture detection than to flame ionization detection. The first studied were a number of substituted benzeneboronic acids.[109,110] In this instance, the introduction of substituents into the benzene ring had two consequences. First, various steric and electronic factors were such that many bifunctional compounds still yielded unstable derivatives, even in instances where stable cyclic benzeneboronates were produced. Those cyclic boronates that were stable, however, displayed good gas chromatographic properties. Second, increases in the molecular weight of the derivatives, compared to the benzeneboronates, led, not unexpectedly, to decreases in volatility. Thus, for the cyclic boronates of the model compound pinacol (2,3-dimethylbutane-2,3-diol), the observed order of elution on a moderately polar OV-17 column was:

TABLE 2. Electron Capture Detector Response toward Pinacol Boronates[a]

Boronic acid	Least detectable amount, pg
Benzene-	150
4-Bromobenzene-	3
2,4-Dichlorobenzene-	2
3,5-Dichlorobenzene-	9
4-Iodobutane-	16
1-Napthalene-	2550
3-Nitrobenzene-	4

[a] The detector temperature has been optimized for maximum sensitivity.[108,113]

benzene-, 4-bromobenzene-, 2,4-dichlorobenzene-, 3,5-dichlorobenzene-, 3-nitrobenzene, 1-napthalene-. All the derivatives showed response to electron capture detection; the 2,4-dichlorobenzeneboronate was detectable in the least amount (Table 2).[110,111]

In situations where the relatively low volatility of the areneboronates may limit their applicability, the use of the more volatile cyclic 4-iodobutaneboronates has been advocated.[111] 4-Iodobutaneboronic acid yielded chemically and thermally stable derivatives with many bifunctional compounds, including some that failed to yield stable boronates when treated with the substituted benzeneboronic acids. The cyclic 4-iodobutaneboronates were approximately threefold more volatile than the corresponding 2,4-dichlorobenzeneboronates, the most useful of the areneboronates. They also gave relatively high electron capture responses, although not as high as those of the 2,4-dichlorobenzeneboronates (Table 2).

Incidentally, some of the areneboronic acids also appear to be of utility in other chromatographic methods. 1-Naphthaleneboronic acid and 9-phenanthreneboronic acid, for example, have been evaluated as reagents for the preparation of derivatives for use in high-pressure and high-performance thin-layer chromatography with selective fluorescence detection.[114]

5. MASS SPECTROMETRIC PROPERTIES OF CYCLIC BORONATES

Approximately four-fifths of the references in the bibliography deal, at last in part, with the mass spectrometric properties of cyclic boronates.

Most data have been obtained by electron impact mass spectrometry. Recently, however, chemical ionization mass spectrometry[60,68,69,79,99,100, 103,104] has received increased attention. The data obtained by both methods are complementary.[60,68] Extensive compilations of mass spectrometric data on the cyclic boronates of cannabinolic acids,[29] catecholamines and related β-hydroxyamines,[52,53] ceramides,[61] corticosteroids,[64–66] the diastereomeric inositols,[51] spingosines,[103,104] polyhydroxyalkylpyrazines (reaction products of sugars and ammonia),[50] sugar acetates,[46] and sugar phosphates[46] have appeared. In addition, Ferrier[107] has reviewed the use of cyclic boronates in the elucidation of carbohydrate structures, and has included not only gas chromatography–mass spectrometric data, but also those obtained by direct insertion mass spectrometry.

The salient features of the mass spectra of cyclic boronates are three. First, the spectra are, in general, relatively simple. Few peaks appear in the high-mass range. In the low-mass range, however, an abundance of ionic fragments often appears. The general mode of fragmentation seems to be little affected by the nature of the alkyl side chain of the boronate moiety. Methane-and benzeneboronates yield simpler mass spectra than the corresponding n-butane and cyclohexaneboronates, which afford spectra that are complicated by fragments due to the stepwise degradation of the alkyl side chain of the boronate moiety. The chemical ionization spectra are even simpler, since the ions generated are of relatively lower energy and, consequently, do not fragment as extensively. The chemical ionization spectra appear to be of particular value in the identification of diastereomers, the electron impact mass spectra of which are similar.[60,68] Second, in contrast to other species, such as, for example, the trimethylsilyl derivatives, cyclic boronates usually yield, upon electron impact, detectable molecular (M) ions, often in high abundance. This behavior is complemented by that exhibited in the chemical ionization mass spectra, where abundant (often the base peak in the spectrum) quasimolecular ($M + 1$) ions appear. Consequently, the molecular weight of the species, and, incidentally, the number of cyclic boronate groups that have been incorporated into the molecule are readily discernible. Third, the identification of the molecular ion or quasimolecular ion and other ionic fragments containing boron is facilitated by the characteristic ratio of abundance of the boron isotopes, $^{10}B : {}^{11}B = 1 : 4.2$.

The mass spectrometric properties of cyclic boronates of 1,2- and 1,3-diols, of which the largest amount of data exists, are, in some ways, reminiscent of another type of diol derivative, the isopropylidene ketal.[9] With both types of derivative, for example, a major fragmentation pathway is carbon–carbon bond cleavage alpha to the heterocyclic ring, with concomitant expulsion of a heterocyclic ion. In the case of the cyclic boronates, these fragments retain the original ring structure, and consequently, indicate wheteher a five- or six-membered cyclic boronate was formed during the original derivatization. Such fragments are of great value in structural analyses, especially of carbohydrates, as has been emphasized by Ferrier.[107]

Another fragmentation pathway common to both isopropylidene ketals and cyclic boronates is loss of the alkyl side chain of the derivative moiety. In the case of an isopropylidene derivative, the loss of a methyl radical produces an ion which is readily stabilized by change delocalization within a 1,3-dioxolane structure. The cyclic boronates, however, show somewhat different behavior. Since boron is an electron-deficient atom, it cannot ordinarily participate in charge delocalization, and, as a result, carbon-boron bond cleavage does not occur. Instead, species like RBO and RHBO are expelled, and the resulting ion is stabilized by charge delocalization within an oxirane structure. In situations where carbon–boron bond cleavage is observed, it is usually indicative of interaction of the boron atom and an electron-rich functionality elsewhere in the molecular ion. Such behavior has been observed with cyclic boronates of sugar acetates,[46] sugar phosphates,[45] and sphingosine N-dimethylaminomethylene derivatives.[103,104] In these instances, the intervening electron-rich functionalities are, respectively, the acetate, phosphate, and N-dimethylaminomethylene groups. Since such phenomena are observed only when certain geometrical conditions exist, they are of some value as structural probes, as has been noted for the sugar boronates.[45,46]

The abundance of relatively intense ions, including the molecular ion, in the high-mass range of the electron impact mass spectra of the cyclic boronates has led to the evaluation of these derivatives for use in mass fragmentography. In this technique, the subject of several informative reviews,[115–119] the intensities of one or more preselected ions in the mass spectrum of the compound of interest are recorded as a function of time, yielding a "mass fragmentogram." This is in contrast to conventional mass spectrometry where the total ionic spectrum is recorded. The technique has the virtues of specificity and sensitivity. It is specific since only when compounds whose spectra contain the preselected ions leave the chromatograph is a mass fragmentogram recorded. Sensitivity is also very good and, under ideal conditions, approaches that of electron capture detection. The ions that are chosen for monitoring are usually those characteristic of, preferably unique to, the compound of interest. In order to obtain maximum sensitivity, it is also important that the preselected ions carry a significant fraction of the total ion current. In many instances, the molecular ion and other high-mass ionic fragments in the electron impact mass spectra of cyclic boronates meet these two requirements.

Qualitatively, mass fragmentography of cyclic boronates has been applied to the detection of catecholamine metabolites[56,58] and the 1,2-diol metabolites of certain tricyclic drugs[70-72] in biological media. A report[66] of a preliminary evaluation of cyclic methaneboronates and other derivatives for the mass fragmentography of corticosteroids has also appeared.

The method can also be used for quantitative analysis. In this instance, internal standards are usually employed. Deuterated analogs of the compound of interest are of particular value in this regard since they show virtually identical chemical and chromatographic behavior. They also display similar mass spectra, although the deuterium-containing ions, of course, appear at slightly higher m/e values. Simultaneous monitoring of the same ionic fragments in the spectra of the compound of interest and its deuterated analog yields pairs of mass fragmentograms, the areas of which are directly proportional to the concentration of these species in the sample. Since the quantity of deuterated internal standard that has been added to the sample is known, the concentration of the compound of interest in the sample is thus readily obtainable.

Quantitatively, mass fragmentography of cyclic boronates has been applied to the analysis of F prostaglandins in samples such as blood plasma and human aortal tissue,[88-90,95] and to the determination of glucose production rates in human children.[48,49] The internal standards that were employed were, respectively, various tetradeuterated F prostaglandins and 6,6-dideuteroglucose.

6. PREPARATION OF CYCLIC BORONATES FOR USE IN GAS CHROMATOGRAPHY–MASS SPECTROMETRY

An impressive feature of the cyclic boronates is the ease with which they can be formed, usually in high yield. The various boronic acids show approximately equal reactivity towards a particular bifunctional compound. Different bifunctional compounds, of course, display different reactivities.[26,27]

In most cases, derivative formation is achieved simply by dissolving equimolar amounts of the bifunctional compound and the boronic acid at room temperature in a suitable organic solvent. Reaction is usually complete within 15 min, although in many instances, derivatization occurs within seconds of dissolution of the reactants. With some compounds, longer reaction times[26,27,52,53] or higher reaction temperatures are necessary, if only to bring about dissolution, or in the case of many sugars,[32,35,36,44–46, 48–51] to effect equilibration of anomers. The failure of an important class of bifunctional compounds, the α-amino acids, to yield cyclic boronates appears due, at least in part, to the lack of a solvent which is able to dissolve the reactants.[27]

Solvents that are suitable for the preparation of cyclic boronates include acetone, chloroform, N,N-dimethylformamide, ethyl acetate, pyridine, and tetrahydrofuran. Of these, pyridine is the most popular. For obvious reasons, alcohols and primary and secondary amines cannot be used as solvents. Acetone is not recommended for use with amino-substituted compounds because of the possibility of Schiff base formation, or, in the case of secondary β-hydroxyamines, oxazolidine formation.[26] The use of catalysts is not necessary, although it has been suggested that pyridine shows catalytic activity in the formation of cyclic boronates of certain ketols.[78]

In some instances, the separate derivative preparation and analysis steps can be combined. Italian workers,[56–58,70–72,105] for example, have demonstrated that cyclic boronates can be formed *within* the gas chromatograph by simultaneous injection of a solution of the boronic acid and a solution of the bifunctional compound.

The derivatization of bifunctional compounds by boronic acids is an equilibrium process in which the equilibrium lies favorably in the direction of cyclic boronate formation. Water is the sole by-product. As mentioned earlier, some cyclic boronates are susceptible to hydrolysis, and, for that reason, these derivatives are not suitable for general use in conventional thin-layer chromatography.[25,114] For purposes of gas chromatography, however the presence of small amounts of water is usually not of consequence, since the cyclic boronates are analyzed shortly after their formation. Work-up of the reaction mixture is usually not necessary, and analyses are performed simply by injecting an aliquot of the reaction mixture into the chromatograph. In situations where hydrolysis in solution is a serious problem, or when it is desired to maintain the integrity of a silane-treated support, the use of water "scavengers" has been recommended. 2,2-Dimethoxypropane (acetone dimethyl acetal) has been used for this purpose,[55,58,59,83, 86,87,91,93,94] usually as the solvent for the derivatization reaction. It reacts with water to form fairly innocuous acetone and methanol. Removal of water in other ways, by azeotropic distillation, for example, is generally not practical owing to the extremely small scale on which the derivatives are prepared.

The boronic acids, in common with many compounds possessing

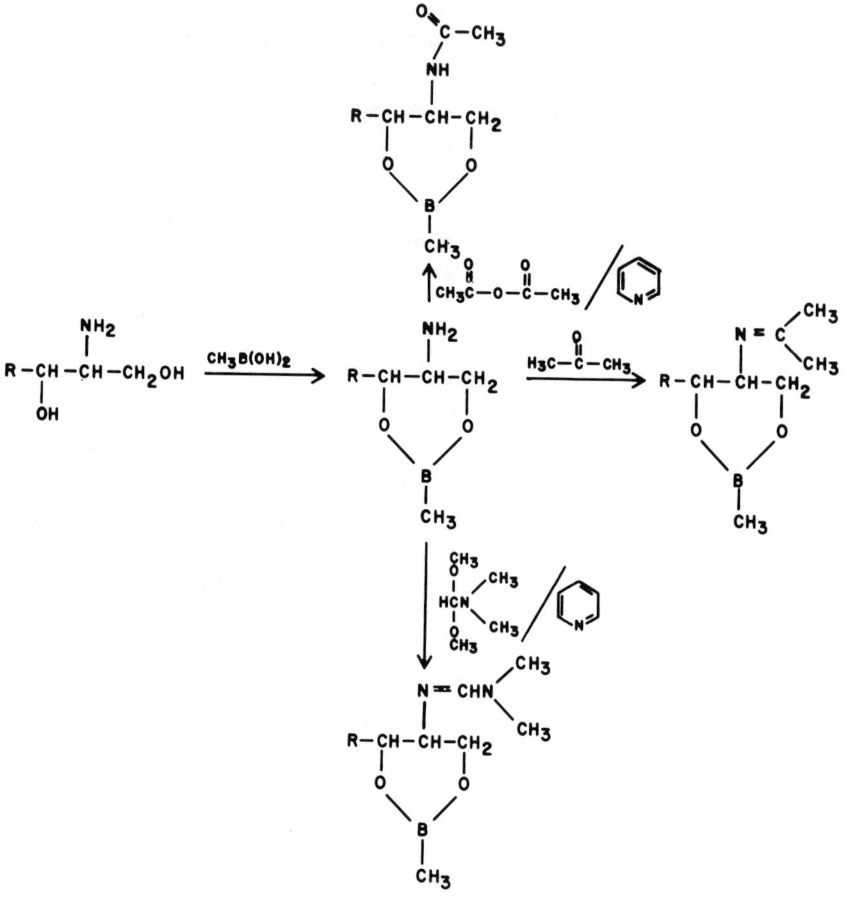

Fig. 3. Alternate methods of derivatization of the cyclic sphingosine methaneboronates.

carbon–boron bonds, are susceptible to autoxidation. Consequently, they are often supplied commercially in the form of aqueous slurries in which the water serves as a "blanket" against atmospheric oxygen. Most of the water can be removed, when necessary, by storage of the slurry on a piece of filter paper or a porous glass plate.[44] The small amount of water that remains in the damp acid does not interfere with subsequent derivatization.

Many cyclic boronates are sufficiently stable to permit the derivatization of isolated amino and hydroxy groups that may be present after boronate formation, thus yielding derivatives with improved gas chromatographic properties. Such boronates can be acylated,[25,26,46,48,49,64,68,103,104] oximated,[26,27,86,94] or trimethylsilylated[24–26,29,32,41,44,50,51,85,90,92,93–98,100, 101] without disturbing the boronate ring. Sphingosine boronates, for example, can be converted either to N-acetyl derivatives, or acetone Schiff bases, or N-dimethylaminomethylene derivatives (Figure 3).[60,68,103,104]

When bifunctional compounds also possess isolated amino or hydroxy groups, it is possible to obtain products other than cyclic boronates. These products are acyclic boronates, and usually arise when the amount of boronic acid used is in excess of that required for derivatization of the proximate functional groups. The presence of such species is made evident with the appearance of broad tailing peaks of dramatically diminished peak height in chromatograms of cyclic boronates (Figure 4).[64] The formation of acyclic boronates has been observed with sphingosines[103] and some corticosteroids.[25,64] They may also be responsible for the poor gas chromatographic behavior of certain carbohydrate cyclic boronates.[31,33]

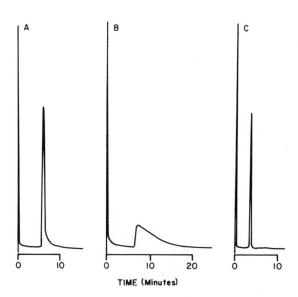

Fig. 4. (A) Chromatogram produced upon injection of an ethyl acetate solution of 5β-pregnan-3α,17α,20β-triol after treatment with an equimolar amount of n-butaneboronic acid; the peak represents the cyclic 17,20-n-butaneboronate of the corticosteroid. (B) Chromatogram produced upon injection of solution A after addition of more n-butaneboronic acid. The peak represents the product of esterification of the 3α-hydroxy group of the cyclic 17,20-n-butaneboronate by n-butaneboronic acid. (C) Chromatogram produced upon injection of solution B after treatment with hexamethyldisilazane–chlorotrimethylsilane. The peak represents the 3-trimethylsilyl ether formed by displacement of the acyclic boronate ester. The cyclic 17,20-n-butaneboronate ring system has not been disturbed. Column: 6 ft × 4 mm i.d. glass packed with 1% OV-17 on 100/120 mesh GAS-CHROM Q. Column temperature: 230°C. Copyright: Elsevier Scientific Publishing Company (*Journal of Chromatography*).

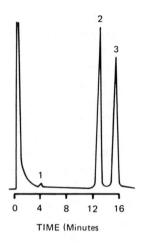

TIME (Minutes

Fig. 5. Compounds separated: (1) tri-*n*-butylboroxine; (2) cyclic 2,3-*n*-butaneboronate of 1-octadecylglycerol; (3) cyclic 1,3-*n*-butaneboronate of 2-octadecylglycerol. Column: 6 ft × 3 mm i.d. glass packed with 3% OV-17 on 100/120 mesh GAS-CHROM Q. Column temperature: 240°C. Copyright: Milton Roy Company.

Fortunately, acyclic boronates are apparently not as stable as cyclic boronates. They can often be destroyed by treatment with certain derivatization reagents,[25,64,103] as illustrated in Figure 4, to yield species with good gas chromatographic properties. Otherwise, the problem can often be obviated simply by using that amount of boronic acid necessary to derivatize only the proximate functional groups.

Boronic acids dehydrate readily to form six-membered cyclic anhydrides (boroxines). Consequently, what is referred to as a "boronic acid" is probably a mixture of the acid itself and its anhydride. For purposes of derivatization, however, the two behave similarly.[27,28,35,84] A peak due to a boroxine is usually observed in chromatograms of cyclic boronates prepared with excess boronic acid (Figure 5).[80,81] Usually, as in the example shown, the presence of an additional peak poses no problem. However, in situations where the boroxine peak interferes with other peaks, it can be eliminated simply by employing either only the required stoichiometric amount of boronic acid or a slight deficiency.

7. EXAMPLES OF THE PREPARATION AND GAS CHROMATOGRAPHY OF CYCLIC BORONATES

The examples that follow are meant to be illustrative rather than comprehensive. Additional information on experimental techniques can be found in the references cited in the bibliography. The examples feature the use of methane- and *n*-butaneboronic acids, the most commonly employed reagents. Since the derivatizations are performed on such a small scale, commercially available small capacity (250 µl–5 ml) screw-capped glass reaction vessels, known variously as Mini-aktors, Reacti-vials, or Silli-vials, can be used to advantage. All chromatograms included in the examples, as well as those that appear elsewhere in the text, were generated with chromatographs equipped with flame ionization detectors.

Example 1—Preparation and Gas Chromatographic Separation of Cyclic n-Butaneboronates of α-Hydroxy Fatty Acids.[80,81] A mixture of the hydroxy fatty acids (∼10 mg) in a Mini-aktor of similar vessel was dissolved in a minimum volume of chloroform. To this solution was added *n*-butaneboronic acid (∼10 mg). The vessel was shaken to dissolve the boronic acid and then was allowed to stand at room temperature for 15 min. An aliquot of the reaction mixture was removed and was injected into the chromatograph. Figure 6 shows the separation of a mixture of the cyclic *n*-butaneboronates of five α-hydroxy fatty acids prepared in this manner.

This example illustrates the experimental simplicity and rapidity of cyclic alkaneboronate preparation. The *n*-butaneboronates of the monoglyceryl ethers shown in Figure 5 were prepared in the same manner.[80,81]

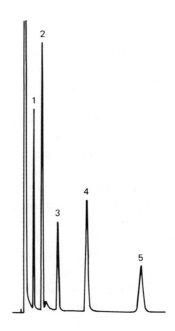

Fig. 6. Compounds separated (as cyclic *n*-butaneboronates): (1) α-hydroxylauric acid (2-hydroxydodecanoic acid); (2) α-hydroxymyristic acid (2-hydroxytetradecanoic acid); (3) α-hydroxypalmitic acid (2-hydroxyhexadecanoic acid); (4) α-hydroxystearic acid (2-hydroxyoctadecanoic acid); (5) α-hydroxyarachidic (2-hydroxyeicosanoic acid). Column: 6 ft × 3 mm i.d. glass packed with 3% OV-17 on 100/120 mesh GAS-CHROM Q. Column temperature: 240°C. Copyright: Milton Roy Company.

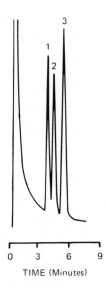

Fig. 7. Compounds separated [as cyclic tris(*n*-butaneboronates)]: (1) D-mannitol; (2) D-sorbitol; (3) D-galactitol. Column: 1.8 m × 5 mm i.d. glass packed with 3% OV-17 on 100/120 mesh GAS-CHROM Q. Column temperature: 200°C. Copyright: Elsevier Scientific Publishing Company (*Carbohydrate Research*).

TIME (Minutes)

Other types of derivative that are useful in the gas chromatography of α-hydroxy fatty acids and similar species include the methyl ester-acetate,[120] the methyl ester-trimethylsilyl ether,[121–123] and the trimethylsilyl ester-ester-trimethylsilyl ether.[123]

Example 2—Preparation and Gas Chromatographic Separation of the Cyclic n-Butaneboronates of the Common Hexitols.[31] A mixture of dry D-galactitol, D-mannitol, and D-sorbitol (1–2 mg total) in a Mini-aktor or similar vessel was dissolved in dry pyridine (∼1.0 ml). To this solution was added *n*-butaneboronic acid (∼10 mg). (Alternatively, a solution of 10 mg of the acid in 1.0 ml pyridine can be added to the mixture of dry sugars.) The vessel was shaken to mix its contents and then was allowed to stand at room temperature for 5 min. Injection of an aliquot of the reaction mixture gave the chromatogram shown in Figure 7.

Treatment of these sugars with *n*-butaneboronic acid yields the cyclic tris(*n*-butaneboronates). This method has been applied to the analysis of mannitol and sorbitol in pharmaceutical products.[39,43]

The common hexitols can also be analyzed by gas chromatography as the hexaacetates. These derivatives, however, require lengthy reaction times for their preparation, and also display long retention times on the columns used for their analysis.[124,125] The hexakis(trimethylsilyl ethers) of these sugars can be prepared somewhat more rapidly, but these derivatives are, at best, only partly resolved on many columns.[125–127]

Example 3—Preparation and Gas Chromatographic Separation of Cyclic Alkaneboronates of Corticosteroids.[64] To a solution of the steroid or steroids (~10 μmol) and ethyl acetate (~1.0 ml) in a Mini-aktor or similar vessel was added methane- or *n*-butaneboronic acid (~10 μmol). The resulting solution was allowed to stand at room temperature for 5 min, at which time an aliquot was withdrawn and was injected into the chromatograph. Figure 8 shows the separation of ten corticosteroid cyclic methaneboronates prepared in this manner.

Under the reaction conditions described, steroids possessing the 17,20-diol and 17,20,21-triol functional groups are completely converted to the corresponding cyclic alkaneboronates. Steroids possessing the 17,21-diol-20-keto functional group are largely converted to the derivatives, but small peaks due to the loss of the side chain and concomitant formation of 17-oxosteroids are also observed. The small peaks between peaks 2 and 3 in Figure 8 are due to 17-oxosteroids formed by thermal decomposition of unreacted 17α,21-dihydroxy-20-oxosteroids. The small peak between peaks 6 and 7 is caused by an unidentified impurity present in steroid 10. With the 17,21-diol-20-oxosteroids, more complete reaction can be obtained by using

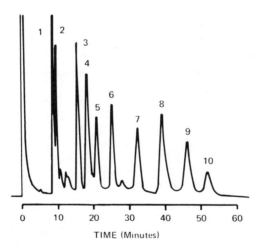

Fig. 8. Compounds separated (as cyclic methaneboronates): (1) 5β-pregnan-3α,17α,20β-triol; (2) 5β-pregnan-3α,17α,20α-triol (pregnanetriol); (3) 5β-pregnan-17α,21-diol-3,20-dione (dihydro-11-desoxycortisol); (4) 5β-pregnan-3α,17α,21-triol-11,20-dione (tetrahydrocortisone, urocortisone); (5) 5-pregnan-17α,21-diol-3,11,20-trione (5β-dihydrocortisone); (6) pregn-4-en-17α,21-diol-3,20-dione (cortexolone); (7) pregn-4-en-17α,21-diol-3,11,20-trione (cortisone); (8) 5β-pregnan-3α,17α,20α,21-tetrol-11-one (cortolone); (9) pregn-4-en-17α,20α,21-triol-3-one; (10) pregn-4-en-11β,17α,21-triol-3,20-dione (cortisol). Column: 12 ft × 4 mm i.e. glass packed with 1% OV-17 on 100/120 mesh GAS-CHROM Q. Column temperature: 230°C. Copyright: Elsevier Scientific Publishing Company (*Journal of Chromatography*).

a slight excess ($\sim$10 mol %) of the alkaneboronic acid. Yields from steroids containing the 20,21-ketol functional group are lower, but can be improved by employing a large excess ($\sim$200 mol %) of the alkaneboronic acid. However, the use of a large excess of reagent is detrimental with steroids possessing hydroxy groups other than those in the side chain, such as single hydroxy groups at the 3 or 11 position. In such instances, acyclic boronates with poor chromatographic properties can be formed.

A comparative gas chromatographic–mass spectrometric study of cyclic methaneboronates and other derivatives of corticosteroids has been reported.[66]

Example 4—Preparation and Gas Chromatographic Separation of Cyclic Alkaneboronates of Catecholamines and Related β-Hydroxyamines.[26] A sample of the β-hydroxyamines ($\sim$1 mg) was dissolved in *N,N*-dimethylformamide ($\sim$1.0 ml). To this solution was added methane- or *n*-butaneboronic acid ($\sim$2 mg). The mixture was then allowed to stand at room temperature overnight, at which time an aliquot was removed and was injected into the chromatograph. Figure 9 shows the separation of four β-hydroxyamine cyclic *n*-butaneboronates (including those of three catecholamines) prepared in this manner.

N,N-dimethylformamide is an excellent solvent for both free β-hydroxyamines and such salts as the hydrochlorides, sulfates, and tartrates.

A variety of derivatives have been employed in the gas chromatography of catecholamines. A useful listing of such species appears in the paper by Anthony, Brooks, and Middleditch.[53]

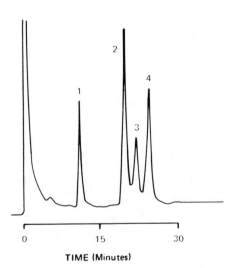

TIME (Minutes)

Fig. 9. Compounds separated: (1) metanephrine cyclic *n*-butaneboronate; (2) epinephrine cyclic bis(*n*-butaneboronate); (3) norepinephrine cyclic bis(*n*-butaneboronate); (4) isoprenaline (isoproterenol) cyclic bis(*n*-butaneboronate). Column: 6 ft × × 1/8 in. glass packed with 1% OV-17 on 100/120 mesh GAS-CHROM Q. Column temperature: 170°C initially, then programmed at 1°C/min. Copyright: Preston Publications, Incorporated (*Journal of Chromatographic Science*).

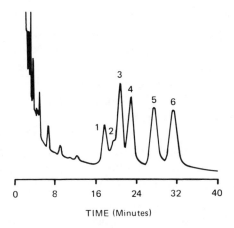

TIME (Minutes)

Fig. 10. Compounds separated: (1) PGE_2 9-methyloxime-11,15-bis(trimethylsilyl ether)-1-trimethylsilyl ester (*syn* or *anti* isomer); (2) PGE_1 9-methyloxime-11,15-bis(trimethylsilyl ether)-1-trimethylsilyl ester (*syn* or *anti* isomer); (3) PGE_2 9-methyloxime-11,15-bis(trimethylsilyl ether)-1-trimethylsilyl ester (*anti* or *syn* isomer); (4) PGE_1 9-methyloxime-11,15-bis(trimethylsilyl ether)-1-trimethylsilyl ester (*anti* or *syn* isomer); (5) $PGF_{2\alpha}$ cyclic 9,11-17-butaneboronate-15-trimethylsilyl ether-1-trimethylsilyl ester; (6) $PGF_{1\alpha}$ cyclic 9,11-*n*-butaneboronate-15-trimethylsilyl ether-1-trimethylsilyl ester. Column: 4-ft glass packed with 3% SE-30 ultraphase on Chromosorb W. Column temperature: 220°C. Copyright: Elsevier Scientific Publishing Company (*Journal of Chromatography*).

Example 5—Preparation and Gas Chromatographic Separation of the Trimethylsilylated Cyclic n-Butaneboronates of the F Prostaglandins: Resolution of Compounds of the E and F Series.[86] To a mixture of prostaglandins E_1, E_2, $F_{1\alpha}$, and $F_{2\alpha}$ (30 µg each) in a small-capacity Mini-aktor or similar vessel was added 20 µl of a 2% (wt/vol) solution of O-methylhydroxylamine hydrochloride in pyridine. The vessel was sealed and was allowed to stand overnight at room temperature. The solvent was then evaporated at room temperature with the help of a fine stream of dry nitrogen. To the residue was added a solution of *n*-butaneboronic acid (2.5 mg) and 2,2-dimethoxypropane (1.0 ml). The vessel was again sealed and was heated at 60°C for 2 min. The reaction mixture was cooled to room temperature and the solvent was again evaporated. Twenty µl of a 1.5 M solution of N-(trimethylsilyl)imidazole in pyridine was then added, and the mixture was heated at 60°C for 5 min. Injection of an aliquot of the cooled reaction mixture gave the chromatogram shown in Figure 10.

This example illustrates the compatibility of cyclic alkaneboronates with other derivatization techniques. Treatment of the prostaglandin mixture with O-methylhydroxylamine converts the 9-keto groups of the E prostaglandins to the corresponding O-methyloximes. The presence of the oxime groups prevents the irreproducible partial formation of enol trimethylsilyl ethers in the subsequent trimethylsilylation step. The two peaks exhibited by each of the E prostaglandins represent resolution of the *syn* and *anti* isomers of the *O*-methyloximes. Treatment with *n*-butaneboronic acid converts the *cis*-9,11-diol functions of the F prostaglandins to the cyclic 9,11-*n*-butaneboronates. Since these cyclic boronates appear to be extremely unstable in the presence of water, the solvent used in this step, 2,2-dimethoxypropane, serves as a water "scavenger." It reacts with water to form acetone and methanol. Finally, treatment with *N*-(trimethylsilyl)imidazole converts the remaining functional groups, that is, the alcoholic 11- and 15- and the carboxylic acid 1-hydroxy groups, to the *O*-trimethylsilyl derivatives.

Most applications of cyclic alkaneboronates to the gas chromatographic–mass spectrometric analysis of prostaglandins have involved compounds of the F series, which possess a cyclopentane-1,3-*cis*-diol structure that geometrically favors cyclic boronate formation.[86–98,101]

Example 6—Preparation and Gas Chromatographic Separation of Trimethylsilylated Cyclic n-Butaneboronates of Pentoses and Hexoses.[32] A mixture of dry D-arabinose, D-fructose, L-fucose, D-galactose, D-glucose, D-lyxose, D-mannose, and D-xylose ($\sim$2 mg each) in a Mini-aktor or similar vessel was dissolved in dry pyridine ($\sim$2.0 ml). To the solution was added *n*-butaneboronic acid ($\sim$80 mg). This mixture was heated at 60°C for 10 min, then it was allowed to stand at room temperature for 48 h. Hexamethyldisilazane (200 μl) and chlorotrimethylsilane (100 μl) were added, the mixture was shaken vigorously for about 30 sec, and was then allowed to stand at room temperature for 5 min. Injection of an aliquot gave the chromatogram shown in Figure 11.

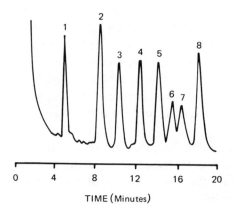

TIME (Minutes)

Fig. 11. Compounds separated (as trimethylsilylated cyclic *n*-butaneboronates): (1) D-lyxose; (2) L-fucose; (3) D-arabinose; (4) D-xylose; (5) D-fructose; (6) D-galactose; (7) D-mannose; (8) D-glucose. Column: 9 ft × 1/4 in. i.d. glass packed with 3% ECNSS-M on 100/120 mesh GAS-CHROM Q. Column temperature: 100°C initially, then programmed from the solvent front at 2°C/min. Copyright: Elsevier Scientific Publishing Company (*Carbohydrate Research*).

This example is another illustration of the compatibility of cyclic alkaneboronates with other derivatization techniques. Pairs of hydroxy groups that are suitably positioned form cyclic *n*-butaneboronates upon treatment with *n*-butaneboronic acid, while those that remain unreacted are subsequently converted to trimethylsilyl ethers. Trimethylsilylation also destroys any acyclic boronates that may have been formed.

Acetylated cyclic methane- and *n*-butaneboronates of pentoses and hexoses have been prepared in a similar two-step manner.[46]

The derivatives that are most often used for the gas chromatography of carbohydrates are alditol acetates and trimethylsilyl ethers. The preparation and chromatographic properties of these and other derivatives of carbohydrates have been reviewed.[128,129]

8. SUMMARY

Cyclic alkane- and areneboronates are thus valuable derivatives for use in the gas chromatography–mass spectrometry of polar, multifunctional organic compounds. Their applicability to a wide variety of organic compounds, ease of preparation, chemical and thermal stability, compatibility with other derivatization techniques, good gas chromatographic properties, and informative mass spectra recommend them for wide use.

REFERENCES

1. M. R. Guerin and W. D. Shults, *J. Chromatogr. Sci.* **7**, 701 (1969).
2. G. Roda and A. Zamorani, *J. Chromatogr.* **46**, 315 (1970).
3. J. E. Attrill, W. C. Butts, W. T. Rainey, and J. W. Holleman, *Anal. Lett.* **3**, 59 (1970).
4. J. J. Pisano, T. J. Bronzert, and H. B. Brewer, *Anal. Biochem.* **45**, 43 (1972).
5. R. W. Kelly, *Tetrahedron Lett.*, 967 (1969).
6. R. W. Kelly, *J. Chromatogr.* **43**, 229 (1969).
7. R. W. Kelly, *Steroids* **13**, 507 (1969).
8. M. A. Kirschner and H. M. Fales, *Anal. Chem.* **34**, 548 (1962).
9. J. A. McCloskey and M. J. McClelland, *J. Am. Chem. Soc.* **87**, 5090 (1965).
10. R. Woods, *Lipids* **2**, 199 (1967).
11. E. Bailey, *Steroids* **10**, 527 (1967).
12. P. Capella and E. C. Horning, *Anal. Chem.* **38**, 316 (1966).
13. A. H. Beckett, G. T. Tucker, and A. C. Moffat, *J. Pharm. Pharmacol.* **19**, 273 (1967).
14. K. Yamasaki, K. Fujita, M. Sakamoto, K. Okada, M. Yoshida, and O. Tanaka, *Chem. Pharm. Bull.* **22**, 2898 (1974).
15. H. G. Kuivila, A. H. Keough, and E. J. Soboczenski, *J. Org. Chem.* **19**, 780 (1954).
16. J. M. Sugihara and C. M. Bowman, *J. Am. Chem. Soc.* **80**, 2443 (1958).

17. R. L. Letsinger and S. B. Hamilton, *J. Org. Chem.* **25**, 592 (1960).
18. A. Finch and J. C. Lockhart, *J. Chem. Soc.*, 3723 (1962).
19. R. L. Letsinger and S. B. Hamilton, *J. Am. Chem. Soc.* **80**, 5411 (1958).
20. E. Nyilas and A. H. Soloway, *J. Am. Chem. Soc.* **81**, 2681 (1959).
21. M. Pailer and W. Fenzl, *Monatsh. Chem.* **92**, 1294 (1961).
22. R. Hemming and D. G. Johnston, *J. Chem. Soc.*, 466 (1964).
23. H. G. Brown and G. Zweifel, *J. Org. Chem.* **27**, 4708 (1962).
24. C. J. W. Brooks and J. Watson, *J. Chem. Soc.*, *Chem. Commun.*, 952 (1967).
25. C. J. W. Brooks and J. Watson, in *Gas Chromatography 1968: Proceedings of the Seventh International Symposium on Gas Chromatography*, C. L. A. Harbourn, ed., Institute of Petroleum, London (1969), pp. 129–141.
26. G. M. Anthony, C. J. W. Brooks, I. Maclean, and I. Sangster, *J. Chromatogr. Sci.* **7**, 623 (1969).
27. C. J. W. Brooks and I. Maclean, *J. Chromatogr. Sci.* **9**, 18 (1971).
28. G. Kresze and F. Schäuffelhut, *Fresenius' Z. Anal. Chem.* **229**, 401 (1967).
29. D. J. Harvey, *Biomed. Mass Spectrom.* **4**, 88 (1977).
30. H. Grote and G. Spiteller, *J. Chromatogr.* **154**, 13 (1978).
31. F. Eisenberg, Jr., *Carbohydr. Res.* **19**, 135 (1971).
32. P. J. Wood and I. R. Siddiqui, *Carbohydr. Res.* **19**, 283 (1971).
33. F. Eisenberg, Jr., in *Methods in Enzymology*, V. Ginsburg, ed., Vol. 28B, Academic Press, New York (1972), pp. 168–178.
34. I. R. McKinley and H. Weigel, *Carbohydr. Res.* **31**, 17 (1973).
35. R. Greenhalgh and P. J. Wood, *J. Chromatogr.* **82**, 410 (1973).
36. P. J. Wood and I. R. Siddiqui, *Carbohydr. Res.* **33**, 97 (1974).
37. F. Eisenberg, Jr., *Anal. Biochem.* **60**, 181 (1974).
38. E. J. Bourne, I. R. McKinley, and H. Weigel, *Carbohydr. Res.* **35**, 141 (1974).
39. M. P. Rabinowitz, P. Reisberg, and J. I. Bodin, *J. Pharm. Sci.* **63**, 1601 (1974).
40. P. J. Wood and I. R. Siddiqui, *Carbohydr. Res.* **36**, 247 (1974).
41. V. N. Reinhold, F. Wirtz-Peitz, and K. Biemann, *Carbohydr. Res.* **37**, 203 (1974).
42. C. J. W. Brooks, *Process Biochem.* **9** (8), 25 (1974).
43. D. L. Sondack, *J. Pharm. Sci.* **64**, 128 (1975).
44. P. J. Wood, I. R. Siddiqui, and J. Weisz, *Carbohydr. Res.* **42**, 1 (1975).
45. J. Wiecko and W. R. Sherman, *Org. Mass Spectrom.* **10**, 1007 (1975).
46. J. Wiecko and W. R. Sherman, *J. Am. Chem. Soc.* **98**, 7631 (1976).
47. T. Kimura, L. A. Sternson, and T. Higuchi, *Clin. Chem.* **22**, 1639 (1976).
48. D. M. Bier, K. J. Arnold, W. R. Sherman, W. H. Holland, W. F. Holmes, and D. Kipnis, *Diabetes* **26**, 1005 (1977).
49. M. D. Bier, R. D. Leake, M. W. Haymond, K. J. Arnold, L. D. Gruenke, M. A. Sperling, and D. M. Kipnis, *Diabetes* **26**, 1016 (1977).
50. H. Tsuchida, K. Kitamura, M. Komoto, and N. Akimori, *Carbohydr. Res.* **67**, 549 (1978).
51. J. Wiecko and W. R. Sherman, *J. Am. Chem. Soc.* **101**, 979 (1979).
52. C. J. W. Brooks, B. S. Middleditch, and G.M. Anthony, *Org. Mass Spectrom.* **2**, 1023 (1969).
53. G. M. Anthony, C. J. W. Brooks, and B. S. Middleditch, *J. Pharm. Pharmacol.* **22**, 205 (1970).
54. G. Munro, J. H. Hunt, L. R. Rowe, and M. B. Evans, *Chromatographia* **11**, 440 (1978).

55. M. Cagnasso and P. Biondi, *Anal. Biochm.* **71**, 597 (1976).
56. P. Biondi and M. Cagnasso, *Anal. Lett.* **9**, 507 (1976).
57. P. Biondi, M. Cagnasso, and C. Secchi, *J. Chromatogr.* **143**, 513 (1977).
58. P. Biondi, C. Secchi, and S. Nicosia, *Anal. Lett.* **11**, 947 (1978).
59. P. Biondi, G. Fedele, A. Motta, and C. Secchi, *Clin. Chim. Acta* **94**, 155 (1979).
60. S. J. Gaskell, C. G. Edmonds, and C. J. W. Brooks, *Anal. Lett.* **9**, 325 (1976).
61. S. J. Gaskell, C. G. Edmonds, and C. J. W. Brooks, *J. Chromatogr.* **126**, 591 (1976).
62. S. J. Gaskell and C. J. W. Brooks, *J. Chromatogr.* **142**, 469 (1977).
63. C. J. W. Brooks and D. J. Harvey, *Biochem. J.* **114**, 15P (1969).
64. C. J. W. Brooks and D. J. Harvey, *J. Chromatogr.* **54**, 193 (1971).
65. C. J. W. Brooks, B. S. Middleditch, and D. J. Harvey, *Org. Mass Spectrom.* **5**, 1429 (1971).
66. T. A. Baillie, C. J. W. Brooks, and B. S. Middleditch, *Anal. Chem.* **44**, 30 (1972).
67. C. F. Poole and E. D. Morgan, *Scan* **6**, 19 (1975); *Chem. Abstr.* **85**, 71807j (1976); *Anal. Abstr.* **32**, 4D64 (1977).
68. C. J. W. Brooks, C. G. Edmonds, and S. J. Gaskell, in *Advances in Mass Spectrometry*, N. R. Daly, ed., Vol. 7B, Institute of Petroleum, London (1978), pp. 1578–1586.
69. S. J. Gaskell and C. J. W. Brooks, *J. Chromatogr.* **158**, 331 (1978).
70. C. Pantarotto, L. Cappelini, P. Negrini, and A. Frigerio, *J. Chromatogr.* **131**, 430 (1977).
71. A. Frigerio, J. Lanzoni, C. Pantarotto, E. Rossi, V. Rovei, and M. Zanol, *J. Chromatogr.* **134**, 299 (1977).
72. C. Pantarotto, L. Cappelini, A. de Pascale, and A. Frigerio, *J. Chromatogr.* **134**, 307 (1977).
73. W. Blum and W. J. Richter, *Helv. Chim. Acta* **57**, 1744 (1974).
74. D. H. Calam, *J. Chromatogr. Sci.* **12**, 613 (1974).
75. M. T. M. Tulp and O. Hutzinger, *J. Chromatogr.* **139**, 51 (1977).
76. S. Y. Chang and C. Grunwald, *J. Lipid Res.* **17**, 7 (1976).
77. I. M. Mutton, *J. Chromatogr.* **172**, 435 (1979).
78. S. J. Shaw, *Tetrahedron Lett.*, 3033 (1968).
79. M. Verzele, E. Vanluchene, and J. van Dyck, *Anal. Chem.* **45**, 1549 (1973).
80. R. F. Kruppa, R. S. Henly, and S. Ramachandran, paper presented at the 44th Fall Meeting of the American Oil Chemists Society, Chicago, Illinois, September 27–October 1, 1970.
81. R. F. Kruppa, R. S. Henly, and S. Ramachandran, paper presented at the Eastern Analytical Symposium, New York, New York, November 20, 1970.
82. W. J. A. VandenHeuvel, *Anal. Lett.* **6**, 51 (1973).
83. P. Bournot, B. F. Maume, and C. Baron, *J. Chromatogr.* **57**, 55 (1971).
84. V. Dommes, F. Wirtz-Peitz, and W. H. Kunau, *J. Chromatogr. Sci.* **14**, 360 (1976).
85. R. L. Jones, P. J. Kerry, N. L. Poyser, I. C. Walker, and N. H. Wilson, *Prostaglandins* **16**, 583 (1978).
86. C. Pace-Asciak and L. S. Wolfe, *J. Chromatogr.* **56**, 129 (1971).
87. M. J. Levitt, J. B. Josimovich, and K. D. Broskin, *Prostaglandins* **1**, 121 (1972).
88. R. W. Kelly, *J. Chromatogr.* **71**, 337 (1972).
89. R. W. Kelly, *Anal. Chem.* **45**, 2079 (1973).

90. R. W. Kelly, in *Advances in Mass Spectrometry*, Vol. 6, A. R. West, ed., Institute of Petroleum, London (1974), pp. 193–198.
91. C. N. Hensby, *Prostaglandins* **8**, 369 (1974).
92. P. L. Taylor and R. W. Kelly, *FEBS Lett.* **57**, 22 (1975).
93. R. W. Kelly and P. L. Taylor, *Anal. Chem.* **48**, 465 (1975).
94. C. Pace-Asciak, *J. Am. Chem. Soc.* **98**, 2348 (1976).
95. A. G. Smith, J. D. Gilbert, W. A. Harland, and C. J. W. Brooks, *Biochem. Soc. Trans.* **4**, 108 (1976).
96. A. G. Smith and C. J. W. Brooks, *Biomed. Mass Spectrom.* **4**, 258 (1977).
97. A. F. Cockerill, D. N. B. Mallen, D. J. Osborne, J. R. Boot, and W. Dawson, *Biomed. Mass Spectrom.* **4**, 358 (1977).
98. L. Fenwick, R. L. Jones, B. Naylor, N. L. Poyser, and N. H. Wilson, *Brit. J. Pharmacol.* **59**, 191 (1977).
99. M. Claeys and B. van Haver, *Org. Mass Spectrom.* **12**, 531 (1977).
100. A. G. Smith, W. A. Harland, and C. J. W. Brooks, *J. Chromatogr.* **142**, 533 (1977).
101. P. L. Taylor, *Prostaglandins* **17**, 259 (1979).
102. S. G. Batrakov, A. G. Panosyan, A. N. Ushakov, B. V. Rozynov, and L. D. Bergel'son, *Izv. Akad. Nauk SSSR Ser. Khim.*, 1739 (1974); *Bull. Acad. Sci. USSR Div. Chem. Sci.* **23**, 1662 (1974).
103. S. J. Gaskell and C. J. W. Brooks, *J. Chromatogr.* **122**, 415 (1976).
104. S. J. Gaskell and C. J. W. Brooks, *Org. Mass Spectrom.* **12**, 651 (1977).
105. G. Belvedere, J. Pachecka, L. Cantoni, E. Mussini, and M. Salmona, *J. Chromatogr.* **118**, 387 (1976).
106. R. J. Ferrier, in *Methods in Carbohydrate Chemistry*, Vol. 6, (R. L. Whistler and J. N. BeMiller, ed., Academic Press, New York (1972), pp. 419–426.
107. R. J. Ferrier, in *Advances in Carbohydrate Chemistry and Biochemistry*, Vol. 35, R. S. Tipson and D. Horton, ed., Academic Press, New York (1978), pp. 31–80.
108. C. F. Poole, *Chem. Ind. (London)*, 479 (1976).
109. C. F. Poole, S. Singhawangcha, and A. Zlatkis, *Chromatographia* **11**, 347 (1978).
110. C. F. Poole, S. Singhawangcha, and A. Zlatkis, *J. Chromatogr.* **158**, 33 (1978).
111. C. F. Poole, S. Singhawangcha, and A. Zlatkis, *Analyst* **104**, 82 (1979).
112. E. J. Sowinski and I. H. Suffet, *J. Chromatogr. Sci.* **9**, 632 (1971).
113. C. F. Poole, *J. Chromatogr.* **118**, 280 (1976).
114. C. F. Poole, S. Singhawangcha, A. Zlatkis, and E. D. Morgan, *J. High Resol. Chromatogr. Chromatogr. Commun.* **1**, 96 (1978).
115. A. E. Gordon and A. Frigerio, *J. Chromatogr.* **73**, 401 (1972).
116. I. Björkhem, R. Blomstrand, O. Lantto, L. Svensson, and G. Öhman, *Clin. Chem.* **22**, 1789 (1976).
117. C. Fenselau, *Anal. Chem.* **49**, 536A (1977).
118. R. Carrington and A. Frigerio, *Drug Metab. Rev.* **6**, 243 (1977).
119. W. D. Lehmann and H. R. Schulten, *Angew. Chem. Int. Ed. Engl.* **17**, 221 (1978).
120. Y. Kishimoto and N. S. Radin, *J. Lipid Res.* **4**, 130 (1963).
121. R. D. Wood, P. K. Raju, and A. Reiser, *J. Am. Oil Chem. Soc.* **42**, 81 (1965).
122. M. G. Horning, K. L. Knox, C. E. Dalgliesh, and E. C. Horning, *Anal. Biochem.* **17**, 244 (1966).
123. M. G. Horning, E. A. Boucher, and A. M. Moss, *J. Gas Chromatogr.* **5**, 297 (1967).
124. J. A. Hause, J. A. Hubicki, and G. G. Hazen, *Anal. Chem.* **34**, 1567 (1962).
125. J. S. Sawardeker, J. H. Sloneker, and A. Jeanes, *Anal. Chem.* **37**, 1604 (1965).

126. F. Loewus, *Carbohydr. Res.* **3**, 130 (1966).
127. W. C. Ellis, *J. Chromatogr.* **41**, 335 (1969).
128. C. C. Sweeley, W. W. Wells, and R. Bentley, in *Methods in Enzymology*, Vol. 8, E. F. Neufeld and V. Ginsburg, ed., Academic Press, New York (1966), pp. 95–108.
129. R. A. Laine, W. J. Esselman, and C. C. Sweeley, in *Methods in Enzymology*, Vol. 28B, V. Ginsburg, ed., Academic Press, New York (1972), pp. 159–167.

Chapter 3

General Aspects of Precolumn Derivatization with Emphasis on Pharmaceutical Analysis

Larry A. Sternson

1. INTRODUCTION

1.1. Purposes for Derivatization

The specific analysis of a wide variety of structurally diverse pharmaceutical preparations has been immeasurably advanced by the development of high-performance liquid chromatography (HPLC). This technique [and its complements, gas–liquid (GLC) and thin-layer (TLC) chromatography] has permitted the quantitation of drugs and often their degradation products (e.g., metabolites) in a diversity of matrices including dosage forms and biological fluids. Under certain circumstances, chemical modification of the analyte can facilitate and improve its analysis. This concept of derivatization has been widely applied to mass spectroscopic, nuclear magnetic resonance, optical spectroscopic, and fluorometric measurements, and is often used in gas–liquid and thin-layer chromatography. Its application to HPLC is,

---

Larry A. Sternson • Department of Pharmaceutical Chemistry, The University of Kansas, Lawrence, Kansas 66045.

therefore, a natural one. Derivatization techniques in HPLC have been the subject of a number of recent reviews.[1–9] Derivatization can either be carried out prior to the chromatographic step or immediately following the elution of the sample band from the column. In the latter case, reaction is carried out on-line, in a flowing stream using a *reaction detector*. This chapter describes preseparation derivatization, which offers the following advantages over postseparation derivatization: greater freedom and flexibility in the selection of reaction conditions (fewer restrictions imposed on reaction kinetics) and usually less need to and difficulty in removing excess reagents prior to read-out. The major disadvantage of the preseparation approach is the possibility of forming multiple products from one analyte or formation of artifacts which may interfere with the determination.

Precolumn derivatization may be introduced into an analytical scheme (a) to enhance the stability of the analyte, (b) to improve the separation of the analyte from the sample matrix, (c) to refine the subsequent chromatographic separation (by improving band shape and/or increasing resolution of adjacent bands), and (d) to improve detectability by increasing response to the detector or by introducing an additional element of specificity into the determination due to the limited reaction possibilities of the reagent.

The HPLC analysis of the antineoplastic agent dianhydrogalactitol (DAG, **1**)[10] includes a preseparation derivatization step that incorporates

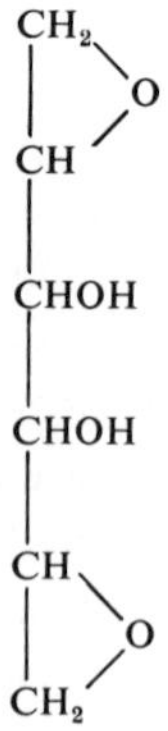

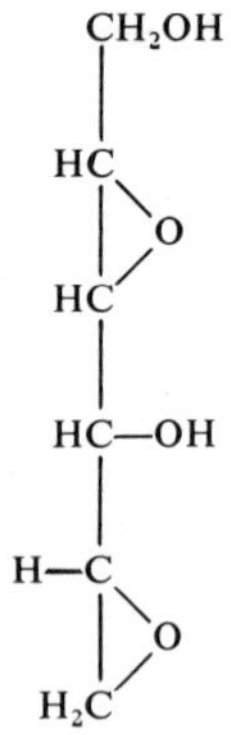

many of the features delineated above. The hydrophilicity of the drug prevents its efficient extraction into water-immiscible solvents, even from salt-saturated solutions. Furthermore, DAG is unstable, binding irreversibly to red blood cells (in the sample containers) through attack by endogenous nucleophiles at the epoxide, and also undergoing intramolecular rearrangement to the thermodynamically more stable 2,3-epoxy isomer, **2**. The lack of chromophoric groups in **1** provides detection limits (>10 µg ml^{-1} of plasma) that are above that necessary for clinical monitoring of the drug. These difficulties were overcome by derivatizing DAG with diethyldithiocarbamate, **3**. Reaction was carried out directly in the blood sample (at room temperature) and conversion to the bisdithiocarbamate, **4**, was quantitative and complete in less than 5 min. Elimination of the epoxides by conversion of **1** to **4** stabilized the analyte from subsequent nucleophilic attack. The derivative is more hydrophobic than the parent drug and can be quantitatively extracted into chloroform. The derivative chromatographs efficiently on a normal phase (CN) column and strongly absorbs uv light ($a_{m,254 \text{ nm}} = 2.8 \times 10^4$), thus providing a route to its clinical monitoring in blood.

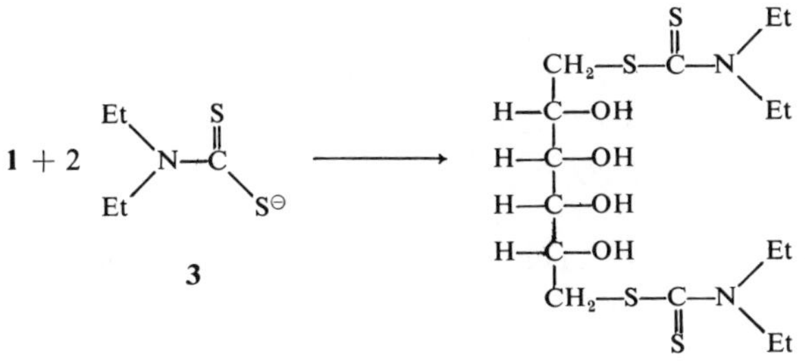

Although this example illustrates the multifaceted utility of precólumn derivatization, the approach has been primarily used to extend the utility of sensitive HPLC detectors to compounds that are relatively insensitive to them, by chemically transforming the molecule of interest into a more readily detectable species. Unlike the situation encountered in GLC, derivatization is rarely employed in HPLC to improve chromatographic separation because of the flexibility in retention characteristics available by mobile phase manipulation and stationary phase selection. An important exception is the resolution of racemic mixtures by preseparation derivatization with chiral reagents, to form diastereomers which are resolvable with common HPLC systems.[11,12] In addition, it has been shown that acetylation improves the liquid chromatographic properties of catecholamines[13] and methylation facilities separation of hydroxyxanthones.[14]

1.2. HPLC Detectors and Compatible Derivatives

Because of the lack of universally sensitive HPLC detectors (comparable to the FID used in GLC), derivatization has been primarily aimed at conferring improved detection properties of analytes. The common (commercially available) detectors are based on refractometric, spectrophotometric, spectrofluorometric, or electrometric (amperometric or coulometric) measurements. Detectors based on measuring refractive index changes are nonspecific and insensitive, and usually not suited to low-level analytical applications. The most widely used detectors are based on spectrophotometric monitoring. These detectors are rugged and generally unaffected by external influences. Their specificity depends on the relative ability of other compounds in the analysis mixture to absorb uv light at the wavelength at which the column effluent is being monitored. Sensitivity is dictated by the molar absorptivity (a_m) of the compound of interest. A compound with $a_m = 10$ can be detected at levels of ~ 1 μg (injected on-column) with most commercial uv detectors. Normally, the introduction of increased conjugation, e.g., (aromaticity) in a molecule increases a_m and produces a bathochromic shift in $\lambda_{\max}$.[15] The chromophoric properties of several substituted aromatic moieties that have been incorporated into derivatizing agents are given in Table 1.

TABLE 1. Functionalities for Enhancing Photometric Detection

Chromophore	$\lambda_{\max}$ nm	$a_{m,254}$	Electroactive ox/red	Reference
Benzyl	254	200	No	16
p-Nitrobenzyl	265	6 200	Yes/red	17, 18
3,5-Dinitrobenzyl		14 000	Yes/red	19
2,4-Dinitrophenyl		$>10^4$	Yes/red	20–22
Benzoyl	230	$<1\,000$	No	23, 24
p-Toluoyl	236	5 400	No	25
p-Chlorobenzoyl	236	6 300	No	25
p-Nitrobenzoyl	254	$>10^4$	Yes/red	24, 26
p-Methoxybenzoyl	262	16 000	No	27, 28
Phenacyl	250	10 000	No	29
p-Bromophenacyl	260	18 000	No	30, 31
p-Nitrophenacyl			Yes/red	32
2-Naphthacyl	248	12 000	No	33
Diethyldithiocarbamoyl	254	12 000	—	10
p-Toluenesulfonyl			No	34
p-Iodobenzenesulfonyl			No	35

Absorbance detectors use either Hg vapor or deuterium lamps as sources. The intensity of the former is greater than that of the latter; however, $>90\%$ of the light from a Hg lamp is emitted as a narrow band at 254 nm. Greatest spectroscopic sensitivity is therefore achieved with a derivative that strongly absorbs radiation at 254 nm; however, because of the variety of structural types of compounds that absorb radiation at this wavelength, detection is relatively nonspecific. Using a deuterium lamp or phosphors in concert with the Hg lamp, wavelengths other than 254 nm are readily monitored, although at a photon output less than that achieved with a Hg lamp at 254 nm. By forming derivatives that absorb light above 300 nm, additional specificity can be introduced into the detection system because of the more limited types of materials that respond at these longer wavelengths. Further sensitivity can be introduced by spectrofluorimetric transducers because they directly measure emitted light rather than the difference between incident and transmitted radiation; and because of the added specificity provided by being able to select both excitation and emission wavelengths. Fluorescent detectors are somewhat more temperamental than absorbance detectors, and respond to a smaller population of compounds (thus improving specificity). Fluorescent molecules are characterized[15] by extended conjugation (polynuclear aromatics) and substituents directly on the aromatic ring that promote $\pi \rightarrow \pi^*$ electronic transitions (electron-releasing substituents, e.g., amines, ethers) in the molecule. Introduction of electron-attracting groups (e.g., nitro, carbonyl, etc.) in conjugation with the aromatic system promote $n \rightarrow \pi^*$ transitions, resulting in increased intersystem crossing and a decrease in fluorescent yield. Fluorescent derivatizing agents can therefore be designed based on structural–electronic relationships. Derivatizing agents that deliver a tag incorporating a 5-dialkylaminonaphthyl- (5) or 4-methylene-7-methoxy-coumarin (6) moiety are two of many groups that impart fluorescence to a target molecule.

5   6

Electrochemical detectors are gaining popularity, but their design is still in a developmental stage. They are based on either amperometric[36] or coulometric[37,38] measurement, the former receiving greater publicity to date.[4,39] Such detectors have been operated primarily in the oxidative

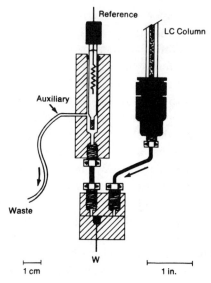

Fig. 1. Thin-layer amperometric detector for HPLC.[39] Reprinted from *Analytical Chemistry.* Copyright (1977) by permission of the American Chemical Society.

(anodic) mode, using either a tubular working electrode or a thin-layer cell. Thin-layer cells (Figure 1) have been employed as the commercial detectors and have been constructed with carbon paste, glassy carbon, platinum, gold, or amalgamated surfaces. To achieve maximum sensitivity the detector is operated in the limiting current plateau for the Faradaic reaction of interest. Selectivity is fulfilled by choosing the lowest potential at which the electrochemical process occurs (i.e., as the potential is increased, an increasing number of compounds respond, and specificity is compromised). Therefore, the operating potential must often be chosen as a compromise between sensitivity and specificity. The maximum potential that can be applied between the electrodes is limited by background current which is influenced by pH, ionic strength, electrochemical reactivity of the solvent (mobile phase), and the presence of electroactive impurities. *Phenols* and *aromatic amines* are two of the major functional groups that are susceptible to electrochemical oxidation.[40] Introduction of electron-donating substituents (e.g., amino, hydroxyl, thio, etc.) in conjugation with (e.g., distributed *ortho-para* to) these moieties facilitates oxidation [$E_{p/2}$ moves to less anodic (positive) potentials]. Alkylation or acylation of arylamines or phenols (to give aryl ethers, amides, or esters) diminishes their oxidizability ($E_{p/2}$ moves to more positive potential) but in the presence of secondary electron-donating substituents in *ortho* and/or *para* positions, such compounds are still amenable to electrochemical detection. Thiols (e.g., phenothiazines)[38] and mercaptans (e.g., cysteine)[39] represent other structural classes of organic compounds that are suitable candidates for electrochemical detection. Electrochemically active derivatives can be prepared from electroinactive molecules prior to chromatographic separation. p-Dimethylaminophenylisocyanate, 7, was designed as an *electrogenic reagent*, i.e., one that is itself electroinactive but reacts with specific substrates to yield electroactive products.[41] It reacts rapidly and quantitatively with arylhydroxylamines, such as phenylhydroxylamines (8), to yield the corresponding hydroxyurea, 9,

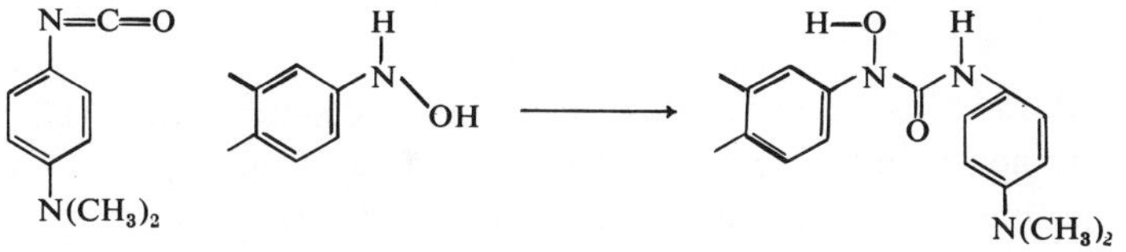

which is readily oxidized ($E_{p/2} = +0.5$ V vs. Ag/AgCl) at a carbon paste surface in a two-electron process (as determined coulometrically) apparently yielding 10. The high signal:noise ratio allows determination of the arylhydroxylamine at levels of 5×10^{-9} M in liver homogenates. Kissinger[4,42] has described the use of alkaline periodate to oxidize biogenic catecholamine derivatives (3-methoxy-4-hydroxybenzyl alcohol, vanillomandelic acid, and 3-methoxy-4hydroxyphenyl glycol) to vanillin, which is readily detected electrochemically after HPLC separation.

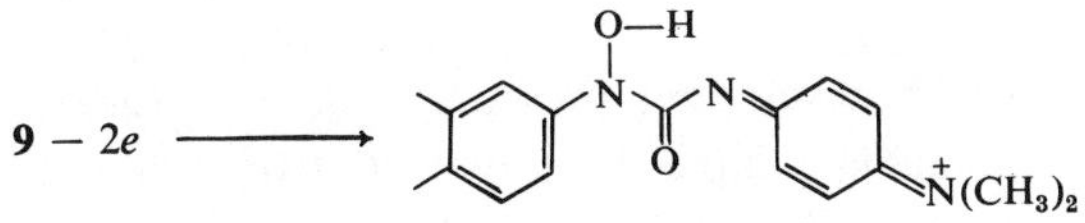

10

Electrochemical detectors based on reduction and measurement of cathodic current have resisted development. Nitro(so) and (thio)carbonyl groups represent examples of reducible functional groups.[43] Among the difficulties associated with such detectors is the contribution to cathodic current caused by dissolved oxygen, reduction of hydronium ion, and trace metal contamination.[39] A detector employing a hanging mercury drop electrode is commercially available from Princeton Applied Research Assoc., and a dropping mercury electrode (DME) based detector has recently been developed by Frei and co-workers.[44] The latter detector has been used both in the dc and differential pulse polarographic mode, and has overcome some

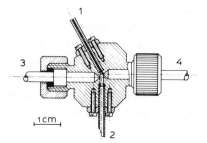

Fig. 2. DME detector for HPLC. (1) Inlet; (2) outlet, counter electrode; (3) reference electrode; (4) DME.[228] Reproduced from the *Journal of Chromatography* by permission of Elsevier Publishing Co.

of the dead volume and response problems associated with DME detectors by using horizontal mercury capillaries (to accelerate drop rates) and conically ground capillaries and reference electrodes (Figure 2). This detector has been applied to the analysis of a series of thiourea-based herbicides[44] as well as to analysis of digoxin species derivatized with 3,5-dinitrobenzoyl chloride.[45] Many of the derivatizing agents that have been designed for improving spectroscopic response have incorporated nitrophenyl moieties (Table 1). Such groups also offer potential for electrochemical detection since the nitro group undergoes facile reduction (4e) at appropriate cathodes. The utility of all of these detectors has been greatly expanded through appropriate precolumn derivatization of compounds that would otherwise be insensitive to them.

1.3. Chemical Reactions Used in Derivatization

The objective in developing preseparation derivatization schemes is to find a reaction that (1) is rapid, (2) is quantitative or at least reproducible, (3) converts the analyte to a single product, (4) utilizes reagents that can be separated from the final product, and (5) forms derivatives that have detection properties different from reactants and preserve the ability to chromatographically separate components. Although few reaction schemes can achieve all of these aims, optimization of reaction conditions is a necessary part of developing derivatization procedures. Reactions should in most cases be run under conditions where reagent concentration remains invariant. Derivatization in trace analysis becomes less efficient as the molecularity of the derivatization reaction increases; the reaction is entropically less favored, i.e., the effective collision probability is reduced. The number of reactions and reagents involved in a derivatization scheme should be minimized and reactions should be carried out under the mildest conditions possible to avoid complex schemes from developing. Reactions need to be optimized with regard to solvent, temperature, reaction time, and susceptibility to catalysis, e.g., acylations can be carried out much more effectively if their propensity for general base catalysis is exploited. Solvent should not *only* be chosen based on most favorable reaction conditions but should also be compatable with subsequent steps in the analysis scheme, e.g., extraction, chromatography. Caution should be exercised in carrying out reactions at elevated temperatures because (1) of the loss in specificity caused by the increased possibility of side reactions occurring, (2) the stability of the derivative may be compromised and therefore reaction times may need to be more strictly defined, (3) components with high vapor pressure (e.g., solvent) may be physically lost from the mixture and (4) because of the inconvenience involved. Similar concerns should be raised before employing any forceful conditions.

Preseparation derivatization can involve oxidation, reduction, displacement, or addition reactions. The latter two categories include attack by either nucleophilic or electrophilic species, depending on the reactivity of the substrate. Most of the reactions described to date involve electrophilic reagents interacting with nucleophilic sites on target molecules. In clinical analysis this approach has the distinct disadvantage that efficient cleanup is often necessary prior to the derivatization step because (1) of the wide variety of endogenous nucleophiles present in biological fluids (at high concentrations relative to the compound of interest) which also react with the tagging agent reducing the specificity of the method, causing consumption of large amounts of reagent, and making separations more difficult, and (2) such reagents (e.g., acylating agents) are often susceptible to degradation by water (the most abundant nucleophile in biological fluid). Nucleophilic reagents are much more specific because of the absence of electrophilic functionalities in endogenous compounds and are usually stable in water. It is these properties that allow the derivatization of dianhydrogalactitol, **1**, with diethyldithiocarbamate, **3**, to be carried out directly in blood. Such derivatization could not be accomplished as readily with electrophilic reagents. Unfortunately, a relatively small percentage of pharmaceuticals is amenable to derivatization with nucleophilic reagents.

Oxidation reactions have been used to a limited extent in preseparation derivatization. Many of these reactions are relatively nonspecific. Several drugs and/or their metabolites, however, have been effectively tagged by this technique. Periodate oxidation of catecholamine metabolites to vanillin, **11**, has provided a useful route to their HPLC analysis.[4,42] The ferricyanide oxidation of morphine **12** produces a fluorescent dimer, **13**, that has been applied to its analysis in urine.[46] Unfortunately, mixed dimers form between morphine-type compounds (i.e., morphine, its metabolites, and the recommended internal standard), and therefore multiple products are formed

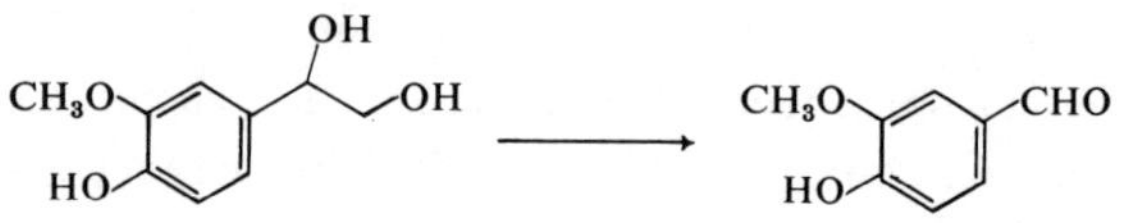

11

in ratios dependent on the relative concentrations of the monomers, diminishing the effectiveness of the method. The antiestrogenic, antineoplastic

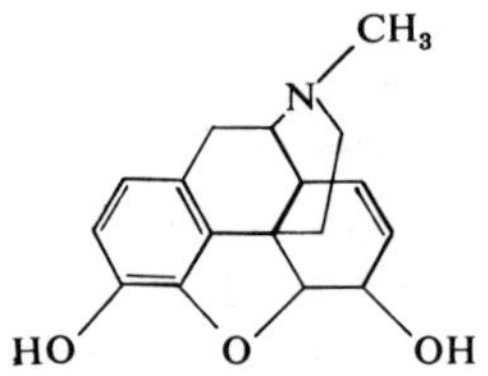

12

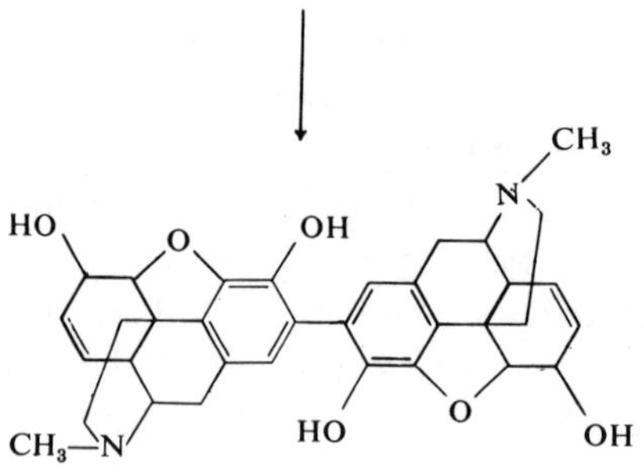

13

agent, tamoxifen (**14**) (used in treatment of metastatic breast cancer) and its two major and potentially therapeutically active metabolites (**14a, 14b**) have been analyzed in whole blood by HPLC with fluorescence detection after photochemically induced oxidation of the drug species to the corresponding phenanthrene, **15**.[47,48] The drug species are first extracted from blood with ether; the ether evaporated and the residue dissolved in HPLC mobile phase (73:27 methanol water, 2.5 mM pentanesulfonate, and 0.5% acetic acid). The mixture is then photolyzed for ~10 min and the final mixture (**15**) chromatographed on an RP-18 column. The acid in the photolysis mixture is necessary to inhibit degradation of **15** by protonating the amine nitrogen, thereby reducing its nucleophilicity, and preventing intra-

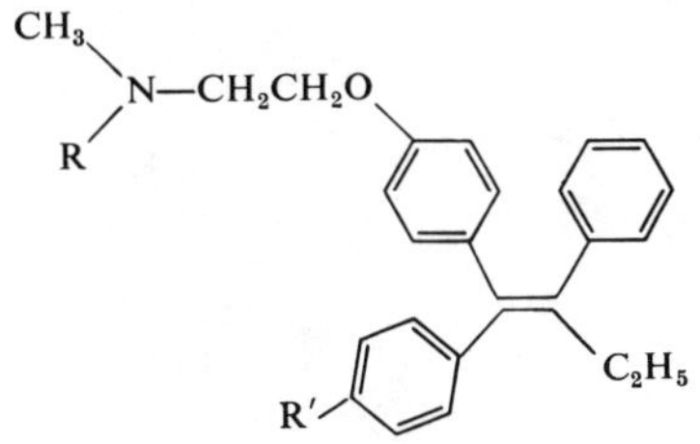

CH3
N—CH2CH2O
R
C2H5
R'

14 R = CH_3; R' = H
14a R = CH_3; R' = OH
14b R = H; R' = H

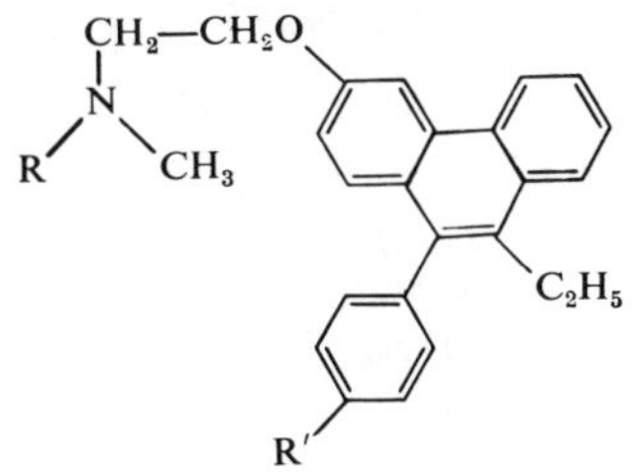

CH2—CH2O
N
R
CH3
C2H5
R'

15

molecularly catalyzed expulsion of the phenol (**16**) and loss in fluorescence intensity.

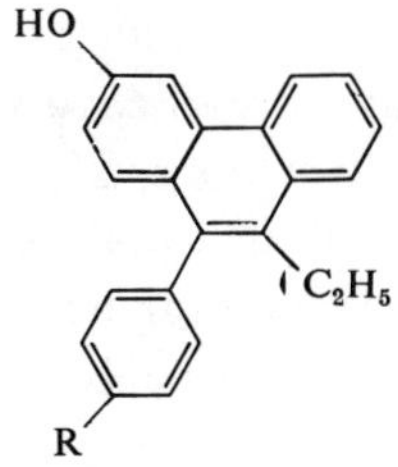

HO
C2H5
R

16

The administration of emetine in cancer chemotherapy has been prevented by the severe myopathy and cardiac arrhythmias associated with its use.[49] By administering the drug as a slow iv infusion, it appears that these side effects can be avoided. To monitor emetine blood levels attained from this route, a sensitive HPLC–fluorescence assay[50] has been developed based on the mercuric acetate oxidation of emetine, **17** (extracted from plasma) to form rubremetine, **18**. Levels approaching 1 ng/ml are readily attained.

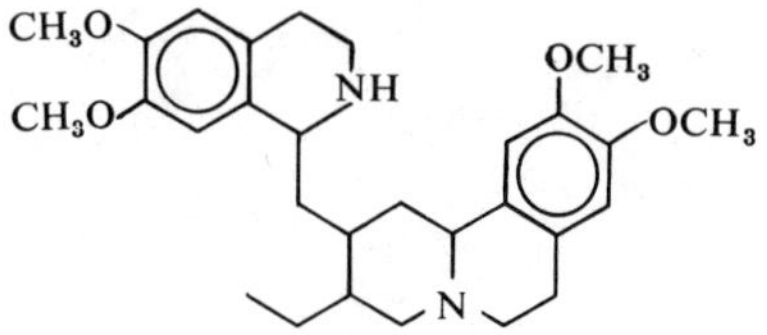

17

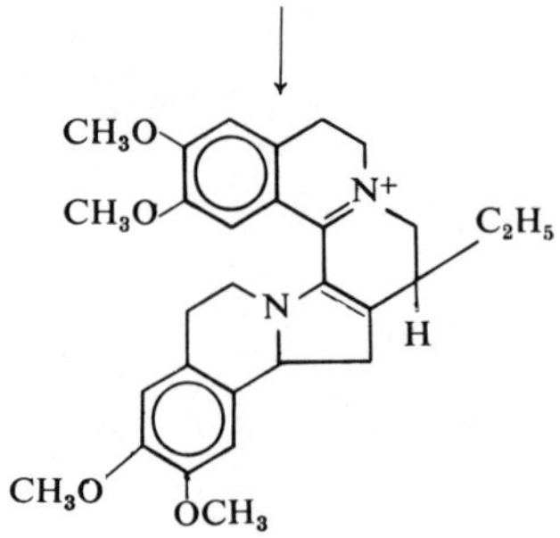

18

The remainder of this chapter is devoted to a discussion of derivatization based on reactions amenable to modification of particular functional groups. The analyte is conceived as a collection of functional groups. The following discussion suggests various approaches as to how these groups can be chemically transformed to facilitate organic analysis by HPLC using commercially available detectors. Advantages and disadvantages of specific reagents and reaction conditions are discussed. A description of the chemistry involved in these transformations is provided to give the reader a more fundamental understanding of variables contributing to the reaction course, thus enabling the reader to optimize both derivative selection and reaction conditions to fit a particular analysis problem.

2. FUNCTIONAL GROUP ANALYSIS

2.1. Amines

By virtue of the lone nonbonded electron pair on nitrogen, amines are subject to electrophilic attack, i.e., acylation and alkylation reactions.

2.1.1. Acylation Reactions

2.1.1.1. Theoretical Aspects. The majority of preseparation derivatization reactions of amines involves acyl transfer reaction. Such reactions proceed in most cases with heterolysis of the C—X bond of RCOX in a two-step mechanism.[51,52] Reaction (1) is initiated by nucleophilic attack (by Y) at the carbonyl carbon to form a tetrahedral intermediate (T.I.) followed by expulsion of a nucleophile to form either product ($k_2 \gg k_{-1}$) or to regenerate starting material ($k_{-1} \gg k_2$):

$$
\underset{\text{R--C--X + Y:}^{\ominus}}{\overset{\text{O}}{\|}} \underset{k_{-1}}{\overset{k_1}{\rightleftharpoons}} \underset{\underset{\text{Y}}{|}}{\overset{\text{O}^{\ominus}}{\underset{|}{\text{R--C--X}}}} \xrightarrow{k_2} \underset{\text{R--C--Y}}{\overset{\text{O}}{\|}} \qquad (1)
$$

$$\text{T.I.}$$

Applying the steady state approximation to the intermediate (T.I.),

$$[\text{T.I.}] = \frac{k_1[\text{ROCX}][\text{Y}]}{k_2 + k_{-1}}$$

the rate of acylation is given by k_2 [T.I.], and the second-order rate constant for acylation (k_a) is

$$k_a = \frac{k_1}{(k_{-1}/k_2) + 1}$$

Product formation is promoted by maximizing k_1 and minimizing the partition ratio for T.I. (k_{-1}/k_2). The effect of both the structural changes in R and of X and Y on k_a can be discussed in terms of their effect on k_1 and k_{-1}/k_2.[53] The kinetics and thermodynamics of the reaction are dictated by the nucleophilicity of the attacking species (Y) and the potential for expulsion of X. Both of these factors can be approximated from the pK_a's of the conjugate acids of Y and X. It must be remembered, however, that this approximation is based on relating kinetic terms for nucleophilicity with the equilibrium affinity for the solvated proton. Furthermore, it neglects steric considerations and the effect of polarizability on nucleophilicity. The stronger the nucleophile (larger pK_a) the more rapidly and tenaciously it will interact with the carbonyl carbon to form T.I. The rate of attack of Y at C=O (k_1) is determined by its nucleophilicity ($pK_{a_{HY}}$ large). The breakdown of T.I. is dictated by the relative magnitude of k_{-1} (return to reactants) and k_2 (formation of products), which in turn is determined by the relative basicities of X and Y, i.e., as Y is made more nucleophilic relative to X, reaction is directed toward product, since Y interacts more strongly with the carbonyl carbon, and X is more easily expelled. Thus derivatization with acylating agents is made more feasible as the substrate increases in nucleophilicity. Reaction can be facilitated by designing derivatizing agents with better leaving groups ($pK_{a_{HX}}$ is low); however, as the reagent RCOX is made more reactive, selectivity is compromised as the reagent becomes less discriminating. It is this difference in leaving group potential (and hence reactivity and specificity of the reagent) that forms the basis for the rational selection of acyl halides, anhydrides, hydroxyimides, or acylimidazoles for use in derivatization of specific functional groups.

Changes in R have an influence on k_1, but only minimal effect on the partition ratio k_{-1}/k_2.[54] Most common acylating agents used for derivatization in HPLC are substituted aromatic acid halides or anhydrides. Hammett plots[55] reveal that substrate reactivity and product stability in acyl transfer reactions are susceptible to the influence of ring substituents. Both acid- and base-catalyzed acylations yield positive Hammett reaction constants (σ), indicating that acylation (k_1) or breakdown of acylated products is promoted by electron-withdrawing substituents (Figure 3) (groups

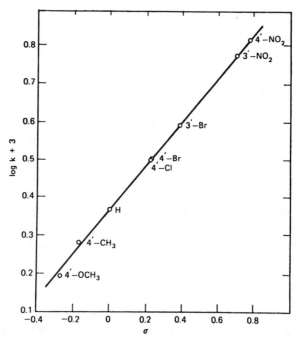

Fig. 3. Hammett plot for the alkaline hydrolysis of substituted ethyl-*p*-biphenyl-carboxylates at 40°C in 88.7% (w/w) ethanol.[229] Reprinted from the *Journal of the American Chemical Society*. Copyright (1953) by permission of the American Chemical Society.

TABLE 2. Hammett Substituent Constants Based on Ionization of Benzoic Acids[a]

Group	Meta position σ_m	Para position σ_p
—CH$_3$	—0.069	—0.170
—CH$_2$CH$_3$	—0.07	—0.151
—C$_6$H$_5$	0.06	—0.01
—CF$_3$	0.43	0.54
—CN	0.56	0.66
—COCH$_3$	0.376	0.502
—CO$_2$C$_2$H$_5$	0.37	0.45
—CO$_2$H	0.37	0.45
—CO$_2^-$	—0.1	0.0
—N$_2^+$	1.76	1.91
—NH$_2$	—0.16	—0.66
—NHCH$_3$	—	—0.84
—N(CH$_3$)$_2$	—	—0.83
—NHCOCH$_3$	0.21	0.00
—$^+$N(CH$_3$)$_3$	>0.88	0.82
—NO$_2$	0.710	0.778
—OCH$_3$	0.115	—0.268
—OC$_6$H$_5$	0.252	—0.320
—OH	0.121	—0.37
—OCOCH$_3$	0.39	0.31
—SCH$_3$	0.15	0.0
—SH	0.25	0.15
—SCOCH$_3$	0.39	0.44
—SCN	—	0.52
—S(O)CH$_3$	0.52	0.49
—SO$_2$CH$_3$	0.60	0.72
—SO$_2$NH$_2$	0.46	0.57
—S(CH$_3$)$_2^+$	1.00	0.90
—SO$_3^-$	0.05	0.09
—F	0.337	0.062
—Cl	0.373	0.227
—Br	0.391	0.232
—I	0.352	0.18

[a] Reference 57. Reprinted from the *Journal of Organic Chemistry*. Copyright (1958) by permission of the American Chemical Society.

with positive Hammett substituent constants, σ; Table 2).[56] Thus, p-nitro-benzoyl chloride ($\sigma_{p,NO_2} = +0.778$)[57] is more reactive than benzoyl chloride ($\sigma_{p,H} = 0.00$), which in turn is more reactive than p-methoxybenzoyl chloride ($\sigma_{p,CH_3} = -0.268$); however, the corresponding p-methoxybenzoyl derivative is the most stable of the three acylated products considered. Such reactions involve a trade-off; more reactive reagents yield less stable products, while less reactive reagents form more stable adducts. Many amine derivatizations in HPLC use p- or m-nitrobenzoyl chloride. Surprisingly, there is relatively little difference in the reactivity of these positional isomers ($\sigma_{p,NO_2} = 0.778$; $\sigma_{m,NO_2} = 0.710$). Acid-catalyzed acylations ($\varrho = 0.106$) are much less susceptible to substrate electronic effects at the reaction site than are the analogous base-catalyzed reactions ($\varrho = 2.2$).

Bases catalyzing acylations can function as nucleophiles by attacking the carbonyl carbon to form a tetrahedral intermediate or may serve as general bases by abstracting a proton from the attacking species. An example of general base catalysis is shown in reaction (2) for the pyridine-cata-

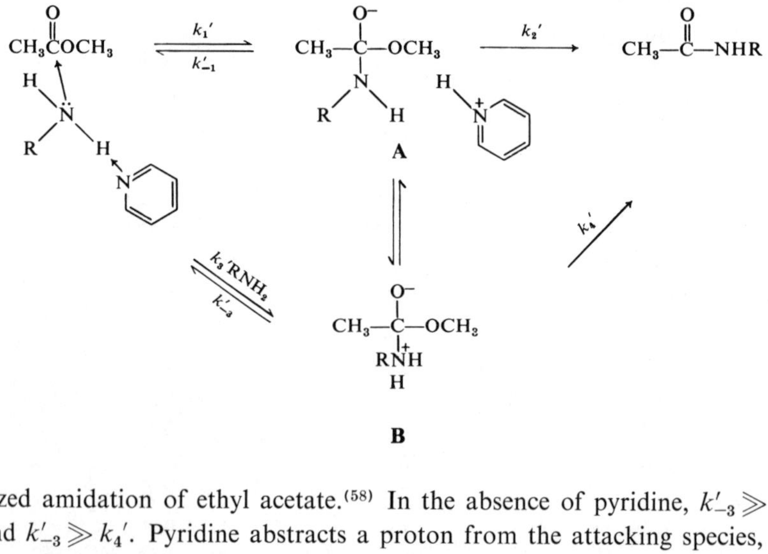

$$(2)$$

lyzed amidation of ethyl acetate.[58] In the absence of pyridine, $k'_{-3} \gg k_3'$ and $k'_{-3} \gg k_4'$. Pyridine abstracts a proton from the attacking species, increasing the nucleophilicity of RNH_2 and the concentration of intermediate **A**, which has less tendency to revert to starting material than does **B** (i.e., $k'_{-1}\mathbf{A} \ll k'_{-3}\mathbf{B}$; $k_1'C_5H_5N > k_3'$). Reaction (3) is an example of nucleophilic catalysis by pyridine.[59] The nucleophilic catalyst provides a highly reactive intermediate at steady state concentrations capable of more rapid reaction with nucleophiles (i.e., $k_3 \ll k_2KC_5H_5N$, where K is the dissociation constant for C to starting materials).

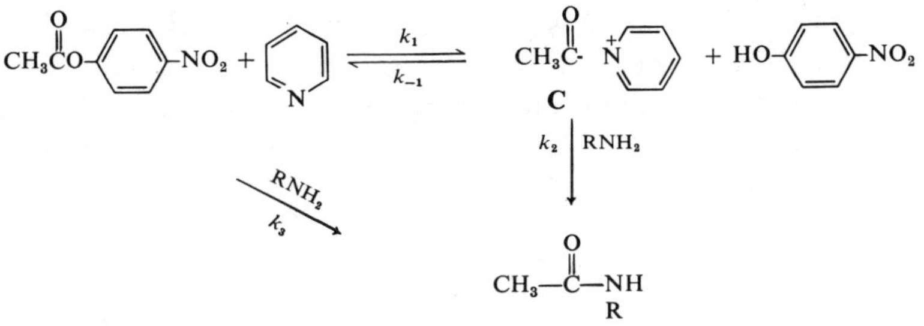

$$v = k(\text{acylating agent})(\text{nucleophile})(\text{catalyst}) \qquad (4)$$

The form of the rate expression [reaction (4)] does not distinguish nucleophilic from general base catalysis:

The type of catalysis depends on the structure of the substrate and the nucleophile. Reactions involving acylating agents with good leaving groups [e.g., p-nitrophenol in reaction (3)] are susceptible to nucleophilic catalysis,[60-62] while poor leaving groups [e.g., methoxide in [reaction (2)] promote general base catalysis. Between these two extremes, both general base and nucleophilic catalysis are observed. Nucleophilic catalysis is more sensitive to changes in the pK_a of the catalyst (β from Brønsted plots of log catalytic rate constant vs. pK_a of catalysis, ~ 0.8) than is general base-catalyzed ($\beta \sim 0.5$) acylation.[63] Common nucleophilic catalysts used to promote acyl transfers include pyridine,[64] 4-dimethylaminopyridine (DMAP),[65] imidazole,[66] and N-methylimidazole (NMI).[67] Although DMAP is 10^4 times more powerful as a catalyst than pyridine, pyridine must still be retained in the system as a proton scavenger and solvent;[68] furthermore, DMAP is prone to decomposition, limiting the concentration that can be used. Imidazole is an effective catalyst for some acyl transfer reactions. Catalytic efficiency is pH dependent.[66] Under acidic conditions, the N-acylimidazolium ion **19** reacts rapidly with nucleophiles ($k_{\text{hydrolysis}}$ for N-acetylimidazolium ion = 5×10^{-2} M^{-1} min^{-1}); at neutral or basic pH, however, the N-acylimidazole is deprotonated to form **20**, which reacts

slowly with nucleophiles ($k_{\text{hydrolysis}}$ for N-acetylimidazole = 9×10^{-5} M^{-1} min^{-1}).[69] Unfortunately, in many instances, in order to facilitate acylation, a neutral or basic environment is necessary, leaving the *N*-acylimidazole in the neutral, relatively unreactive form. In reactions with acid anhydrides, imidazole will in fact inhibit the rate of acyl transfer.[70] *N*-Methylimidazole is superior as a catalyst for acylation reactions because the corresponding 1-acyl-3-methylimidazolium ion, **21**, maintains a positive charge on nitrogen

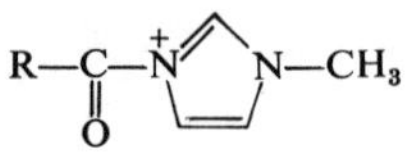

$$R-\overset{\underset{\parallel}{O}}{C}-\overset{+}{N}\diagup\diagdown N-CH_3$$

21

(regardless of pH) until dissipated by acyl transfer.[67,69] It should be noted that the nature of the counterion exerts major kinetic effects on the rate of reaction of 1-acyl-3-methylimidazolium ions with amines.[71] Although *N*-acylpyridinium ions are more reactive than 1-acyl-3-methylimidazolium ions, *N*-methylimidazole reacts faster and more effectively with acylating agents than does pyridine. *N*-Methylimidazole is an efficient catalyst of acylation reactions.

Acyl transfer reactions involving amines, although in some cases, acid catalyzed, are normally not carried out under acid conditions.

In general, the reactivity of carboxylic acid derivatives toward nucleophiles decreases in the order acid chloride > acid anhydride > ester > amide. This loss in reactivity correlates well with the double-bond character existing between the leaving group and acyl group[51]:

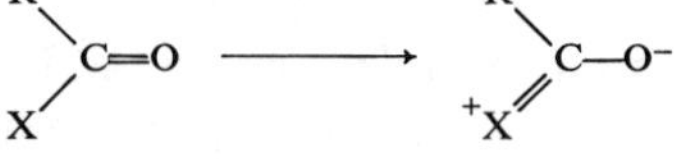

$$\overset{R}{\underset{X}{\diagdown}}C=O \longrightarrow \overset{R}{\underset{\overset{+}{X}}{\diagdown}}C-O^-$$

Thus, because the interaction between the acyl function in an anhydride is greater than its interaction with the chloride of an acid chloride, the latter is more reactive toward nucleophiles:

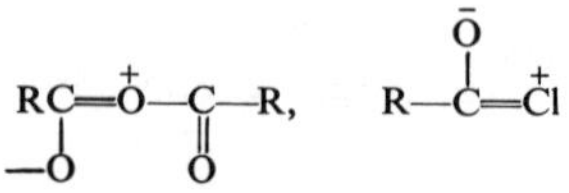

$$R\overset{+}{C}=\overset{+}{\underset{-O}{O}}-\overset{\underset{\parallel}{O}}{C}-R, \quad R-\overset{\overset{\bar{O}}{|}}{C}=\overset{+}{C}l$$

(assuming the R groups in the two compounds are similar). Since this higher bond order imparts stability to acylated products, amides are quite stable, relatively resistant to hydrolysis and transacylations.

Whereas primary and secondary amines react with acylating agents to form amides, tertiary amines are in most cases unreactive toward such functionalities owing to the instability of the products that would be formed ($RCONR_3^{\oplus}$):

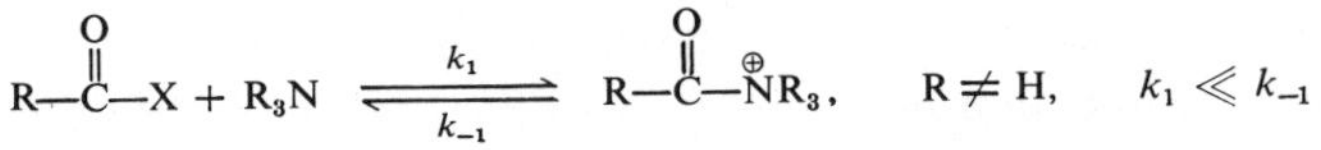

$$R-\overset{O}{\underset{\|}{C}}-X + R_3N \underset{k_{-1}}{\overset{k_1}{\rightleftharpoons}} R-\overset{O}{\underset{\|}{C}}-\overset{\oplus}{N}R_3, \quad R \neq H, \quad k_1 \ll k_{-1}$$

Aliphatic amines are generally more reactive than the corresponding aromatic amines because of the reduction in nucleophilicity of aromatic amines caused by delocalization of the electron pair on nitrogen about the ring:

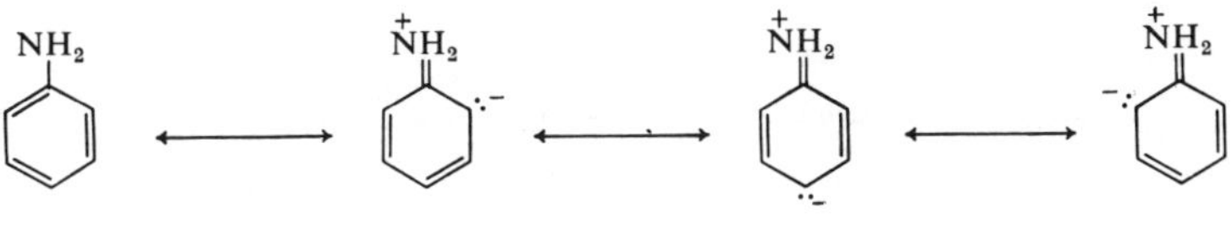

Further perturbation in aromatic amine nucleophilicity is observed as substituents are introduced onto the arene, as previously discussed. As substituents are introduced on the amine nitrogen which sterically impede approach of an acylating agent to the nucleophilic electron pair, acylation is also made more difficult:

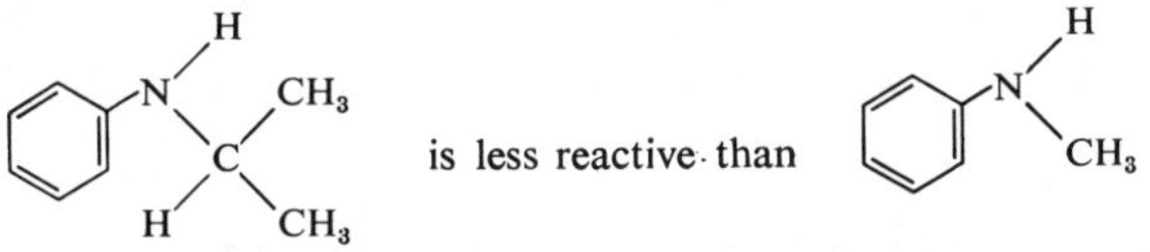

A rule of thumb cannot be invoked to compare relative reactivity of primary vs. secondary amines toward electrophiles because competing effects of increased steric bulk in secondary amines reduce reactivity, while nucleophilicity of nitrogen is increased by the inductive electron donation by the additional alkyl group in the secondary amine.

Acylation of tertiary amines (in which one of the substituents on nitrogen is a methyl group) has been accomplished by reaction with pentafluorobenzylchloroformate (**22**) [72-75] [reaction (5)]. Reaction proceeds with the concerted elimination of chloromethane to form the corresponding disubstituted amide. This derivatization reaction has been applied to the conversion of tertiary amines to electron-capture-active derivatives for sensitive GLC analysis. Although this chemical transformation has not as yet been applied to HPLC analysis, by using appropriately substituted chloroformates (i.e., replacing the $C_6F_5CH_2$ group in **22** with strong chromo-phoric, fluorophoric, or electroactive groups), the methodology should serve a useful role in improving detectability for monitoring various tertiary amines. The reaction is not suitable for derivatization of tertiary amines containing a pyridine nucleus (anywhere in the molecule) because the re-agent appears to attack pyridine resulting in ring opening and generation of a variety of degradation products.[76] A further potential problem encoun-tered with this reaction is in metabolic studies of tertiary amines where a principle pathway often involves N-demethylation. Under these circum-stances, this derivatization sequence would be incapable of differentiating between parent and metabolite (since both would be converted to the same product).

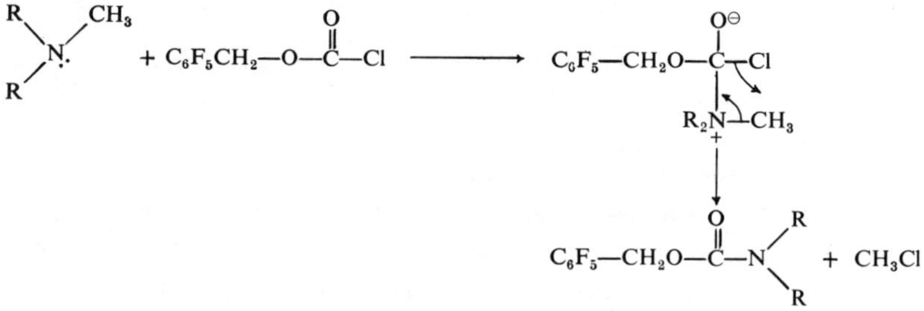

$$(5)$$

Sulfonyl chlorides, **23** (acid chlorides of sulfonic acids), represent an-other class of electrophilic reagents used to derivatize primary and sec-ondary amines [reaction (6)]; tertiary amines do not react. The resulting

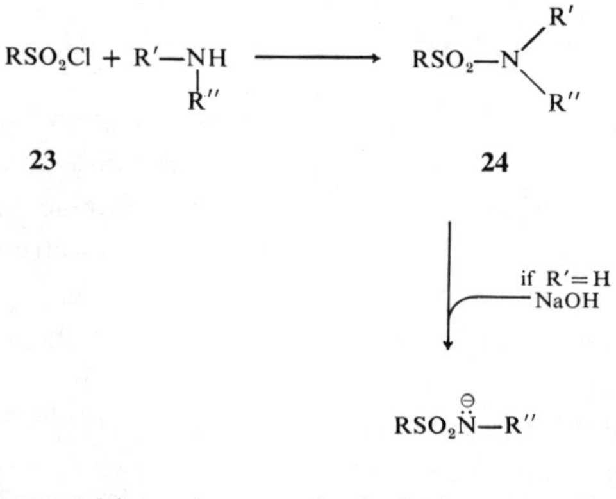

$$(6)$$

sulfonamides, **24**, are stable, resistant to hydrolysis or reaction with alternate nucleophiles. Sulfonamides offer the advantages over carboxyamides in that sulfonamides derived from primary amines can be separated from secondary amines by pH adjustment. The amide hydrogens in sulfonamides derived from primary amines are relatively acidic ($pK \sim 8.5$) and therefore at $pH \geq 10$, these compounds are ionized, allowing them to partition favorably from an organic phase into water, thereby affecting their separation from sulfonamides derived from secondary amines whose partition characteristics are unaffected by pH alterations. Such crude separations may facilitate later HPLC analysis by eliminating potential interferences which may be present in the analysis mixture. Similarly, the retention characteristics of sulfonamides derived from primary amines can be modified by pH adjustment. The stability of most commercial chromatographic supports, however, limits the upward adjustment of pH to ~ 8.5.

Sulfonyl halides have been used primarily to convert primary and secondary amines to fluorescent sulfonamides (fluorescence being controlled by the organic component of the reagent). Corresponding carboxamides are usually not suitable fluorescent derivatives because the carbonyl group (in conjugation with the aromatic nucleus) promotes intersystem crossing and therefore a loss in fluorescence yield.[15]

2.1.1.2. Applications—Fluorogenic Reagents. The most commonly employed group of reagents used to convert primary and secondary amines to fluorescent products are 5-dialkylaminonaphthalenesulfonyl chlorides, **25**, which yield fluorescent sulfonamides which exhibit emission bands between 470 and 510 nm. Reagents **25a–25e** differ in the chain length of the alkyl substituent on nitrogen. The reagents (and resulting derivatives) increase in

a, R = CH_3,	dansyl chloride
b, R = C_2H_5,	ethansyl chloride
c, R = n—C_3H_7,	propansyl chloride
d, R = n—C_4H_9,	bansyl chloride
e, R = n—C_5H_{11},	pentansyl chloride

25

hydrophobicity progressing from **25a** to **25e**; however, they have similar reactivities toward amines. Thus, reagents can be selected for partition chromatography based on desired retention characteristics, i.e., on a reverse phase column, the corresponding derivatives exhibit increasing capacity factors going from **25a** to **25e**. The major advantage in using the more hydrophobic reagents occurs in instances where the product is also being analyzed by mass spectrometry, since the longer alkyl groups offer a more favorable fragmentation for specific ion-current monitoring.[77] Dansyl

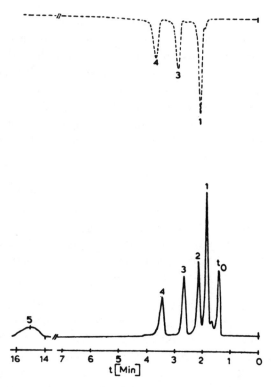

Fig. 4. Chromatogram (HPLC) of a cough syrup after derivatization. ———, uv detection; – – –, fluorescence detection. Peaks: (1) Dns–ephedrine; (2) narcotine; (3) Dns–cephaeline; (4) Dns–emetine; and (5) codeine; and t_0 dodecylbenzene.[80] Reproduced from the *Journal of Chromatography* by permission of Elsevier Publishing Co.

derivatives are the most commonly used of this group of fluorescent reagents. The sulfonyl chloride reacts with both amines (primary and secondary) and phenols; however, by using sodium acetate as a base, simultaneous derivatization of the phenolic group can be prevented. Reactions are routinely carried out in aprotic solvents such as acetone, acetonitrile, dimethylformamide or ethyl acetate. More forceful conditions (e.g., employing potassium carbonate[78] or potassium fluoride[79] solubilized with the crown ether, 18-crown-6) accelerate the reaction, often giving higher yields of fully substituted products, but are less selective (e.g., react with phenols). The reagent itself and reaction by-products fluoresce causing potential interferences at high sensitivity. Dansylation has been applied to the pre-HPLC derivatization of a number of drugs in pharmaceutical dosage forms. Frei and co-workers[80] describe the dansylation of cephaeline, emetine, ephedrine, and morphine in capsules and syrups. Codeine and narcotine (lacking phenolic groups), however, failed to react with dansyl chloride. Syrups could be derivatized directly without prior extraction of the active substance (after dilution with an equal volume of distilled water) with an acetone solution of dansyl chloride (Figure 4). Emetine capsules composed of a complex excipient and drug formulation were slurried in water, agitated in an ultrasonic bath (for 10 min) and the inert, insoluble ingredients then removed by centrifugation. The resulting solution was dansylated as described above and then analyzed by HPLC. In most cases derivatization took place at all available amino and phenolic sites to yield completely substituted derivatives. In several instances, the sensitivity reached by fluorimetric monitoring of the dansylated amine was similar to that obtained by uv detection; however, fewer potentially interfering peaks

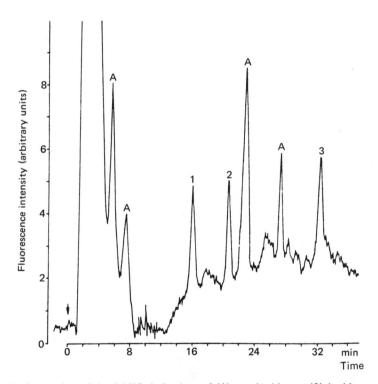

Fig. 5. Separation of the DANS derivatives of (1) aprobarbitone, (2) barbitone, and (3) heptabarbitone, extracted from 20 µl of blood. Each peak corresponds to 50 ng of barbiturate in the blood. A: DANS derivatives of unidentified blood constituents, or by-products of the dansylation reactions. Separation conditions: During the first 8 min, water, then a linear water–methanol gradient (100% → 35%) 20 min at a flow rate of 1 ml/min. Column: 500 × 2.5 mm; reversed phase packing based on pellicular silica with octadecyl residues bonded to the surface.[81] Reproduced from the *Journal of Chromatographic Science* by permission of Preston Publications, Inc.

were observed in the fluorescence tracing. The improved selectivity and sensitivity provided by dansylation has permitted monitoring these substances at levels of 1–10 ng/ml with reproducibility of $\sim 2\%$ (limited by the efficiency of the derivatization step).

Dansylation has also been applied to the pre-HPLC derivatization of amine-containing drugs for their analysis in biological fluids. In these cases the drug must be separated from the biological matrix prior to derivatization. This approach has been applied to the analysis of barbiturates (including aprobarbital, barbital, heptabarbital, secobarbital, and butabarbital) (Figure 5),[81] tocainide, (26), an antiarrhythmic drug,[82] a variety of polyamines[83–86] of interest in cancer research, tranylcypromine,[87] L-alanosine,[88] and free valine in plasma.[89]

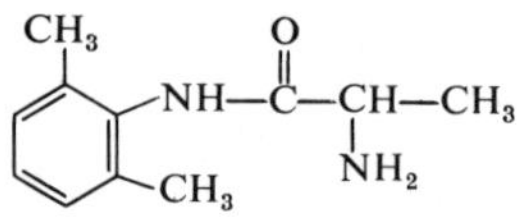

26

An assay for epinephrine, 27, in blood has been described based on its initial conversion to the tridansylated derivative, 28 (involving substitution at amine and phenolic positions),[90] prior to chromatographic analysis. Reaction proceeds to completion in 10 min, with 3 mol of DnsCl reacting per mole of epinephrine. Because of steric hindrance, 28 is relatively

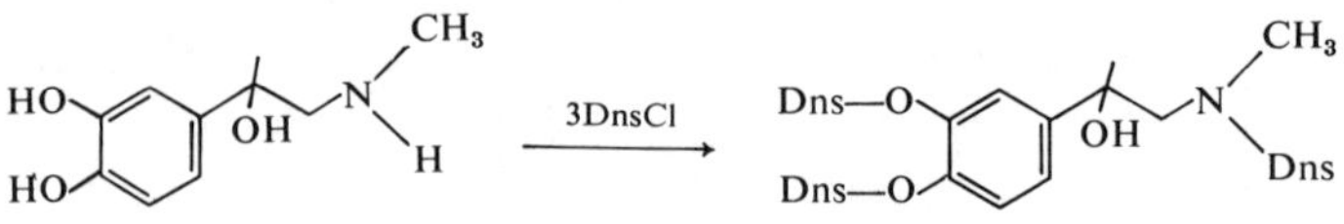

27   28

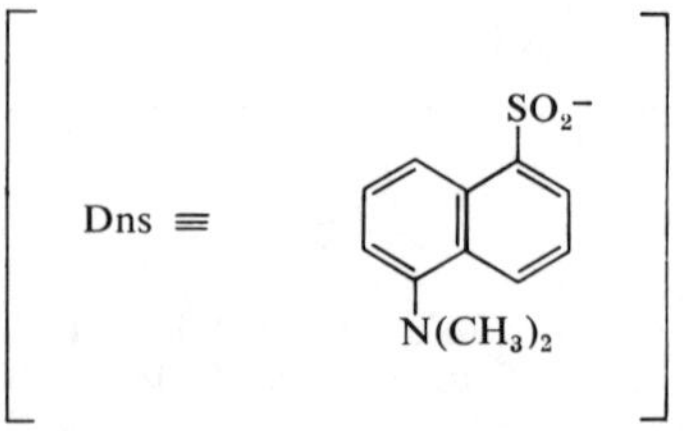

unstable, undergoing degradation with expulsion of one of the DNS residues to form dimers, such as 29. This lability of the tridansylated adduct limits

its potential as a practical derivative for a clinically useful epinephrine assay. Chlorpromazine and its metabolites[91,92] and cannabanoids[93] have also been analyzed in biological fluids by prechromatographic dansylation of the analytes. In these cases, however, TLC was used to separate the derivatives. The adaptation of this procedure to HPLC should be easily accomplished.

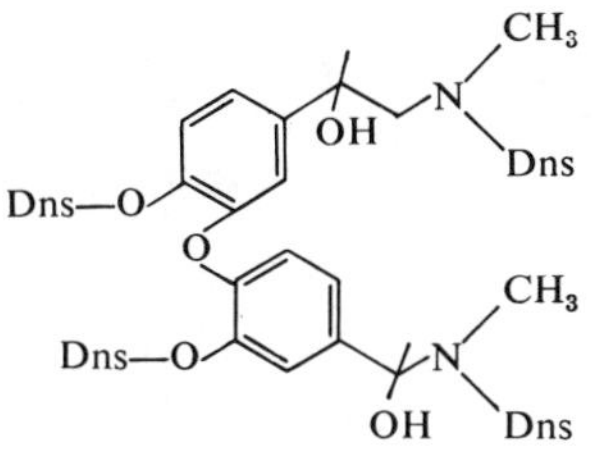

29

Karger and co-workers[94] found that the addition of the chiral metal chelate L-2-R-4-octyldiethylenetriamine-M(II), **30** (R = ethyl, isopropyl, or isobutyl; M = Zn or Cd), to the mobile phase of a bonded reverse phase

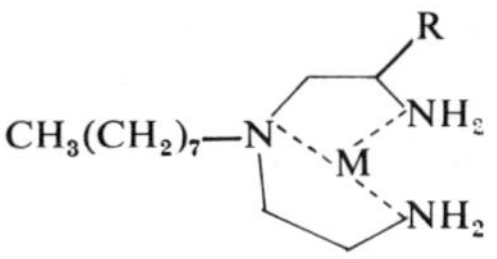

30

column results in the resolution of the dansyl derivatives of a wide variety of D,L-amino acid pairs (Figure 6; Table 3). The metal center imposes a rigid conformation at the exchange site and competes with the sulfonamide acidic hydrogen to permit the formation of a multidentate ligand capable of interaction through both the sulfonamide and carboxylate groups. The enantiomeric amino acids are thus separated as "diastereomeric" complexes, which are increased in stability and detectability by the sulfonamide moiety.

Two other acylating agents commonly used to produce fluorescent derivatives of *primary* amines are fluorescamine and *o*-phthalaldehyde. These reagents are fluorogenic, i.e., they form fluorescent products but are not fluorescent themselves.[95] This, of course, minimizes interferences and reduces the need for separation of the derivative from the reagent. Fluo-

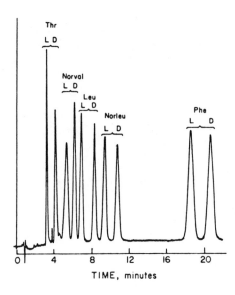

Fig. 6. Separation of d,l-dansyl amino acids. Conditions: 0.65 mM 1-2-isopropyl-dien-Zn(II); 0.17 M NH$_4$Ac to pH 9.0 with aqueous NH$_3$; 35/65 AN/H$_2$O; T = 30°C; flow rate, 2 ml/min; column: 15 cm by 4.6 mm i.d. Hypersil 5 μm C$_8$; solutes: Thr = threonine; Norval = norvaline; Leu = leucine; Norleu = norleucine; Phe = phenylalanine.[94] Reprinted from *Analytical Chemistry.* Copyright (1979) by permission of the American Chemical Society.

rescamine, **31**, reacts instantaneously with primary amines (at pH 8–9) in water to yield **32** [reaction (7)]. Although the reagent degrades rapidly in water, degradation products are not fluorescent.[96] The product, **32**, has a relatively low quantum yield in comparison with *o*-phthalaldehyde.[95] *o*-Phthalaldehyde, **33**, reacts rapidly with primary amines in the presence of a mercaptan (RSH) in aqueous solution (pH 9–11) to yield an isoindole,

$$ \tag{7} $$

31   **32**

34 [reaction (8)], with maximum fluorescence at λ_{ex} = 340 nm, λ_{em} = 455 nm [97]: The reagent is soluble and stable in water, unlike fluorescamine. It offers about tenfold greater sensitivity than fluorescamine and is considerably less expensive. Although the adduct (**34**) is relatively stable at pH 9–11 ($t_{1/2}$ ∼ 10 hr), it does degrade to a nonfluorescent species **35**, decomposition occurring more rapidly as pH is lowered.[98] Therefore, if used for precolumn

TABLE 3. Retention and Selectivity Parameters for Dansyl Amino Acid Derivatives[a,b]

Solute[d]	R[c] = ethyl		R = isopropyl		R = isobutyl	
	k'	α^e	k'	α^e	k'	α^e
α-Ala	3.4		3.65		6.5	
		1.00		0.79		0.8
	3.4		4.6		7.6	
α-NH$_2$ butyric	2.5		2.85		4.8	
		1.00		0.88		0.89
	2.5		3.25		5.4	
Norleu	8.25		10.0		18.0	
		0.96		0.86		0.92
	8.60		11.6		19.5	
Leu	6.15		7.3		14.2	
		0.94		0.84		1.08
	6.50		8.7		13.1	
Thr	2.55		2.7		5.1	
		1.19		0.71		0.82
	2.15		3.8		6.2	
Ser	4.0		4.1		7.8	
		1.25		0.67		0.74
	3.2		6.1		10.5	
Asp	1.9		2.9		5.4	
		1.00		1.16		1.00
	1.9		2.5		5.4	

[a] Common conditions: 0.65 mM L-2-R-dien-Zn; 0.17 M NH$_4$Ac to pH 9.0 with aqueous NH$_3$; 35/65 AN/H$_2$O; T = 30°C; flow rate 2 ml/min; column: 15 cm × 4.6 mm i.d. Hypersil 5 μm C$_8$.

[b] Reference 94. Reprinted from *Analytical Chemistry*. Copyright (1979) by permission of the American Chemical Society.

[c] R = alkyl substituent on metal chelate.

[d] Solutes are dansyl amino acids.

[e] $\alpha = k_L'/k_D'$.

derivatization, mobile phases designed to separate phthalaldehyde adducts should be maintained at relatively high pH.

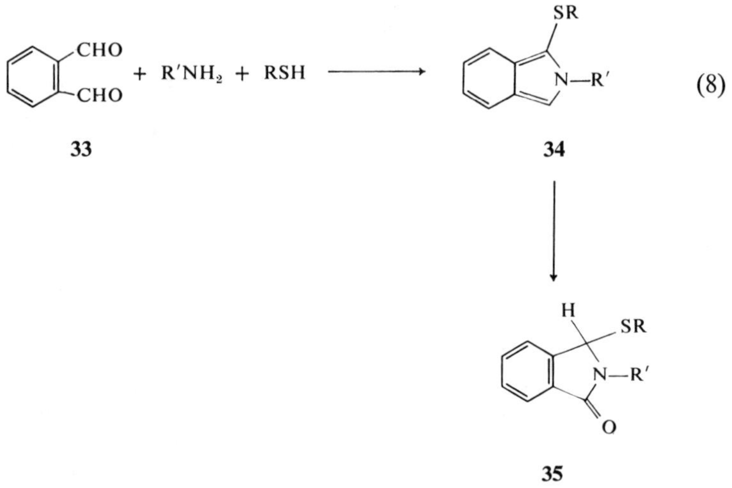

$$(8)$$

Fluorescamine is not widely used for preseparation derivatization, although its use in postcolumn reactors is well documented.[99,100] Fluorescamine has been used for preseparation derivatization of dopamine and norepinephrine isolated from rat brain.[101] A detection limit of 1 nmol of catecholamine/ml of biological fluid was reported; however, derivatization can involve formation of two products. Similarly, McHugh[102] found that amino acids derivatized with fluorescamine gave two products which were in equilibrium with one another: the expected adduct and a lactone formed between the free amino acid carboxyl group and proximal hydroxyl:

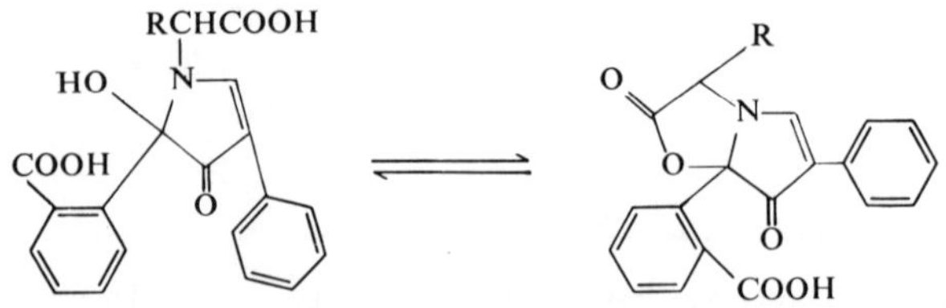

Stewart[103] described a fluorometric method for the determination of chlordiazepoxide, **36**, in dosage forms and biological fluids (urine and plasma) in which the drug is initially subjected to acid hydrolysis and the liberated amines, methylamine and 3-amino-5-chlorobenzophenone, are reacted with fluorescamine to yield an intensely fluorescent product at excitation and emission wavelengths of 390 and 486 nm, respectively. The method described in the literature measures bulk fluorescence and accordingly suffers from the disadvantages of (1) insufficient sensitivity for analysis of biological samples obtained after single-dose administration of chlordiazepoxide and (2) being subject to interference from other amine-containing drugs. By subjecting the adduct to HPLC analysis, it would appear that such difficulties could be overcome.

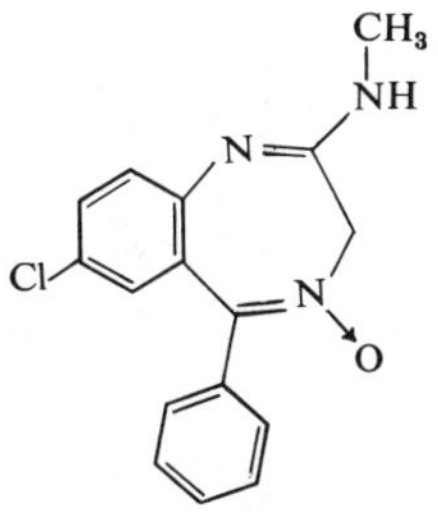

36

Precolumn o-phthalaldehyde derivatization has been applied to the analysis of amino acids.[104,105] Derivatization is complete in 1 min and products can be detected fluorometrically in column effluent at levels of ~40 pg (Figure 7). The only disadvantages of this procedure are the low response to cysteine and lysine and the inability to directly derivatize proline and hydroxyproline. This procedure has also been applied to analysis of catecholamines and their metabolites in urine[106] and brain tissue,[107] and histamine[108] in urine. After separation of the phthalaldehyde adducts on an RP-18 column, fluorometric detection in the low-picogram range is possible. A clinically useful method for determination of the aminoglycoside antibiotic gentamycin in human serum at levels of 500 ng/ml has been developed[109] based on its initial separation from the biological matrix by column chromatography. After washing the column with water, an aqueous solution of o-phthalaldehyde was added to the column and the derivatized gentamycin was eluted from the column with ethanol. An aliquot of the eluent was then chromatographed on an RP-18 column and the derivative detected fluorometrically.

2.1.1.3. Applications—Carboxylic Acid Chlorides. In some instances uv derivatization may be preferred to fluorescence labelling, in spite of loss in the inherent sensitivity provided by fluorescence. The spectrophotometric techniques are often simpler and more predictable.

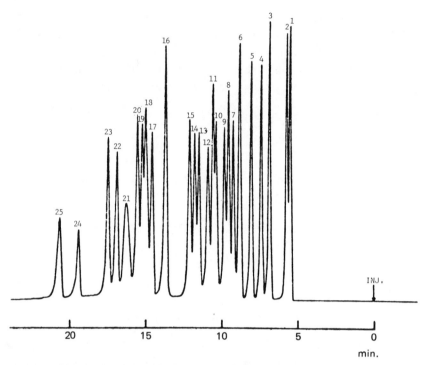

Fig. 7. Separation of *o*-phthalaldehyde derivatized amino acids. Gradient run: column: 200 mm × 4.6 mm i.d. Nucleosil RP-18, 5 μm. Mobile phase: solvent A—citrate–phosphate buffer, pH 7.7; solvent B—Methanol. Convex solvent gradient (0.5) from 20% B to 70% B in 15 min. Peaks: (1) cysteic acid; (2) Asp; (3) Glu; (4) *S*-carboxymethyl cysteine; (5) Asn; (6) Ser; (7) Gln; (8) His; (9) methionine sulfone; (10) Thr; (11) Gly; (12) Arg; (13) *β*-Ala; (14) Tyr; (15) Tyr; (16) *α*-aminobutyric acid; (17) Trp; (18) Met; (19) Val; (20) Phe; (21) NH_4^+; (22) Ile; (23) Leu; (24) Orn; (25) Lys.[105] Reprinted from *Analytical Chemistry*. Copyright (1979) by permission of the American Cancer Society.

Clark and Wells[110] examined the effect of ring substitution of a series of benzoyl chlorides and benzene sulfonyl chlorides as derivatizing agents for primary and secondary amines (Tables 4 and 5). The five *para* substituents studied were H, NO_2, OCH_3, CH_3, and Cl. In general, the benzamides showed more intense uv absorption (given as the extinction coefficient, ε) than the corresponding sulfonamide. The 4-methoxybenzamides had the highest absorptivity at 254 nm, and were thus concluded to offer the lowest detection limits. The *p*-methoxybenzoyl chloride derivatization of low levels (6 ng) of amphetamine in aqueous solution (reagent dissolved in tetrahydrofuran, THF) was carried out in 95% yield when the acid chloride to amine ratio was ≥500. 4-Methoxybenzoyl chloride also has the greatest stability in water of the reagents investigated, allowing the reaction to be carried

TABLE 4. Ultraviolet Absorption Properties of Benzamides[a]

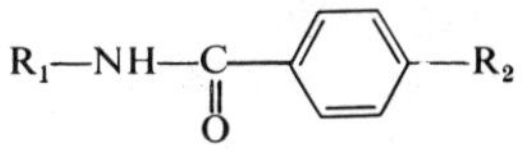

$$R_1—NH—\underset{O}{\overset{\|}{C}}—\text{⟨⟩}—R_2$$

R_1	R_2	λ_{max}	ε_{max} [b]	ε_{254} [b]
$C_6H_5CH(CH_3)$	H	225.5	12 400 ± 267	3 290 ± 154
n-C_3H_7	H	225.5	10 500 ± 140	2 470 ± 38
$C_6H_5CH(CH_3)$	NO$_2$	265.0	12 000 ± 246	10 600 ± 246
n-C_3H_7	NO$_2$	265.0	11 300 ± 232	10 100 ± 169
$C_6H_5CH(CH_3)$	OCH$_3$	252.0	17 500 ± 130	17 400 ± 130
n-C_3H_7	OCH$_3$	252.0	15 100 ± 196	14 900 ± 163
$C_6H_5CH(CH_3)$	CH$_3$	236.0	14 900 ± 242	7 440 ± 293
n-C_3H_7	CH$_3$	236.0	13 200 ± 268	5 390 ± 136
$C_6H_5CH(CH_3)$	Cl	236.0	16 500 ± 270	8 290 ± 138
n-C_3H_7	Cl	236.0	14 000 ± 130	6 260 ± 54

[a] Reference 110. Reproduced from the *Journal of Chromatographic Science* by permission of Preston Publications, Inc.
[b] Values reported are 95% confidence limits.

TABLE 5. Ultraviolet Absorption Properties of Benzenesulfonamides[a]

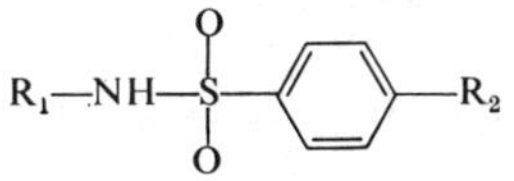

$$R_1—NH—\underset{O}{\overset{O}{\underset{\|}{\overset{\|}{S}}}}—\text{⟨⟩}—R_2$$

R_1	R_2	λ_{max}	ε_{max}	ε_{254} [b]
$C_6H_5CH(CH_3)$	H	—	—	635 ± 28
n-C_3H_7	H	221.0	8 430 ± 177	508 ± 91
$C_6H_5CH(CH_3)$	NO$_2$	268.5	10 800 ± 173	8500 ± 168
n-C_3H_7	NO$_2$	266.5	10 100 ± 222	8430 ± 194
$C_6H_5CH(CH_3)$	OCH$_3$	241.0	15 300 ± 282	4660 ± 100
n-C_3H_7	OCH$_3$	239.0	16 300 ± 178	3330 ± 60
$C_6H_5CH(CH_3)$	CH$_3$	229.0	11 600 ± 318	815 ± 140
n-C_3H_7	CH$_3$	227.0	11 700 ± 130	541 ± 76
$C_6H_5CH(CH_3)$	Cl	233.5	12 600 ± 270	1630 ± 75
n-C_3H_7	Cl	231.5	14 700 ± 315	1060 ± 55

[a] Reference 110. Reproduced from the *Journal of Chromatographic Sciences* by permission of Preston Publications, Inc.
[b] Values reported are 95% confidence limits.

out directly in the aqueous media with minimum hydrolysis, using THF to solubilize the reagent. Examples of amidation of drugs with p-toluenesulfonyl chloride,[111] m-toluoyl chloride,[112] m-nitrobenzoyl chloride,[113] p-nitrobenzoyl chloride,[114] and 4,N,N-dimethylaminoazobenzene sulfonyl chloride[115] for preseparation derivatization are also presented in the literature. Nitrophenylacetate derivatives of amines have been prepared by reaction with N-succinimidyl-p-nitrophenylacetate, **37**,[116] for use in improving sensitivity in HPLC analysis (with uv detection) of primary amines [reaction 9].

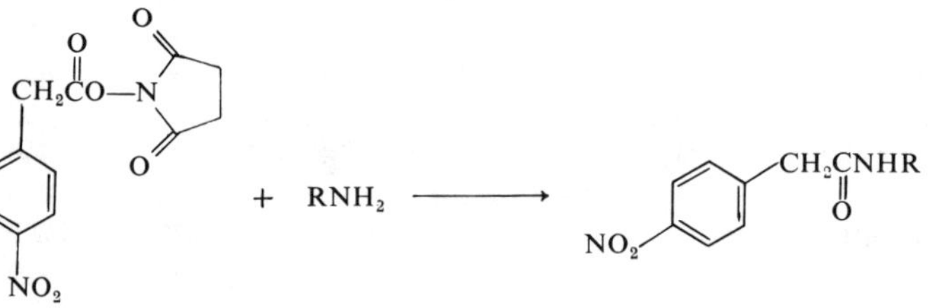

$$\text{37} \quad + \quad RNH_2 \quad \longrightarrow \quad (9)$$

Mixtures of optical isomers have been resolved to allow the monitoring of individual enantiomers by formation of appropriate diastereomeric mixtures using an optically active resolving agent, and subsequent chromatographic separation.[112] Optically active amines have been resolved chromatographically after reaction with chiral acylating agents. (+)-α-Methoxy-α-trifluoromethylphenylacetyl chloride, **38**, will form diastereomers with

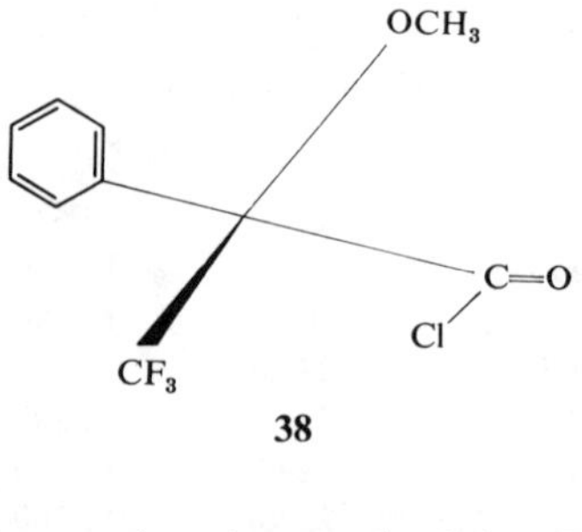

38

most amines,[112] which can then be resolved by GLC, TLC, or HPLC. Racemic amino acids were resolved by HPLC as a diastereomeric mixture after reaction with N-d-10-camphorsulfonyl p-nitrobenzoate[117] (Table 6).

Pirkle[118] has ingeniously developed a method for resolution of enantiomeric mixtures of amines derivatized with 3,5-dinitrobenzoyl chloride

TABLE 6. Retention Times (min) of D- and L-Amino Acid Derivatives[a] of N-d-10 Camphorsulfonyl p-Nitrobenzoate

Eluent A: isooctane–dichloromethane–isopropanol (79:16:5)
Eluent B: isooctane–dichloromethane–isopropanol (63:32:5)
Flow rate, 0.4 ml/min. Column, MicroPak–NH$_2$

Amino acid	Eluent A			Eluent B		
	L	D	D/L	L	D	D/L
Leucine	3.9	4.4	1.1	2.7	2.8	1.0
Isoleucine	4.4	5.0	1.1	2.9	3.1	1.1
Phenylalanine	6.2	8.5	1.4	3.3	4.1	1.2
Methionine	7.4	10.0	1.4	3.6	4.6	1.3
Alanine	7.2	9.3	1.3	3.7	4.4	1.2
Glutamic acid	12.8	16.8	1.3	4.2	5.2	1.2
Tryptophan	29.2	49.6	1.7	9.0	14.9	1.7
Tyrosine	33.2	47.2	1.4	11.6	16.2	1.4

[a] Reference 117. Reproduced from the *Journal of Chromatography* by permission of Elsevier Publishing Co.

by HPLC on a rationally devised chiral column. He suggests that if a chiral stationary phase is to differ in its affinity for enantiomers, there must be a minimum of three points of interaction with at least one of the enantiomers, and at least one of these sites of interactions must be stereospecific.

Chiral 2,2,2-trifluoro-1-(9-anthryl)ethanol, **39**, was covalently bound to a silica support to form the stationary phase. The trifluoroethanol portion

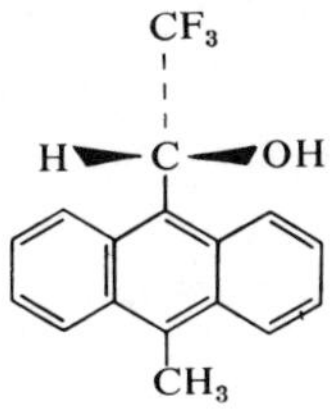

39

of the stationary phase affords two-point binding to molecules through hydrogen bonding of electron-rich atoms on the solutes with the hydroxylic and methine hydrogens. This interaction of **39** with appropriate enantiomers affords "diastereomeric chelatelike" complexes, **40a** and **40b**. These diastereomeric chelates will differ only in stability (and therefore differ in chromatographic retention characteristics) if differences exist in the free energy of interaction of the individual enantiomers at the third interaction site; i.e., a stereochemically dependent third interaction site must be included. The anthryl substituent is a π base; by conversion of the amine

40a

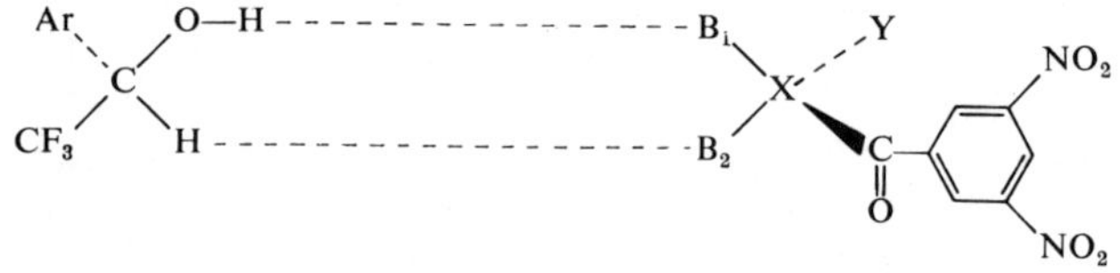

40b

solute to a 3,5-dinitrobenzamide derivative, a π-acid function is generated that can potentially interact with the anthryl moiety through a π–π donor–acceptor complex. The efficiency of this interaction will depend on the stereochemical orientation of the 3,5-dinitrobenzoyl group relative to the anthryl moiety. As these groups become more proximal to one another, interaction is enhanced and retention on-column is increased (i.e., **40b** is eluted before **40a** because maximum distance is maintained between the two groups in **40b**). Therefore, elution order can also be predicted. This approach has not only been applied to resolution of enantiomeric mixtures of amines, but also *alcohols, amino acids, hydroxy acids, lactones, thiols,* and *sulfoxides* (Figure 8). The tremendous potential for this approach is evidenced by the wide variety of functionalities responsive to this technique and the excellent separation of enantiomers that is achieved (Table 7).

2.1.2. Arylation

2.1.2.1. Theoretical Aspects. Although amines are most often derivatized for HPLC analysis through acylation reactions (pathway *A*), they have also been subjected to nucleophilic displacement reactions with aro-

TABLE 7. Separation of Derivatized Enantiomers upon Chiral Stationary Phase[a]

Q	B	R	α	k_1'	k_2'
NH	C_6H_5	CH_3	1.19	3.01	3.57
NH	C_6H_5	C_2H_5	1.29	2.63	3.40
NH	C_6H_5	i-C_3H_7	1.26	2.29	2.89
NH	α-Thienyl	CH_3	1.14	3.04	3.45
NH	p-$CH_3OC_6H_4$	CH_3	1.18	5.19	6.13
NH	p-$CF_3C_6H_4$	CH_3	1.11	2.03	2.25
NH	CO_2CH_3	CH_3	1.08	4.16	4.48
NH	CO_2CH_3	$CH_2C_6H_5$	1.10	4.00	4.41
NH	CO_2CH_3	i-C_3H_7	1.05	2.75	2.89
NH	CO_2CH_3	$CH_3S(CH_2)_2$	1.04	4.81	5.03
NH	CO_2CH_3	C_6H_5	1.19	4.17	4.48
NH	CONH n-C_4H_9	CH_3	1.30	2.10	2.61
NH	CONH n-C_4H_9	i-C_3H_7	1.33	0.82	1.10
NH	CONH n-C_4H_9	$CH_2C_6H_5$	1.56	1.22	1.90
NH	CONH n-C_4H_9	$(CH_2)_2SCH_3$	1.37	1.66	2.30
NH	CONH n-C_4H_9	C_6H_5	1.78	0.75	1.33
NH	CH_2OH	C_6H_5	1.38	2.36	3.30
NH	CH_2OOCCH_3	C_6H_5	1.18	5.30	6.20
O	CO_2CH_3	$2,5$-$(CH_3)_2C_6H_3$	1.05	2.51	2.63
O	CO_2CH_3	CCl_3	1.06	1.88	2.00
O	CO_2CH_3	CH_3	1.05	3.39	3.57
O	CO_2CH_3	$CH_3O_2CCH_2$	1.12	6.95	7.81
O	C_6H_5	CH_3	1.08	1.56	1.68
O	C_6H_5	C_2H_5	1.10	1.23	1.35
O	C_6H_5	C_3H_7	1.12	1.09	1.23
O	C_6H_5	i-C_3H_7	1.14	0.81	0.93
O	C_6H_5	t-C_4H_9	1.13	0.81	0.92
O	α-Naphthyl	CH_3	1.04	2.21	2.29
O	9-Anthryl	CH_3	1.06	2.53	2.67
O	2-Phenanthryl	CH_3	1.07	3.13	3.36
O	p-$CH_3SC_6H_4$	C_3H_7	1.10	1.56	1.72
O	p-$CH_3OC_6H_4$	C_3H_7	1.12	1.60	1.79
S	C_6H_5	CH_3	1.07	1.47	1.57
S	α-Naphthyl	i-C_3H_7	1.16	1.05	1.21

[a] Reference 118. Reprinted from the *Journal of Organic Chemistry*. Copyright (1979) by permission of the American Chemical Society.

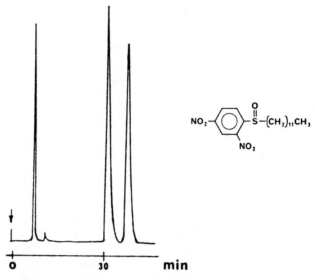

Fig. 8. Separation of the enantiomers of racemic *n*-dodecyl 2,4-dinitrophenyl sulfoxide. The first peak is durene, added as a retention marker.[118] Reprinted from the *Journal of Organic Chemistry*. Copyright (1979) by permission of the American Chemical Society.

matic (pathway *B*) or aliphatic substrates (pathway *C*) as a means of derivatization. Reaction of amines with aromatic reagents proceeds by nucleophilic addition to form a tetrahedral intermediate[119] **(41)**.

Proton transfer occurs before or concomitantly with expulsion of the leaving group (X) to give the corresponding aniline, **42** [reaction (10)]. The reaction is more applicable to amines with strong nucleophilic character (R_1R_2NH), and is facilitated by reagents which incorporate good leaving groups (i.e., where X in **41** is a poor nucleophile) and by substituents on the aromatic reagent (G in **41**) that stabilize the transient benzanion intermediate, **41**. The leaving group potential for X in 1-X-2,4-dinitrobenzenes decreases in the order $F > NO_2 > OSO_2C_6H_4CH_3\text{-}p > SO_6CH_5 \sim Br \sim Cl > I$.[120] The intermediate is stabilized by electron-withdrawing substituents that promote delocalization of charge. The reaction of amino acids

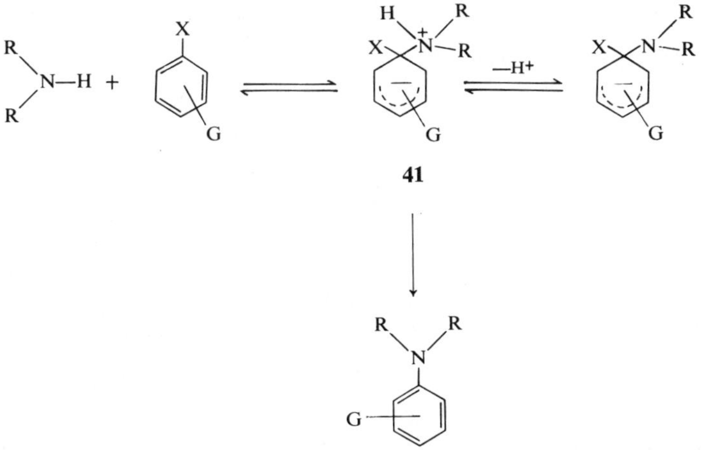

41

$$(10)$$

42

with 2,4-dinitrofluorobenzene (Sanger's reagent),[121] **43**, illustrates this type of derivatization. Reaction is initiated by nucleophilic attack by the amine

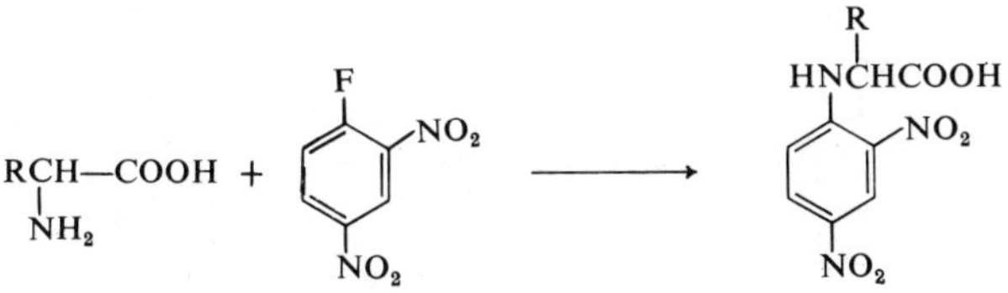

43   **45**

on **43** to yield the intermediate, **44**; its negative charge is delocalized about the ring (**44a–44c**) and further stabilized through resonance interaction with the nitro groups (**44d, 44e**). Expulsion of fluoride yields the product, **45**.

2.1.2.2. Applications. Reaction with 2,4-dinitrofluorobenzene has been applied to the preseparation derivatization of amino acids,[121,122] dipeptides,[123] and aromatic amines[124] to facilitate their separation and spectrophotometric detection. This reagent also reacts with phenols[125] in-

troducing a potential interferent; however, the corresponding aromatic
ether (from phenols) and amines can be separated by an acid wash.

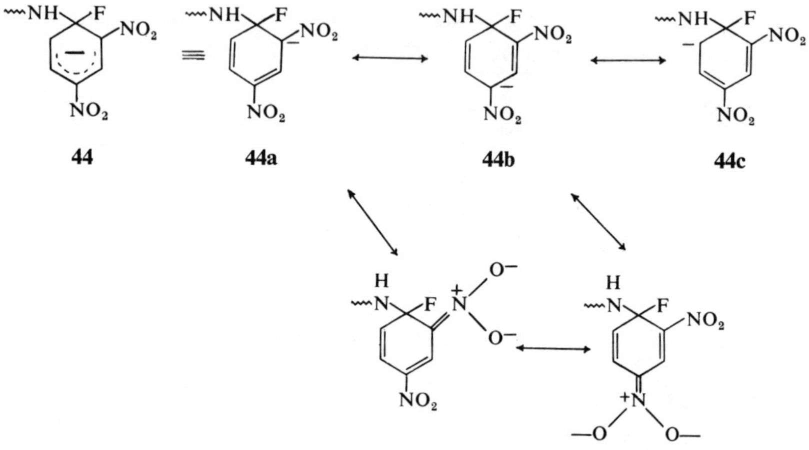

4-Chloro-7-nitrobenz-2,1,3-oxadiazole, **46**, reacts with primary and
secondary *aliphatic* amines under mildly basic conditions to form intensely
fluorescent products [reaction (11)]. The reagent does not form fluorescent
products with aromatic amines or phenols and therefore offers additional
selectivity as a derivatizing agent. Compound **46** reacts with amphetamines

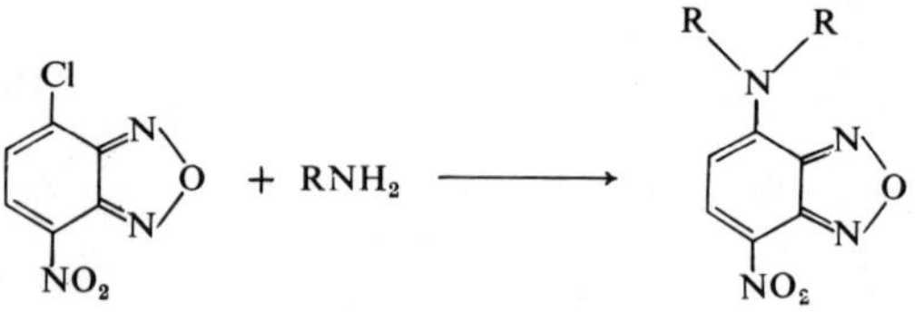

in urine to yield fluorescent products ($\lambda_{ex} = 482$ nm, $\lambda_{em} = 484$ nm) detectable at nanogram levels.[126,127] The reagent has been applied to the
analysis of nitrosamines after reduction to the primary amine.[128] The 2,1,3-
oxadiazole-derivatives are then separated chromatographically and detected
fluorimetrically. This reagent has recently been used to monitor the levels
of pareptide (L-prolyl-*N*-methyl-D-leucyl-glycinamide), a pharmacologi-
cally important synthetic peptide,[129] to quantitate the levels of other aliphatic amines[130] as well as some hydroxyphenylalkylamines.[131]

2.1.3. Alkylation

Alkylation has not been widely used for derivatization of amines. Mikkelson *et al.*[132] has determined pilocarpine, **47**, in ocular aqueous humor by quarternization of the methylimidazole tertiary amine with *p*-nitrobenzylbromide to form **48** [reaction (12)]. The resulting derivative is sep-

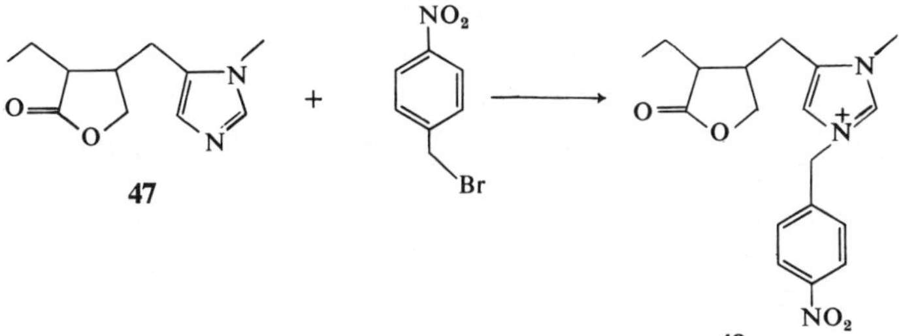

$$(12)$$

arated on an RP-18 column with a methanol:water mobile phase containing sodium octanesulfonate as an ion-pairing agent, permitting its resolution from isopilocarpine, a major decomposition product and impurity. The derivative strongly absorbs uv light ($\varepsilon \sim 12\,000$) at 254 nm. The method has also been applied successfully to a number of aromatic heterocyclic and tertiary alkyl amines.

2.2. Alcohols and Phenols

2.2.1. Acylations

2.2.1.1. Theoretical Aspects. Alcohols, by virtue of the available lone pair electrons on oxygen, are also susceptible to attack by electrophilic reagents. They participate in acylation reactions to form esters with acid chlorides, acid anhydrides, and sulfonyl chlorides in a manner similar to amide formation with amines. Reactions are also subject to similar catalytic influences. In general, esterification reactions occur somewhat more readily than the corresponding amidation; however, the resulting esters are more prone to hydrolytic (solvolytic) degradation than the corresponding amides. Sulfonate esters, in particular, are often quite labile toward water (the corresponding sulfonamides are less easily hydrolyzed).

Whereas preseparation derivatization of alcohols is often necessary in GLC to avoid analyte decomposition and improve chromatographic properties, its use in HPLC has been primarily to improve detection prop-

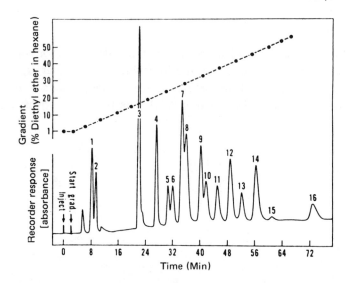

Fig. 9. Separation of 16 perbenzoylated hydroxy compounds on Corasil II (37–50 μm; Waters Assoc.) by HPLC. Temperature, ambient. Benzoates of: (1) 1-butanol; (2) methanol; (3) ethylene glycol; (4) glycerol; (5) α-D-xylose; (6) β-D-xylose; (7) α-D-Mannose; (8) α-D-glucose, α-D-galactose; (9) β-D-glucose, D-galactose; (10) D-galatose; (11) β-D-mannose; (12) sucrose; (13) α- and β-maltose; (14) lactose; (15) maltotriose; (16) lactose oligosaccharides impurities.[134] Reproduced from the *Journal of Chromatography* by permission of Elsevier Publishing Co.

erties of analytes. Accordingly, derivatization has involved the conversion of alcohols to aromatic esters with chromophoric or fluorophoric properties.

2.2.1.2. Applications. Benzoylation has been applied to the analysis of hydroxysteroids[133] and mono-, di-, and trisaccharides (Figure 9).[134] The resulting esters strongly absorb uv light ($a_m \geq 10^4$) at the λ_{max} of 230 nm. By using the corresponding p-nitrobenzoate esters, the extinction coefficient of the ester remains about the same as the corresponding benzoate ($\sim 10^4$), but a bathochromic shift in λ_{max} to 254 nm is observed.[133] This results in a tenfold improvement in sensitivity to common spectrophotometric HPLC detectors. p-Nitrobenzoylation has been applied to the analysis of digitalis glycosides (Table 8) [in water (Figure 10),[135] pharmaceutical dosage forms,[136] and plant extracts[137]], polyhydric alcohols,[138] steroids,[133] and mono-, di-, and trisaccharides.[139] Digitalis glycosides[136] and other polyhydric alcohols[138] form polyesterified products. All hydroxyl groups on the sugar moieties are *quantitatively* derivatized. Only the 14-HO group of digitoxigenin was resistant to p-nitrobenzoylation. Conversion of sugars to benzoyl or p-nitrobenzoyl esters not only improves detectability but also allows for separation of the anomers of slowly mutorotating sug-

TABLE 8. Chemical Structures of the Digitalis Glycosides and Aglycones Investigated[a]

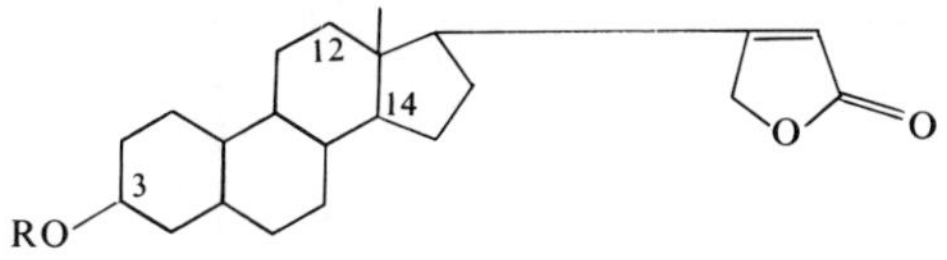

Compound	12–	14–	16–	R[b]
Digitoxigenin	H	OH	H	H
Digitoxigenin monodigitoxoside	H	OH	H	D–
Digitoxigenin bisdigitoxoside	H	OH	H	D–D–
Digitoxin	H	OH	H	D–D–D–
Acetyldigitoxin	H	OH	H	AcD–D–D–
Lanatoside A	H	OH	H	G–AcD–D–D–
Desacetyl lanatoside A	H	OH	H	G–D–D–D–
Gitoxigenin	H	OH	OH	H
Lanatoside B	H	OH	OH	G–AcD–D–D–
Desacetyl lanatoside B	H	OH	OH	G–D–D–D–
Digoxigenin	OH	OH	H	H
Digoxigenin monodigitoxoside	OH	OH	H	D–
Digoxigenin bisdigitoxoside	OH	OH	H	D–D–
Digoxin	OH	OH	H	D–D–D–
Acetyldigoxin	OH	OH	H	AcD–D–D–
Lanatoside C	OH	OH	H	G–AcD–D–D–
Desacetyl lanatoside C	OH	OH	H	G–D–D–D–
Lanatoside D	OH	OH	OH	G–AcD–D–D–
Diginatigenin	OH	OH	OH	H
Diginatin	OH	OH	OH	D–D–D–
Lanatoside E	H	OH	$-O-\overset{\overset{\displaystyle O}{\|}}{C}-H$	G–AcD–D–D–
Gitaloxigenin	H	OH	$-O-\overset{\overset{\displaystyle O}{\|}}{C}-H$	H
Gitaloxin	H	OH	$-O-\overset{\overset{\displaystyle O}{\|}}{C}-H$	D–D–D–

[a] Reference 136. Reproduced from the *Journal of Chromatography* by permission of Elsevier Publishing Co.
[b] D = digitoxose; AcD = acetyldigitoxose; G = glucose.

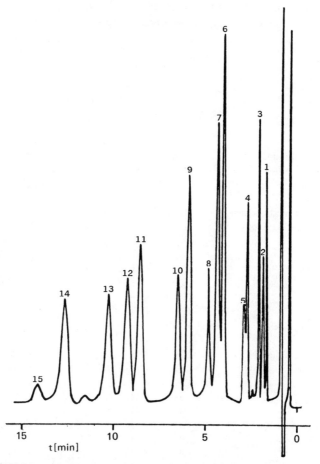

Fig. 10. HPLC of three 4-NB derivatives each of the A, B, C, D, and E series of digitalis glycosides. (1) Gitoxigenin; (2) digitoxigenin; (3) diginatigenin; (4) digoxigenin; (5) gitaloxigenin; (6) gitoxin; (7) digitoxin; (8) diginatin; (9) digoxin; (10) gitaloxin; (11) lanatoside B; (12) lanatoside A; (13) lanatoside D; (14) lanatoside C; (15) lanatoside E. Solvent system: n-hexane–chloroform–acetonitrile (30:10:9). Column: SI-60, 5 μm; 15 cm × 3 mm i.d. Flow-rate: 1.5 ml/min; p = 120 atm. Apparatus: Hewlett-Packard UFC 1000 equipped with a DuPont 842 uv dectector (λ = 254 nm). Injection: Rheodyne 7105 injection system: 20-μl injection volume in chloroform.[136] Reproduced from the *Journal of Chromatography* by permission of Elsevier Publishing Co.

ars[134,139] if the sugar is not dissolved in pyridine before derivatization is carried out. If, however, sugars are kept in pyridine, other anomerization reactions take place. This procedure has been applied to the determination of mono- and disaccharide composition in pharmaceutical syrups.[139] Compounds were separated on a micro silica column.

To improve sensitivity, esterification of alcohols using 3,5-dinitro-benzoyl chloride was attempted. Although the molar absorptivity of the resulting ester is higher than the mononitrobenzoyl ester, the derivative is more rapidly hydrolyzed to the original alcohol.[133] 3,5-Dinitrobenzoyl chloride has, however, been used for the preseparation derivatization of ppm levels of ethylene glycol in a series of polyethyleneglycols.[140]

Anisyl chloride (p-methoxybenzoyl chloride) has been used to derivatize solutions of hexachlorophene[141] and for the determination of ppm levels of pentaerythritol tetranitrate (used to prevent anginal attack) in plasma samples.[142] Again, derivatization is used to improve detectability. A similar rationale was given for using azobenzene-4-sulfonyl chloride in the derivatization of estrogen for visualization after TLC separation.[143]

2.2.2. Reactions Specific to Phenols

Phenols represent a special class of alcohols. The most important chemical difference between phenols and aliphatic alcohols is that the former ($pK_{a,\phi OH} = 10$) is about a million times stronger an acid that the latter ($pK_{a,ETOH} \sim 16$). Thus, by appropriate pH adjustment, phenols can exist as strongly nucleophilic anionic species in aqueous solution (pH $\geq$ 11), while this is not possible for alcohols. Accordingly, in alkaline solution, phenols are susceptible to productive reaction with weaker electrophiles than are alcohols. In basic solution, phenols (but not aliphatic alcohols) react with dansyl chloride (dissolved in acetone) to yield fluorescent sulfonate esters (dansyl phenols). Attempts should be made to eliminate water from the environment in which the derivative is stored, because of its propensity for undergoing hydrolysis. The fluorescence characteristics of dansyl phenol derivatives vary greatly with the dielectric constant of the media in which emission is measured.[5] Nonpolar, low dielectric solvents increase fluorescence intensity with a hypsochromic shift in emission wavelength maxima (λ_{em}), while more polar solvents cause bathochromic shifts in λ_{em} and a reduction in fluorescence intensity. Roos has recently described[144] a method for the analysis of estrogens in pharmaceutical tablet, suspension, and injectable dosage forms using this reagent. After separation from inert materials, the estrogens are converted to sulfonate esters with dansyl chloride (e.g., estradiol dansyl derivative, **49**) and the derivatized solutions analyzed by HPLC with fluorescence detection. Note only the phenolic hydroxyl reacts. The method can simultaneously detect estradiol valerate, estrone, ethinyl estradiol, α-estradiol, and estradiol at 50-pg levels. This method is simpler, more reliable, and more sensitive than similar procedures in which dansylated estrone, estradiol, and estriol are separated by TLC and detected by fluorodensitometry[145,146] or by measuring fluorescence of the eluted TLC spot.[147] Urinary estrogen levels have also been measured

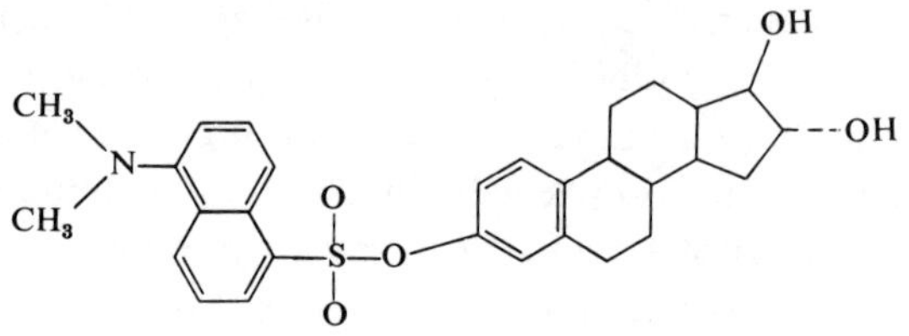

49

as dansylated derivatives.[147,148] Dansylation has also been used as a fluorogenic label for chlorophenols[149] and hydroxylated chlorobiphenyls[150] prior to their chromatography.[144] Fluorescent intensity of the adduct decreased and emission characteristics changed as the number of halogens in the phenol increased (Table 9). Another application of dansylation prior to chromatography involved the HPLC analysis of hydroxybiphenyls in urine[151] and determination of cannabanoids (utilizing TLC) by mass spectrometry[152] with potential application to HPLC analysis.

TABLE 9. Effect of Chloro Substituents on Fluorescence of Dansyl-Phenol[a]

Dansyl product	Relative fluorescence intensity		Wavelength maxima, nm	
	Hexane	Benzene	Hexane ex, em	Benzene ex, em
Phenol	55	42	350 466	355 500
p-Chlorophenol	49	36	352 470	361 508
m-Chlorophenol	48	26	355 474	361 515
O-Chlorophenol	46	17	355 476	362 515
2,4-Dichlorophenol	41	6	356 482	365 525
2,4,5-Trichlorophenol	11	1	357 485	368 530

[a] Reference 5. Reproduced from the *Journal of Chromatographic Science* by permission of Preston Publications, Inc.

2.2.3. 1-Ethoxy-4-(dichloro-5-triazinyl) naphthalene– Fluorescent Reagent for Alcohols

Whereas primary aliphatic alcohols fail to form fluorescent dansyl derivatives, the more reactive reagent, 1-ethoxy-4-(dichloro-s-triazinyl) naphthalene, **50**, was shown to react with the aliphatic C_{21} hydroxyl of corticosteroids to yield fluorescent derivatives, suitable for TLC analysis with fluorescence detection.[153] This reagent has not, unfortunately, been studied further.

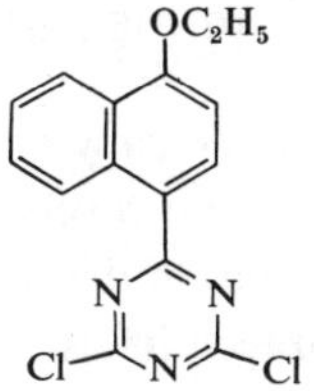

50

2.3. Carboxylic Acids

2.3.1. Introduction

The major groups of carboxylic-acid-containing compounds of bio-medical interest are fatty acids, prostaglandins, and penicillins. The pre-separation derivatization of carboxylic acids has been carried out to improve their chromatographic properties as well as to enhance detectability. The classically used derivatives have been esters and to a lesser extent amides.

Carboxylic acids can be converted to esters by three different synthetic routes.

(1) Direct reaction with an alcohol under acid conditions:

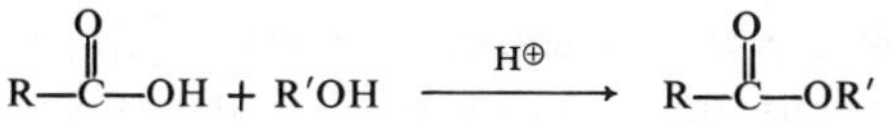

(2) Conversion to an acid chloride or acid anhydride and then reaction with an alcohol:

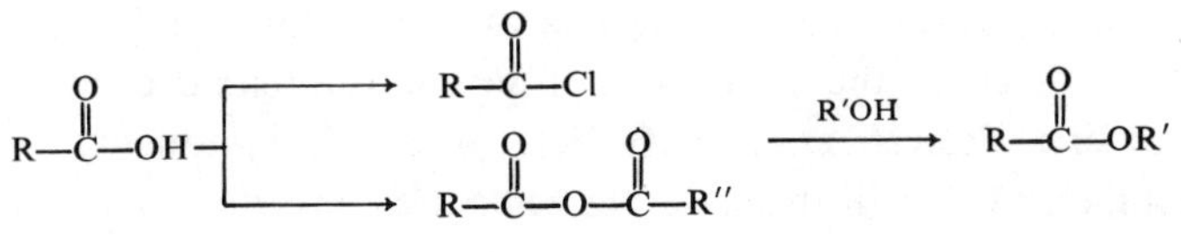

(3) Ionization of the carboxylic acid followed by alkylation:

$$\underset{\overset{\|}{R-C-OH}}{\overset{O}{}} \longrightarrow \underset{\overset{\|}{R-C-O^{\ominus}}}{\overset{O}{}} \xrightarrow{\text{R'X}} \underset{\overset{\|}{R-C-OR'}}{\overset{O}{}}$$

Route (2) is also applicable to amide formation. Because of the harsh conditions and low yields associated with esterification via route (1), it has not been widely used as an analytical reaction.

2.3.2. Use of Acid Halide Intermediates

Route (2), involving the intermediacy of a reactive acylating agent, has been employed to a limited extent; however, the lability of the intermediate toward water and other nucleophiles in the reaction media and the multistep nature of the process has discouraged its use as well. It has, however, proved effective in the amidation of fatty acids with *p*-methoxyaniline.[154] Acids (C_6–C_{24}) are converted to the corresponding acid chloride with either a

$$RCOOH + \phi_3P + CCl_4 \longrightarrow \phi_3PO + CHCl_3 + \underset{\overset{\|}{R-C-Cl}}{\overset{O}{}} \xrightarrow{RNH_2} \underset{\overset{\|}{R-C-NHR}}{\overset{O}{}}$$

$$(13)$$

mixture of triphenylphosphine and carbon tetrachloride [reaction (13)], or by using polymeric polystyryl-diphenylphosphine/CCl_4 [reaction (14)] as chlorinating agent:

$$RCOOH + \text{polymer}-C_6H_4-P-\phi_2 + CCl_4 \longrightarrow \text{polymer}-C_6H_4P-O(\phi_2)$$

$$+ \underset{\overset{\|}{R-C-Cl}}{\overset{O}{}} \xrightarrow{RNH_2} \underset{\overset{\|}{RC-NHR}}{\overset{O}{}} \quad (14)$$

The advantage of the latter approach is that no organophosphine or phosphine oxide is present in solution after anilide formation and therefore they cannot interfere with subsequent chromatographic separations. Furthermore, the polymeric material chlorinates the acid more rapidly than ϕ_3P/CCl_4. The acid chlorides are not isolated but reacted with *p*-methoxyaniline to form amides with molar absorptivities of $>2.4 \times 10^4$ at 254 nm, which are then chromatographed on an RP-18 column with methanol–water or acetonitrile–water mobile phases. This approach [route (2)] has also been applied to the separation of acyclic isoprenoid enantiomers by diastereomeric derivatization with R(+)- or S(−)-α-methyl (*p*-nitrobenzylamine), **51**.[155] In this case the acids are converted to acid chlorides with oxalyl chloride, **52**, and then reacted with the chiral amine to yield diastereomeric amides [reaction (15)]. The mixture is then chromatographed on a silica column and components eluted with a mobile phase of 20% tetra-

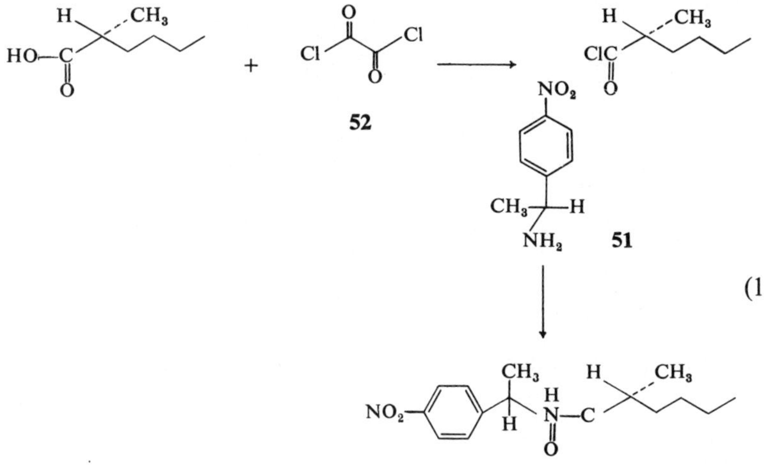

$$(15)$$

hydrofuran in *n*-heptane. Separation factors (α) for enantiomers varied from 1.03 to 2.21 (Table 10).

2.3.3. Alkylation of Carboxylate Anions–Theory

The carboxylic acid esterification for analytical application has primarily been carried out via alkylation of the carboxylate anion [route (3)]. It must be pointed out that carboxylate anions are relatively poor nucleophiles because of their weak basicity [charge delocalization between the two

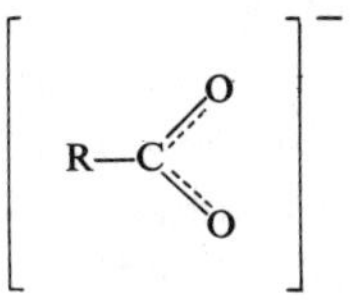

53

oxygen atoms (**53**)] and their tendency to be efficiently solvated by most protic solvents.[156,157] The solvation sphere of nucleophiles must be dis-

TABLE 10. Results for the Liquid Chromatographic Separation of Diastereomeric Acyclic Isoprenoid Acids[a]

Acid number	Chiral position	k'			First eluted pair	
		1st eluted	2nd eluted	α	Reagent	Acid
I	2	3.13	6.92	2.21	S	R
II	3	7.79	9.22	1.18	R	R
III	3, 7	6.52	8.04	1.23	R	R
IV	4, 8	7.83	8.25	1.05	R	S
V	5, 9	7.67	7.67	1.00		
VI	3	9.00	11.00	1.22	R	R
VII	3	8.71	12.80	1.47	R	S
VIII	3	9.25	9.50	1.03		
IX	3, 7	7.58	8.71	1.15	R	R

[a] Reference 155. Reproduced from the *Journal of Chromatographic Science* by permission of Preston Publications, Inc.

rupted when they engage in substitution at carbon, so more tightly solvated anions are less nucleophilic. Thus, carboxylate anions which are strongly solvated by means of hydrogen bonds are weakly nucleophilic in protic solvents:

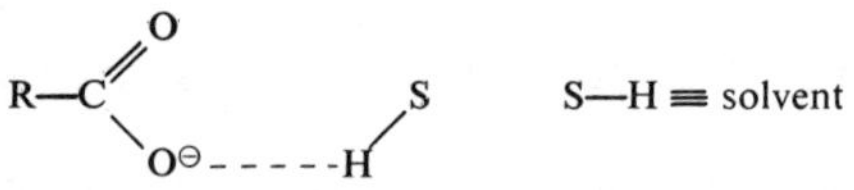

The rate of reaction between an anion (carboxylate) and neutral molecules (alkyl halide) increases by several orders of magnitude going from a protic solvent (e.g., water, alcohol, acetic acid) to polar aprotic solvents (e.g., acetone, acetonitrile, dimethylacetamide, and dimethylsulfoxide).[158] In aprotic solvents, solvation of the carboxylate portion of the molecule is less extensive (no possibility for hydrogen bonding), increasing the activity coefficient of the anion (i.e., its inherent nucleophilicity), thus accelerating reaction. This rate enhancement is more pronounced for small anions than for larger ones, which may be more effectively solvated through other types of intermolecular interactions. Therefore, to facilitate nucleophilic displacement, esterification reactions involving carboxylate anions and alkyl halides are carried out in polar aprotic solvents.

The nucleophilicity of the carboxylate anion is also compromised by coulombic interaction with its companion cation. By using a hindered organic base (e.g., diisopropylethylamine) compatable with aprotic solvents to abstract the acidic proton from the carboxylic acid, the electrostatic stabilization of the ion pair is diminished (over that observed with more traditional bases, e.g., potassium hydroxide) and its reactivity enhanced. A much more effective means of increasing the nucleophilicity of carboxylate anions is to complex their companion cations with a crown ether, creating "naked anions"[159] which are extremely reactive. The cavity of the crown ether can be designed to accommodate cations of different sizes. Their application to catalysis of carboxylate anion alkylation has involved derivatives of the crown ether 18-crown-6, which selectively complexes potassium ion.[160] Thus, the carboxylic acid is neutralized with potassium hydroxide and a small amount of crown ether is added [reaction (16)]. Stoichiometric concentrations of the crown ethers are not necessary. One may use crown ethers in molar ratios of 1:20 to 1:100 to catalyze alkylation of carboxylate salts. Reaction with an alkyl halide then affords the corresponding ester in high yield. Durst[159] achieved better than 90% yield of phenacyl esters of a series of aliphatic acids (C_1–C_7) and a variety of aromatic acids using crown ether catalysis.

$$RCO_2^{\ominus}K^{\oplus} + \quad \text{[crown ether]} \quad \longrightarrow \quad \left[\text{[crown ether·}K^+\text{]} \ RCO_2^{\ominus} \right] \equiv RCO_2^{\ominus} \, \text{(}K^+\text{)} \tag{16}$$

$$RCO_2^{\ominus} \, \text{(}K^+\text{)} + R'Br \longrightarrow RCO_2R' + \text{(}K^+\text{)} \, Br^{\ominus}$$

$$\text{(}K^+\text{)} \, Br^{\ominus} + RCO_2^{\ominus}K^{\oplus} \longrightarrow RCO_2^{\ominus} \, \text{(}K^+\text{)} + KBr \downarrow$$

2.3.4. Alkylation of Carboxylate Anions—Applications

The preparation of simple benzyl esters of carboxylic acids for analytical purposed by reaction of carboxylic ions with the corresponding alkyl halides or diazo compounds has been somewhat unsuccessful owing to the toxicity, instability, and poor reaction yields obtained with such reagents. Reagents with better leaving groups and greater stability were sought to produce such benzyl esters in high yield, in processes applicable to analytical systems.

2.3.4.1. Triazenes. 1-Benzyl-3-*p*-tolyltriazene, **54a**, a diazolike precursor, reacts with free fatty acids to give the corresponding benzyl esters [reaction (17)].[161] The acid is dissolved in ether; the triazene is added and maintained at 35°C for 3 hr. The mixture is then washed successively with hydrochloric acid (to remove *p*-toluidine and decompose excess reagent)

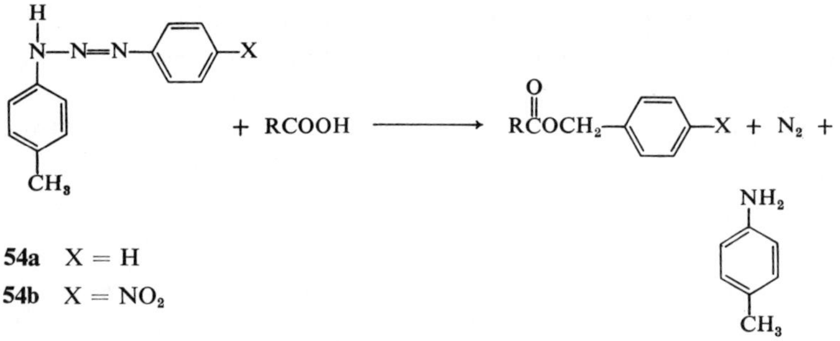

54a X = H
54b X = NO$_2$

(17)

and base (to remove excess acid) and the final ether solution subjected to HPLC. Unfortunately, in this instance, upon derivatization, differences in physical properties among the fatty acids decreased, thereby rendering HPLC separation more difficult. More recently *p*-nitrobenzyl-3-*p*-tolyltriazene (**54b**) was introduced[162] as a reagent for the preparation of esters with enhanced absorptivity at 254 nm. The disadvantages of esterification by triazenes are (a) the reagents are carcinogenic,[163] (b) provisions must be made to vent the nitrogen evolved during reaction, (c) the derivatizing agent is expensive and must be used in large molar excess, and (d) the reaction produces by-products which may interfere with subsequent chromatographic analysis.

2.3.4.2. O-Substituted Isoureas. A second approach to the selection of a reactive alkylating agent capable of benzylating carboxylic acids involves the use of O-substituted isoureas. Isoureas are powerful alkylating agents because of their propensity for expelling neutral ureas [reaction (18)]. *O-p*-Nitrobenzyl-*N*,*N'*-diisopropylisourea, **55**, reacts with carboxylic acids (dissolved in methylene chloride) to give the corresponding *p*-nitrobenzyl ester [reaction (18)].[164] Reaction is carried out at 80°C for 2 hr; after cooling, an aliquot of the reaction mixture is subjected to chromatographic analysis. The reagent, **55**, is prepared by condensation of *p*-nitrobenzyl alcohol with diisopropylcarbodiimide, **56**, [reaction (19)].[165] This reagent has been applied to analysis of picomole quantities of fatty acids[164] and is also used for monitoring bile acids (e.g., lithocholic acids) and their metabolites in incubation mixtures containing intestinal bacteria.[166] In both cases, esters were separated on silica columns.

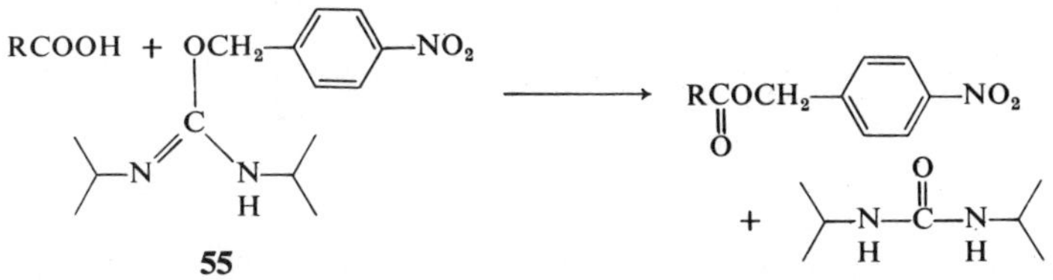

$$RCOOH + OCH_2\!-\!\!\!\left\langle\!\!\!\bigcirc\!\!\!\right\rangle\!\!\!-NO_2$$

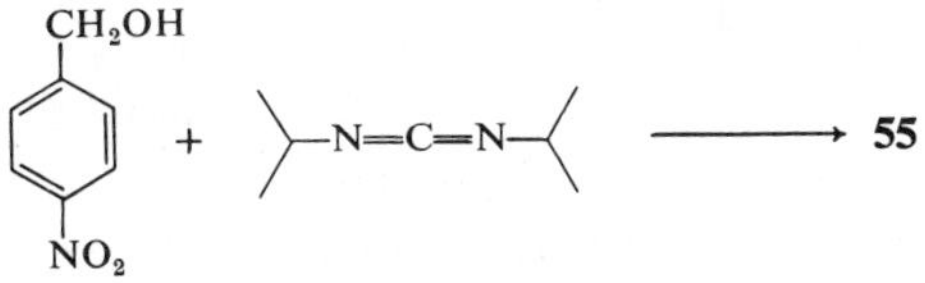

(structure **55**)

$$\longrightarrow \quad RCOCH_2\!-\!\!\!\left\langle\!\!\!\bigcirc\!\!\!\right\rangle\!\!\!-NO_2 \qquad (18)$$

$$+ \quad \rangle\!-\!N\!-\!\overset{O}{\underset{H}{C}}\!-\!N\!-\!\langle$$

$$\underset{NO_2}{\left\langle\!\!\!\bigcirc\!\!\!\right\rangle}\!\!\!-CH_2OH \quad + \quad \rangle\!-\!N\!=\!C\!=\!N\!-\!\langle \quad \longrightarrow \quad 55 \qquad (19)$$

(structure **56**)

2.3.4.3. Phenacyl Halides. Whereas most alkyl halides lack sufficient reactivity to esterify carboxylate anions in high yield, α-haloketones, **57**, have been employed effectively to derivatize carboxylic acids. Specifically,

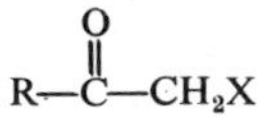

$$R\!-\!\overset{O}{\underset{}{C}}\!-\!CH_2X$$

57

phenacyl bromides (**57**; $R = C_6H_5$, $X = Br$) have been used as reagents to form esters with strong uv absorbance. The alkyl bromide is used rather than the corresponding alkyl chloride because of its stronger alkylating ability. Since the bond between carbon and the leaving group has undergone partial cleavage in the transition state, reaction rates depend upon the leaving group. The reaction order for halide leaving groups is $I^\ominus > Br^- > Cl^- \gg F^-$. This order is determined by the strength of the C–X bond which increases in the order $I < Br < Cl < F$. The carbonyl group adjacent to the site of substitution retards reactions by S_N1 mechanisms and greatly enhances S_N2 reactivity. Rate enhancement is not due solely to inductive electron withdrawal by the carbonyl, because other electronegative groups in the α position do not produce comparable effects. Direct participation by the electron-deficient carbonyl group in attracting the electron-rich re-agent has been proposed,[167] i.e., there should be significant overlap between the bond-forming p orbital of the attacking atom and the electron-deficient π orbital of the carbonyl carbon. This interaction constitutes partial bond formation between the attacking atom and the carbonyl carbon and helps stabilize the activated complex, thus facilitating substitution:

The only major disadvantage in working with phenacyl halides is their strong lacrymating properties, requiring that they be handled in a fume hood.

Phenacyl esters of fatty acids (C_{12}–C_{24}) have been prepared as derivatives to facilitate analysis of the fatty acid composition in chick fibroblast phospholipids and free fatty acids, phospholipids in platlets,[168] and in plasma and blood.[169] The derivatives strongly absorb uv light ($a_{m,254\text{ nm}}$ = 14 000) and are separated on an RP-18 column. Fatty acids have been similarly derivatized as p-bromophenacyl, p-nitrophenacyl, p-chlorophenacyl, and 2-naphthacyl esters.[170,171] Reactivity is not greatly affected by these substituents; however, the substituents do alter the spectrophotometric properties (λ_{max}, a_m) of the derivatives. The carboxylate anion is formed with lithium carbonate or N,N-diisopropylethylamine in dimethylformamide and the phenacyl halide then added [reaction (20)]. The esters

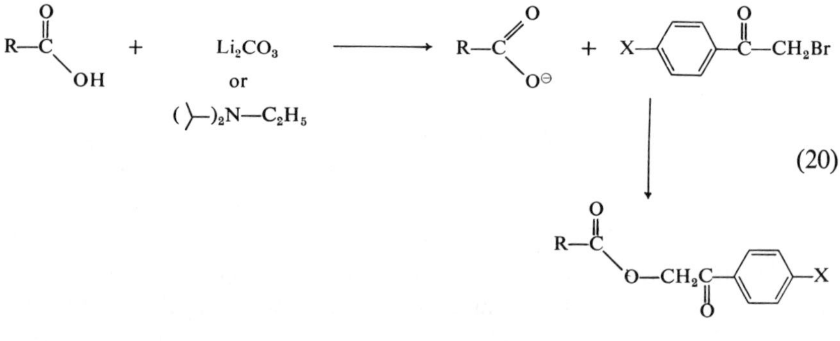

$$(20)$$

are again separated on RP-18 columns. Phenacyl esters of fatty acids[160] and dicarboxylic acids[172] have been prepared with crown ether catalysts, combining the potassium salt of the acid with 18-crown-6 or dicyclohexyl-18-crown-6 and then adding the phenacyl halides. In both cases, liquid chromatographic separation of these esters was obtained on a column packed with C_9-bonded phase.

Prostaglandins represent a special class of fatty acids. HPLC analysis of such compounds has involved ionization of the carboxylic acid moiety with a hindered amine (e.g., diisopropylethylamine) followed by reaction with p-nitrophenacyl[173,174] or p-bromophenacyl[175] bromide. The former

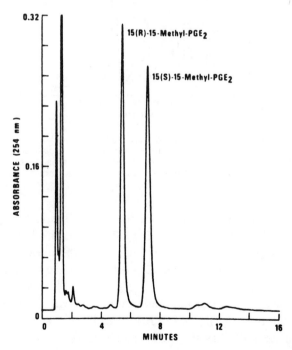

has been used to separate ten F series prostaglandin analogs, eight series E analogs, isomeric 15R and 15S methyl prostaglandins (Figure 11) of the E

Fig. 11. Separation of the p-nitrophenacyl esters of 15-methyl-PGE$_2$ on one 2.1 mm (i.d.) × 25 cm microparticulate silica gel column. Conditions: mobile phase, methylene chloride–acetonitrile–dimethylformamide (160:40:1); 2000 psi; 0.65 ml/min.[173] Reproduced from *Prostaglandins* by permission of Geron-X Publications, Inc.

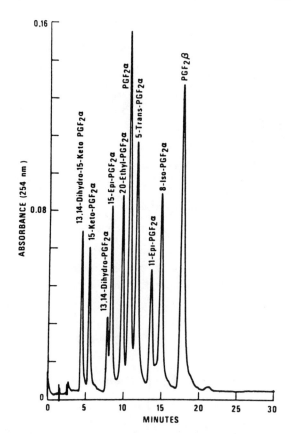

Fig. 12. HPLC separation of a mixture of F series prostaglandin *p*-nitrophenacyl esters on two series coupled 2.1 mm (i.d.) × 25 cm microparticulate silica gel columns. Conditions: mobile phase, methylene chloride–hexane–methanol (55:45:5); 3000 psi; 0.3 ml/min.[173] Reproduced from *Prostaglandins* by permission of Geron-X Publications, Inc.

and F series (Figure 12), PGA_2 and PGB_2,[173] and the 15 epimers, 15-epi-$PGF_{2\alpha}$ and $PGF_{2\alpha}$.[174] Separations have been carried out both on silica and on RP-18 columns. The latter reagent has also been used to monitor *in vitro* enzymatic degradation of several prostaglandins to the corresponding 15-keto metabolites.[175]

A final class of biomedically important carboxylic acids are the penicillins. Penicillins, V, F, K, and dihydropenicillin F, have been analyzed by reverse phase HPLC as their *p*-bromophenacyl esters.[176] Esters were formed either by crown-ether-catalyzed alkylation of the potassium carboxylate or by reaction of the penicillin with triethylamine followed by alkylation. Both approaches gave similar kinetic and thermodynamic responses. Derivatization with naphthacyl bromide has also been applied to the analysis of barbiturates (phenobarbital, hexobarbital, heptobarbital, butobarbital, amobarbital, and pentobarbital).[177] Adducts are formed in acetone at 30°C in 30 min in a quantitative manner using cesium carbonate as a catalyst. Derivatives are separated on a microparticulate reverse phase (C-18) column using a methanol:water (80:20) mobile phase.

2.3.4.4. 4-Bromomethyl-7-methoxycoumarin. Most of the reagents designed for derivatization of acids for HPLC analysis impart strong uv absorbance to the adduct. 4-Bromomethyl-7-methoxycoumarin, **58**, has been used to form fluorescent esters of carboxylic acids (with $\lambda_{ex} = 365$ nm,

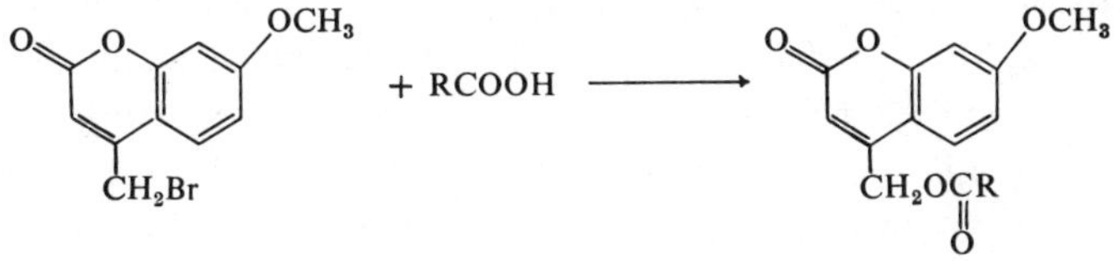

$$\text{(21)}$$

58

$\lambda_{em} = 420$ nm) [178–183] [reaction (21)]. The reaction has been applied to the analysis of fatty acids as well as acidic herbicides.[182] Derivatization can be carried out in the presence of solid potassium carbonate or with a crown ether phase transfer catalyst to facilitate esterification[181] and offer potential for trace level analysis of suitable substrates.

2.3.4.5. N-Chloromethylphthalimides. N-Chloromethylphthalimides, **59**, represent a new class of uv-sensitive reagents for the formation of derivatives suitable for HPLC.[184] The reactive chlorine is displaced by alkali metal or

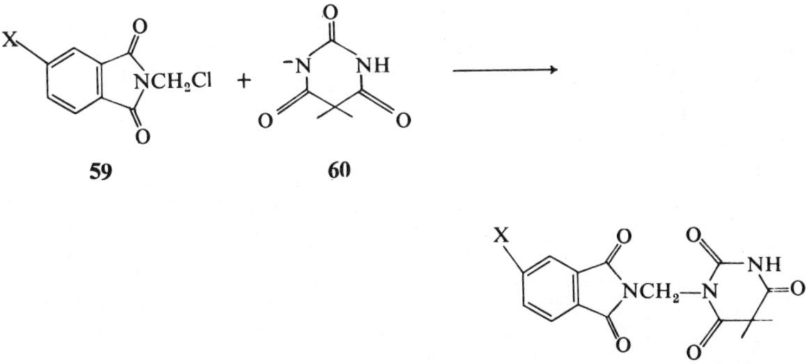

$$\text{(22)}$$

ammonium salts of CH–, –OH, and –NH-acid compounds. The reagent does not react efficiently with free amines, alcohols, or carboxylic acids.

Generating the anionic form of the analyte to be derivatized is a prerequisite for reaction. Thus, this reagent offers specificity not found with other alkylating agents, i.e., only anionic forms of substrates efficiently react. This scheme has been applied to the derivatization of fatty acids (C_2–C_{18}), dicarboxylic acids (C_2–C_6), penicillins, and barbiturates (Figure 13) (**60**) [reaction (22)]. Reaction is carried out in aprotic solvents (acetonitrile is preferred by these authors) and is facilitated by amine bases (e.g., triethylamine or 4-dimethylaminopyridine) or the phase transfer catalyst, 18-crown-6, when alkali metal (particularly potassium) salts of the acid are used.

The molar extinction coefficient is the same for all phthalimidomethyl monocarboxylic acid esters ($\varepsilon_{254} = 1390$, $\lambda_{max} = 224$, $\varepsilon_{\lambda_{max}} = 18\,700$). The 4-nitro analog of **59** (X$=$NO$_2$) has a λ_{max} at 235 nm and an extinction coefficient ($\varepsilon_{\lambda_{max}} = 23\,900$) similar to that of the unsubstituted phthalimidomethyl derivative at its λ_{max}. However at 254 nm, the wavelength most commonly monitored by fixed-wavelength detectors, the absorptivity of the nitro compound ($\varepsilon_{254} = 9500$) is seven times greater than that of the unsubstituted derivative ($\varepsilon_{254} = 1390$), offering greater sensitivity. All derivatives were chromatographed on LiChrosorb RP-8 (7-μm particle size) using acetonitrile–water mobile phases.

2.3.4.6. 1-Chloromethylisatin. Similarly, 1-chloromethylisatin, **61**, has been used to derivatize carboxylic acids.[185] The reaction has been applied to the TLC analysis of a series of saturated acids (C_1–C_{18}), unsaturated

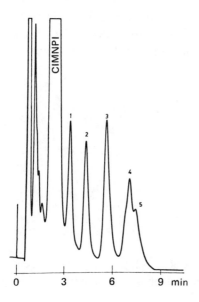

Fig. 13. Separation of a 4-nitrophthalimidomethyl-substituted barbiturate mixture. Separation system: Column, RP-18; mobile phase, acetonitrile–water (3:2); flow rate, 1.5 ml/min; detection uv at 254 nm. Peak designation: (1) methylphenobarbital derivative; (2) phenobarbital derivative; (3) cyclobarbital derivative; (4) amobarbital derivative; (5) secobarbital derivative.[184] Reproduced from the *Journal Chromatography* by permission of Elsevier Publishing Co.

acids, and aromatic acids (substituted benzoic acids). Reaction is carried out for 10 min at 50°C in dimethylformamide, with potassium bicarbonate and crown ether (18-crown-6) catalysis [reaction (23)]. At the completion

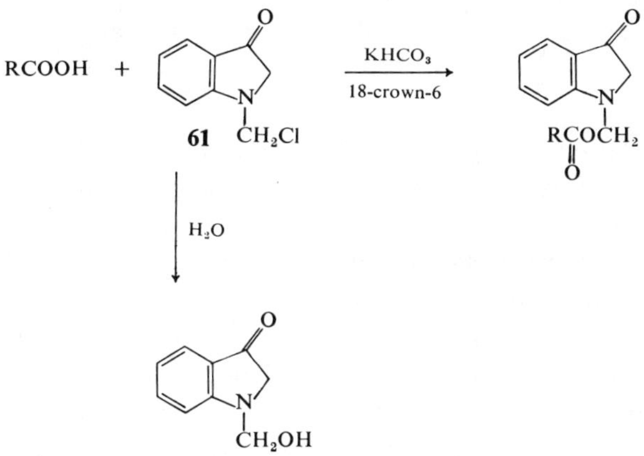

$$(23)$$

of reaction, excess reagent is destroyed with water and the mixture subjected to TLC separation.

2.4. Aldehydes and Ketones

2.4.1. Theoretical Considerations

Carbonyl compounds (aldehydes and ketones) undergo nucleophilic addition, with the nucleophile attacking carbon and the electrophile reacting with oxygen. The products of such addition reactions, **62**, often undergo β elimination to yield an unsaturated product, **63** [reaction (24)]. These

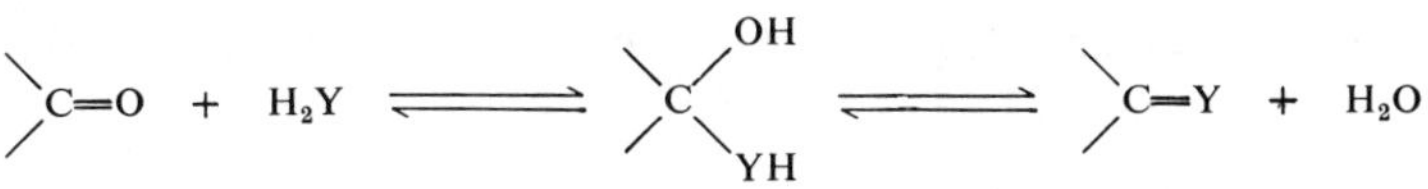

$$(24)$$

two-step reactions are all reversible, i.e., the products are susceptible to hydrolysis. Nucleophilic addition at carbonyl groups is both acid and base catalyzed: acids promote polarization of the carbonyl group (**64**) thus increasing the electrophilicity at carbon; bases exert their effect by increasing the fraction of nucleophile in its reactive conjugate base form.[186]

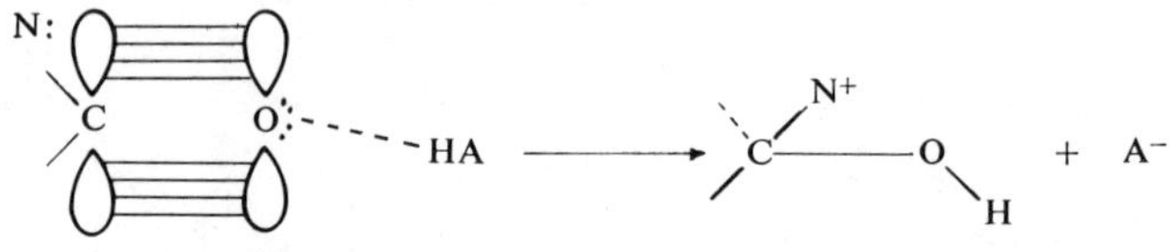

64

The application of chemical derivatization of aldehydes and ketones for HPLC analysis has primarily involved the use of nitrogen nucleophiles (e.g., oximes, hydrazines, semicarbazide, etc.). Derivatization is carried out primarily to improve detectability but it will also stabilize aldehydes, interfering with their oxidation to carboxylic acids. The rate of reaction of such nucleophiles with carbonyl compounds in two-step addition–elimination processes is pH dependent, with a change in rate-determining step occurring with pH variation.[187] This pH dependence is illustrated by a discussion of the addition of hydroxylamine, N, to a carbonyl compound (C) to yield the oxime, P [reaction (25)]. The reaction proceeds through a

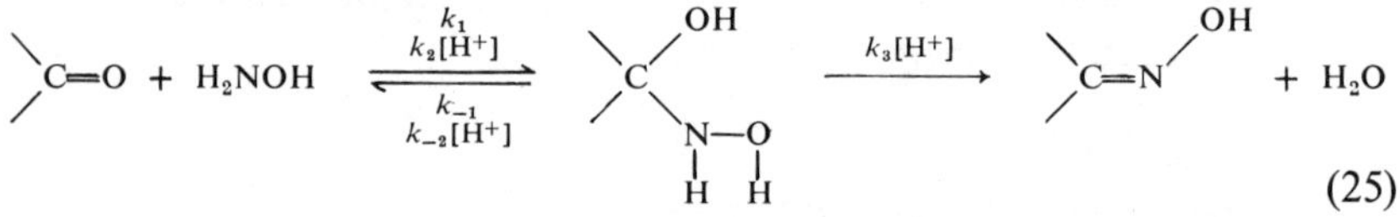

$$C \qquad N \qquad\qquad I \qquad\qquad P \tag{25}$$

tetrahedral intermediate, I, in at least two steps, either of which may be rate limiting. The rate of product formation is described by the rate of breakdown of the tetrahedral intermediate, I:

$$v = \frac{dP}{dt} = k_3 a_H \mathbf{I} \tag{26}$$

The value of I can be determined by applying the steady state approximation to it:

$$\frac{d\mathbf{I}}{dt} = (k_1 + k_2 a_H)\mathbf{NC} - (k_{-1} - k_{-2}a_H + k_3 a_H)\mathbf{I} = \mathbf{O}$$

$$[\mathbf{I}] = \frac{(k_1 + k_2 a_H)\mathbf{NC}}{k_{-1} + k_{-2}a_H + k_3 a_H} \tag{27}$$

Substitution of equation (27) into equation (26) gives an expression for the rate of oxime formation:

$$v = \frac{k_3 a_H (k_1 + k_2 a_H) \mathbf{NC}}{k_{-1} + k_{-2} a_H + k_3 a_H} \tag{28}$$

Since the apparent first-order rate constant $k_{obs} = v/C$,

$$k_{obs} = \frac{k_3 a_H (k_1 + k_2 a_H) \mathbf{N}}{k_{-1} + k_{-2} a_H + k_3 a_H} \tag{29}$$

At higher pH (near neutrality) the terms in parentheses and in the denominator of equation (29) containing a_H become insignificant relative to other terms and may be neglected. Consequently, under these conditions, equation (29) simplifies to

$$k_{obs} = \frac{k_1}{k_{-1}} k_3 a_H \mathbf{N} \tag{30}$$

At high pH, nucleophilic addition occurs in a rapid equilibrium step followed by a slow (rate-determining) dehydration of the tetrahedral intermediate, i.e., the rate depends on the equilibrium concentration of the intermediate (k_1/k_{-1}) and its rate of dehydration $(k_3 a_H)$ [reaction (25)].

Dehydration is an acid-catalyzed process, so the rate increases as pH is decreased (Figure 14). However, as the pH is further decreased, a greater fraction of the nucleophile exists in the unreactive protonated form. Thus, two effects are acting in opposition to one another—dehydration is accelerated by the increased acidity, but the concentration of effective nucleophile is reduced. As a result, at intermediate pH's, the rate of reaction reaches a limiting, pH-independent value. In this region, $k_1 \gg k_2 a_H$, but $k_3 a_H \gg k_{-1}$ and $k_3 a_H \gg k_{-2} a_H$, and equation (29) reduces to the pH-independent expression

$$k_{obs} = \frac{k_3 a_H k_1 \mathbf{N}}{k_3 a_H} = k_1 \mathbf{N} \tag{31}$$

Further reduction, to pH's well below the pK_a of the nucleophile causes a decrease in rate, in apparent contradiction to previous statements that the rate determining step, dehydration of $\mathbf{I}$ is acid catalyzed (Figure 14). At such pH's, the fraction of unprotonated nucleophile is so low that the rate of addition is decreased and eventually falls below the dehydration step. At this point, the rate of addition becomes rate determining $(k_2 a_H \gg k_1)$; the rate of dehydration of the intermediate is again much faster than its return to reactants, i.e., $k_3 a_H \gg k_{-1}$ and $k_3 a_H \gg k_{-2} a_H$. Equation (29) thus reduces to

$$k_{obs} = k_2 a_H \mathbf{N}$$

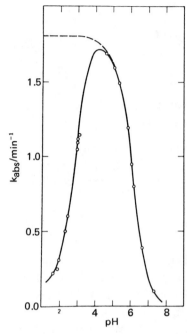

Fig. 14. pH-rate profile for the reaction of hydroxylamine with acetone in water at 25°C. Dashed line: rate of the acid-catalyzed dehydration step; full line: observed rate.[187] Reprinted from the *Journal of the American Chemical Society.* Copyright (1959) by permission of the American Chemical Society.

The rate of reaction depends on the fraction of total nucleophile present in the free unprotonated form. Since $\mathbf{N} = [K_a/(a_H + K_a)]\mathbf{N}_t$ (where K_a is the ionization constant for the nucleophile, N), nucleophiles that are weaker bases exist to a greater extent in the unprotonated form at lower pH, and therefore in these cases the acid-catalyzed process makes a greater contribution to the overall rate of reaction than is observed with more basic nucleophiles.[188,189]

Although base catalysis has been presumed to be specific base catalysis,[190] general base catalysis of thiosemicarbazone formation has also been observed.[189] The following two mechanisms for general-base-catalyzed dehydration of the tetrahedral intermediate have been suggested:

In reaction (33), the general base facilitates proton abstraction followed by hydroxide liberation. In the kinetically equivalent case [reaction (34)], the base abstracts a proton in a fast preequilibrium step to give the carbinolamine anion, which then undergoes a slow general-acid-assisted loss of hydroxide with formation of the imine. The mechanism depicted by reaction (33) seems more reasonable because of the instability of the carbinolamine anion [in reaction (34)].

Nucleophiles such as hydroxylamines, hydrazines, and semicarbazides which contain an electronegative atom with a free electron pair adjacent to the nucleophilic atom display an unusually high nucleophilic reactivity, which has been termed the "α effect".[60,157] The electron pair on the adjacent atom increases the effective electron density at the reaction center in the transition state by utilization of electrons and orbitals on adjacent atoms, as seen for the addition of X–Y to a ketone:

$$\ddot{X}-\overset{\delta+}{Y}------ \overset{\displaystyle\diagdown}{\underset{\diagup}{C}}====O^{\delta-} \longleftrightarrow \overset{\delta+}{X}=Y----- \overset{\displaystyle\diagdown}{\underset{\diagup}{C}}====O^{\delta-}$$

$$\textbf{65a} \qquad\qquad\qquad \textbf{65b}$$

Attack of X–Y at the carbonyl carbon results in a partial removal of electrons from Y (**65a**), which can be stabilized by electron donation from X (**65b**). Through hybridization of atomic orbitals, molecular orbitals are generated that can accomodate the extra bonds required for such stabilization. In addition, polarizability, steric effects, and intramolecular catalysis contribute to the α effect.

The reaction of carbonyl compounds with nitrogen nucleophiles results in the formation of imines ($\diagdown$C$=$N—R) which can exist as either of two geometric isomers:

$$\overset{A}{\underset{B}{\diagdown}}C=O + H_2N-R \longrightarrow \overset{A}{\underset{B}{\diagdown}}C=\overset{R}{\underset{\ddot{\,}}{N}} + \overset{A}{\underset{B}{\diagdown}}C=\overset{\ddot{\,}}{N}\diagdown_R \qquad (35)$$

In many instances these isomers are readily interconvertible. However, certain imino isomers are separable by chromatographic techniques. Therefore, caution should be taken in preparing imino derivatives for analytical purposes; one should determine that the derivative is eluted as a single component, not as two peaks arising from a resolvable mixture of geometric isomers.

2.4.2. Applications

2.4.2.1. 2,4-Dinitrophenylhydrazine. The primary reagent that has been used to improve the uv detectability of aldehydes and ketones is 2,4-dinitrophenylhydrazine, **66**, which forms chromophoric hydrazones, **67** [reaction (36)]. The reaction is most commonly carried out by dissolving the aldehyde

$$
\begin{array}{c}
R \\
\diagdown \\
C{=}O + H_2NN{-} \\
\diagup \\
R
\end{array}
\quad\quad
\begin{array}{c}
R \\
\diagdown \\
C{=}N{-}N{-} \\
\diagup \\
R
\end{array}
\tag{36}
$$

         66   **67**

or ketone in methanol, acidifying the solution with concentrated hydrochloric acid, and then adding a methanolic solution of dinitrophenylhydrazine. Reaction is carried out for $\sim$2 hr at 50–60°C. This reaction has been applied to the analysis of simple aldehydes (C_1–C_2, benzaldehyde, and substituted benzaldehydes, glyoxal, glutaraldehyde)[191–193] and ketones (C_3–C_7, all isomers),[192,193] prior to chromatographic analysis, as well as to the determination of formaldehyde in tobacco smoke.[194]

    Pyruvic acid and α-ketoglutarate have been analyzed in serum by conversion to the corresponding 2,4-dinitrophenylhydrazones which are then separated on a Zipax Permaphase (AAX) anion exchange column.[195] The derivatized pyruvate appeared as two peaks (Figure 15) which were subsequently shown to represent syn–anti geometric isomers (**68a–68b**).

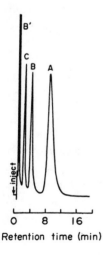

Fig. 15. Liquid chromatogram of a standard mixture of the DNPH derivatives of pyruvic acid and a ketoglutaric acid, and trimellitic acid (internal standard). Operating conditions: column, 50-cm Zipax Permaphase AAX (particle size, 30–50 μm); mobile phase, 11.1 g of KH_2PO_4 + 0.9 g of K_2HPO_4 per 1 of water; column temperature, 55°C; flow rate, 1.2 ml/min (pressure, 50 kg/cm²); detector, uv photometer (254 nm). Peaks: (*A*) DNPH-α-ketoglutaric acid; (*B,B'*) DNPH-pyruvic acid; (*C*) trimellitic acid.[195] Reproduced from the *Journal of Chromatography* by permission of Elsevier Publishing Co.

Retention time (min)

Isomerization was dependent on water concentration and could be minimized by cooling the solution after derivatization was complete. 2,4-Dinitrophenylhydrazones of a series of quinones (benzoquinones, naphthoquinones, anthraquinones, and phenanthrenquinones) were also prepared

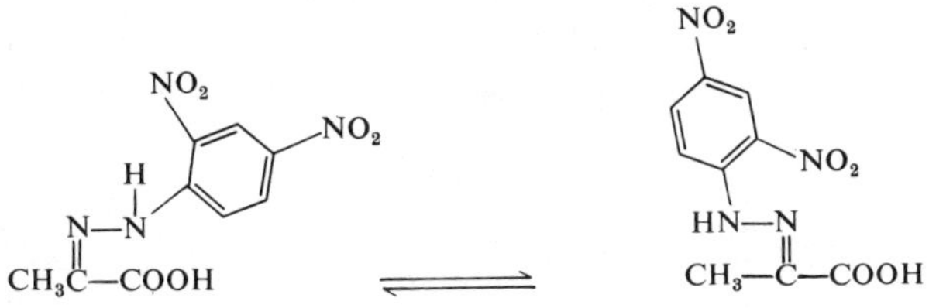

68a   68b

as derivatives prior to HPLC separation on silica or normal phase (CN) columns using petroleum–isopropanol mobile phase.[196] This derivatization technique has been applied to the analysis of a variety of steroids including 3-keto steroids (progesterones, testosterones, corticosterones).[197] and 17-keto steroids.[198-201] This includes the analysis of 17-keto steroids[198] and their 11-desoxy metabolites[199] in urine by TLC procedures, and the analysis of 17-keto steroids in blood and urine by HPLC on RP-18 or β,β'-oxydipropionitrile (ODPN) columns.[200] This latter method has been applied to the separation of androsterone, etiocholanolone, and their epimers dehydroepiandrosterone, 11β-etiocholanolone, and 11β-OH androsterone. Conjugated and esterified estrogens (estrone, equiline, and equilenin) were determined in pharmaceutical tablet dosage forms by similar derivatization techniques followed by HPLC separation on a chemically bonded ether (ETH-Permaphase) column and elution with n-heptane containing 1.1% isopropanol.[201]

   2.4.2.2.   4-Amino-3-hydrazino-5-mercapto-1,2,4   triazole.   4-Amino-3-hydrazino-5-mercapto-1,2,4 triazole, **69**, is a hydrazine that reacts selectively

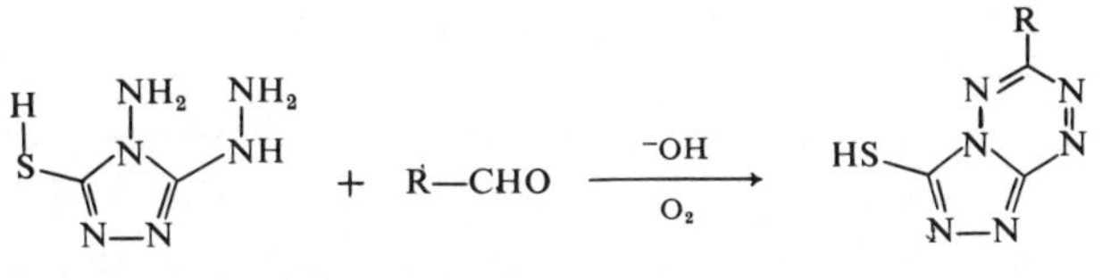

69   70   (37)

with aldehydes to form magenta- and violet-colored 6-mercapto-3-substituted-5-triazolo (4,3-b) 5-tetrazines, **70** [reaction (37)].[202] Although this reaction has not been applied to chromatographic analysis, it offers advantages as a preseparation derivatization reagent in HPLC in that the reaction is rapid, specific, and produces a strong chromophore.

2.4.2.3. O-p-Nitrobenzylhydroxylamine. An alternative reagent for derivatization of carbonyl compounds prior to HPLC analysis is *O-p*-nitrobenzylhydroxylamine, **71**, which reacts with aldehydes and ketones to form oxime ethers, **72**, that strongly absorb uv light [reaction (38)].[203] This

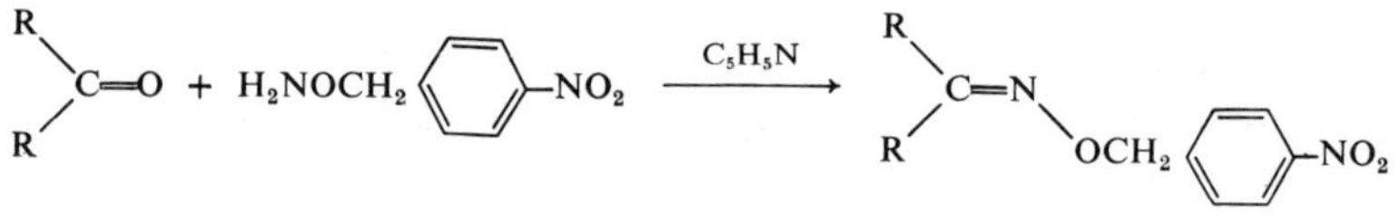

$$\begin{array}{c} R \\ {} \\ R \end{array} C{=}O \ + \ H_2NOCH_2 \text{—}\langle\text{—}\rangle\text{—}NO_2 \ \xrightarrow{\ C_5H_5N\ } \ \begin{array}{c} R \\ {} \\ R \end{array} C{=}N \diagdown OCH_2\text{—}\langle\text{—}\rangle\text{—}NO_2 \tag{38}$$

71   **72**

reaction has been applied to the analysis of a variety of prostaglandins. Derivatization requires the initial esterification of the prostaglandin with diazomethane followed by oximation in pyridine. This approach has been applied to the analysis of PGE$_1$ [reaction (39)], PGE$_2$, PGD$_1$, PGD$_2$,

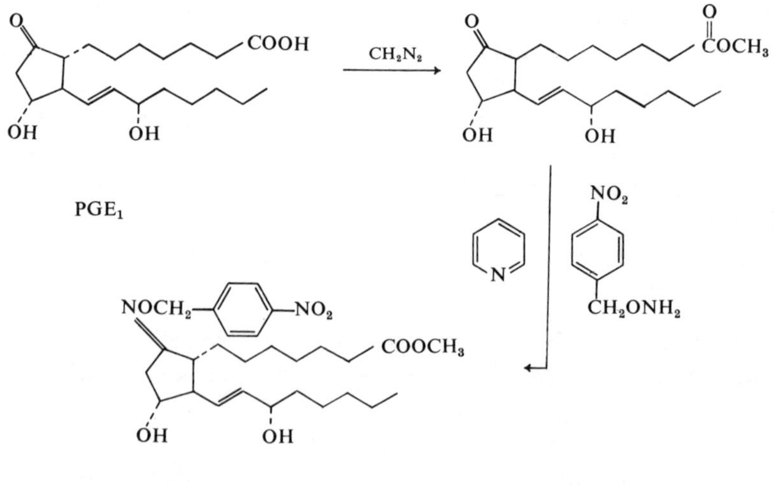

$$\tag{39}$$

PGE$_1$

PG.   PGB$_2$, PGA$_1$, PGA$_2$, 15-keto-PGF$_{2\alpha}$, and thromboxane B$_2$ (Table 11). Mixtures are separated on an RP-18 column. The detection limits attainable were strongly influenced by the position and number of double bonds relative to the carbonyl group.

2.4.2.4. Dansyl Hydrazine. A more sensitive reagent for carbonyl compounds is dansyl hydrazine, **73**, which forms fluorescent hydrazones,

TABLE 11. Retention Data for Reversed Phase Liquid Chromatography[a] of Esterified Prostaglandin Oximes[b]

Compound	k' [c]	(I) Detection limit[d,e]	k'	(II) Detection limit[e]
PGE_2	1.24	95	1.83	70
PGE_1	1.42	95	2.08	110
PGD_2	1.20	50	1.83	100
PGD_1	1.08	50	2.08	120
	1.20			
PGA_2	1.88	90	2.71	90
	2.06		2.87	
PGA_1	2.20	40	3.25	70
	2.40		3.42	
PGB_2	4.32	225	3.60	40
PGB_1	5.00	120	4.08	35
15-keto-$PGF_2\alpha$	1.46	40	2.00	10
15-keto-PGE_2	2.16	25	3.92	35
	2.32		4.12	
15-Methyl-PGB_2	4.84	65	4.04	120

[a] Column: 60 cm µBondapak C_{18}; mobile phase: 85/15 acetonitrile/water; flow rate: 0.45 ml/min; temperature: $25 \pm 3°C$.
[b] Reference 203. Reprinted from *Analytical Chemistry*. Copyright (1977) by permission of the American Chemical Society.
[c] $k' = (V_r - V_0)/V_0$.
[d] Nanograms on-column required to produce a signal:noise ratio of 5:1 at 0.04 AUFS setting of detector.
[e] I = p-Nitrobenzyloximes of prostaglandin methyl esters. II = p-Nitrobenzyloximes of prostaglandin pentafluorobenzyl esters.

74, ($\lambda_{ex} = 350$ nm, $\lambda_{em} = 505$ nm)[204] [reaction (40)]. This reagent has been used to derivatize keto steroids (androsterone, etiocholanolone, dehy-

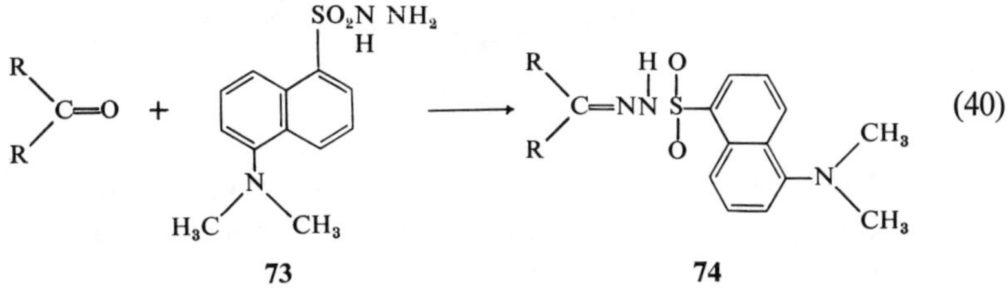

(40)

73   74

droepiandrosterone, pregnanolone in pregnancy urine and keto estrogens) prior to TLC analysis.[205] In this case, excess dansyl hydrazine had to be removed prior to chromatographic analysis. Pyruvate was added after derivatization was complete and the dansyl pyruvate (**75**) was washed out of the reaction media with pH 5 buffer, to avoid interferences during chromatography. Hydrocortisone[230] and cortisol[204] have been monitored in

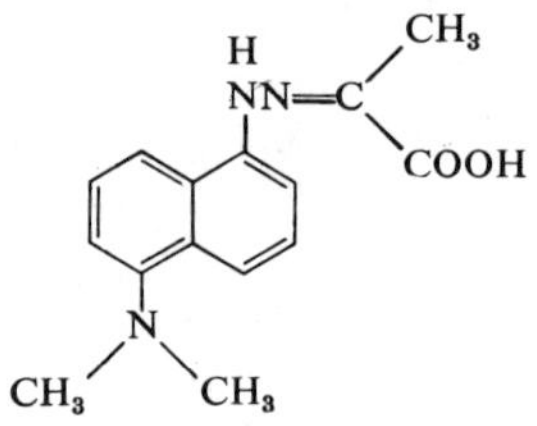

75

human plasma and urine by initially extracting the drug with dichloromethane, evaporating the solvent to dryness, reconstituting the residue in ethanolic hydrochloric acid solution, and then adding dansyl hydrazine (as a 0.02% ethanol solution). The derivative is separated on a silica column (Hitachi gel No. 3042) using dichloromethane:ethanol:water (948:35:17) mobile phase and detected fluorimetrically. The detection limits are ~0.2 ng of cortisol. A similar dansylation procedure has been carried out on reducing sugars prior to TLC analysis and fluorometric detection.[206]

2.4.2.5. Phenylenediamines and naphthalene-2,3-diamines. Several specific derivatizing agents have been designed to improve the detectability of α-keto acids. o-Phenylenediamines[207] and naphthalene-2,3-diamines [208] form quinoxalones **76** and benzoquinoxalones, respectively, with phenylpyruvate and other α-keto acids, which strongly absorb uv light:

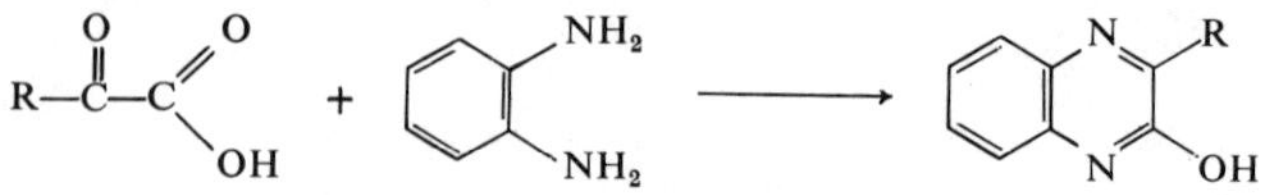

$$\text{(41)}$$

76

2.4.2.6. N-Methylnicotinamide. N-Methylnicotinamide hydrochloride, **77**, is a fluorogenic reagent for α-keto acids[209–211] which has been applied to the HPLC analysis of pyruvic and α-ketoglutaric acid. The resulting fluorophore, **78**, has been formed in postcolumn reactors but the reaction scheme appears equally well suited for preseparation derivatization of α-keto acids. The first step in the reaction is highly temperature dependent and the final fluorescence produced was almost proportional to NaOH concentration and increased with increasing concentrations of N-methylnicotinamide up to 50 mM. Increasing the reaction time of the first step causes only modest increase in fluorescence because higher temperatures promote hydrolysis of the amide in **77** and in the α adduct. In the second step, acidification of the

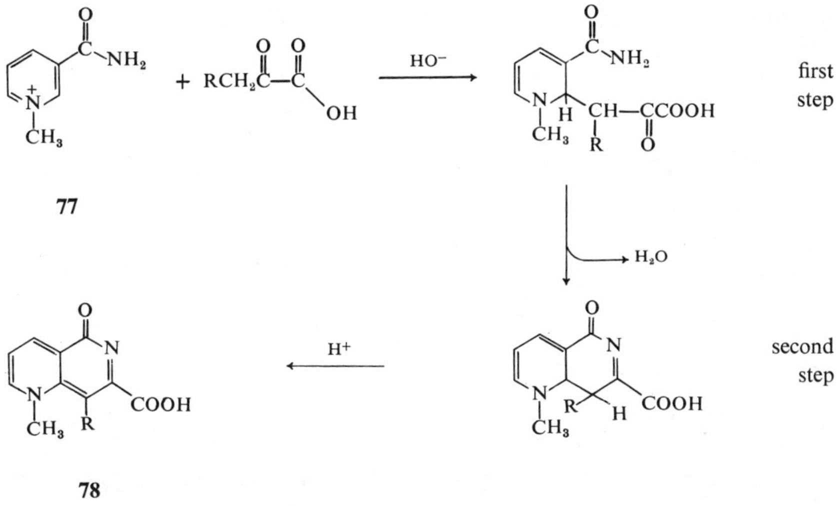

alkaline solution was essential for transformation of α adduct into fluorophore **78**, and the reaction proceeded proportionally with increasing temperature, acidity, and reaction time.

2.5. Miscellaneous Nitrogen-Containing Compounds

2.5.1. α-Amino Acids

In addition to responding to the reactions described for amines and carboxylic acids, α-amino acids react specifically with phenylisothiocyanate, **79**, to form the corresponding phenylthiohydantoin (PTH), **80** [reaction (42)].[212–214] The amino acid is dissolved in 60% aqueous pyridine containing phenylisothiocyanate and the mixture heated at 40°C for 1 hr. The resulting thiohydantoin can then be separated by TLC [212] or HPLC on an ether (ETH-permaphase) or C-18 [213] column. PTH amino acids have also been separated by HPLC on silica columns where the derivatives are divided into three groups based on their polarity.[214] The nonpolar PTH

amino acids are eluted with heptane: chloroform (50:50), moderately polar
derivatives with pure chloroform, and the polar compounds with 3% methanol in chloroform.

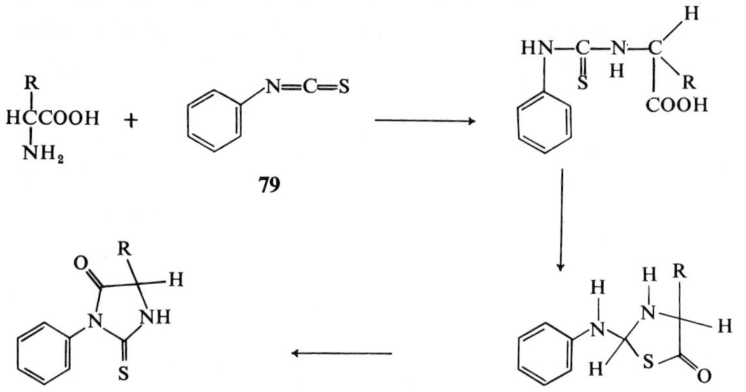

(42)

80   (PTH amino acid)

2.5.2. Guanidines

Guanidines represent a large class of antihypertensive agents used in
treatment of cardiovascular disease. Although chemical derivatization has
not been applied to HPLC analysis of guanidines, guanethidine, **81**,[215] and

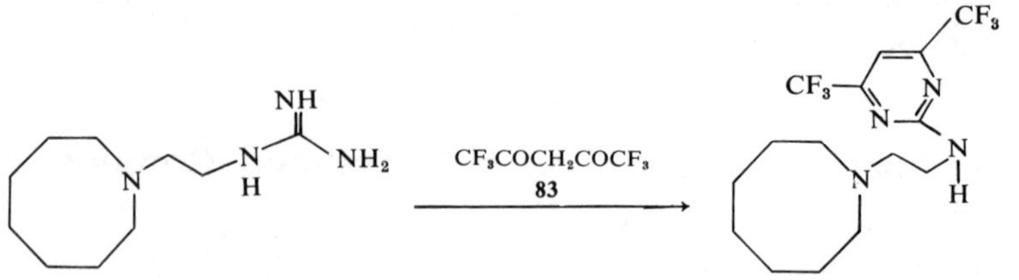

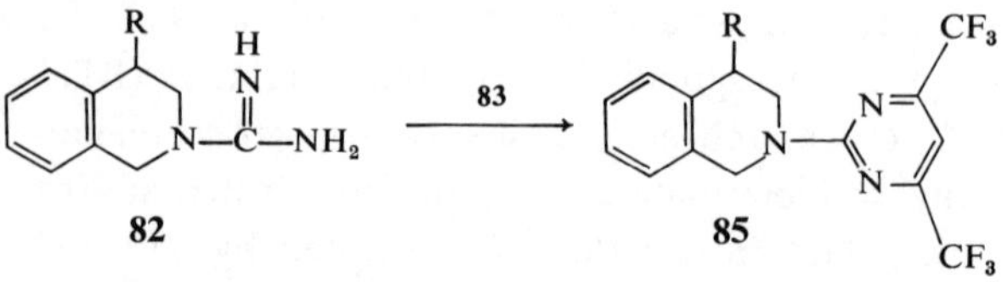

debrisoquin (and its 4-hydroxy metabolite) **82**[216] have been derivatizec
with hexafluoroacetylacetone, **83**, prior to GLC analysis. The resulting

pyrimidines, **84** and **85**, strongly absorb uv light, so this procedure should be readily adaptable to HPLC analysis.

Plasma or urine samples containing **81** or **82** are extracted with benzene containing hexafluoroacetylacetone (**83**). The mixtures are heated at 100°C for 2 hr and after alkaline hydrolysis of excess reagent, the samples are subjected to GLC analysis. This procedure has also been applied to the GLC analysis of guanidine and methylguanidine,[217] which are found in fresh beef and fish and reportedly degraded to the carcinogenic nitrosocyanamide and nitrosourea,[218] and again should be readily adaptable to HPLC.

2.5.3. Biguanides

A number of oral hypoglycemic acids used in the management of diabetes are biguanides. Metformin, **86**, can be analyzed by HPLC after derivatization with p-nitrobenzoyl chloride.[219] The resulting s-triazine, **87**, strongly absorbs uv light, is separated on a μ-Bondapak phenyl column, and eluted with 40% methanol in water.

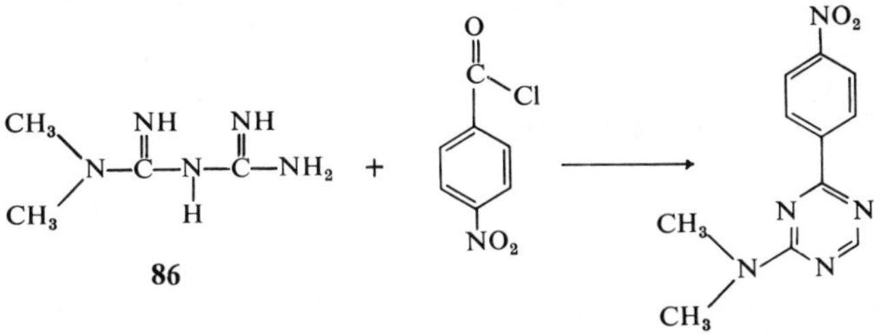

Metaformin was extracted from salt-saturated urine samples with acetonitrile containing the derivatizing agent. Extractive alkylation followed by HPLC permits detection of the adduct to levels of 200 ng/ml of urine.

2.5.4. Hydrazines

Unsymmetrical hydrazines react with 2-hydroxy-5-nitrobenzaldehyde, **88**, to form hydrazones, **89**, which have been analyzed by GLC.[220] This procedure has been applied to the analysis of 25 different acyclic and cyclic hydrazines. The derivative strongly absorbs uv light and should also be readily reduced at a mercury electrode. Therefore, this procedure should be applicable to HPLC analysis of hydrazines with uv or polarographic detection.

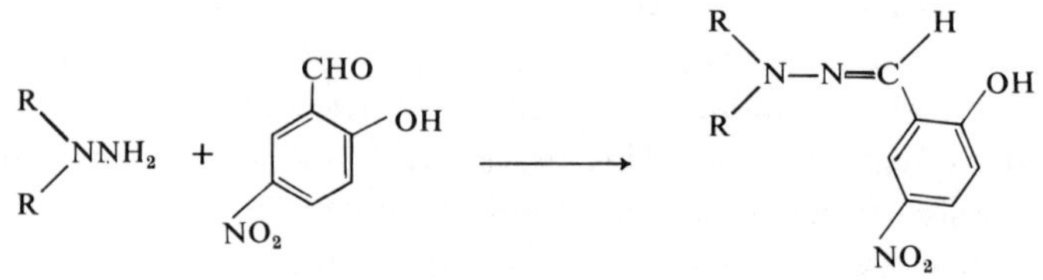

88   89

2.5.5. Arylhydroxylamines

Arylhydroxylamines, **90**, are metabolites of primary aromatic amines that induce cancer.[221] These compounds have been analyzed directly by HPLC on RP-18 columns[222]; however, the instability of the analyte[223,224] compromises precision for quantitative analysis. By derivatization with methyl isocyanate, arylhydroxylamines are converted to hydroxyureas (**91**) which are stable and suitable for HPLC analysis with uv detection.[225]

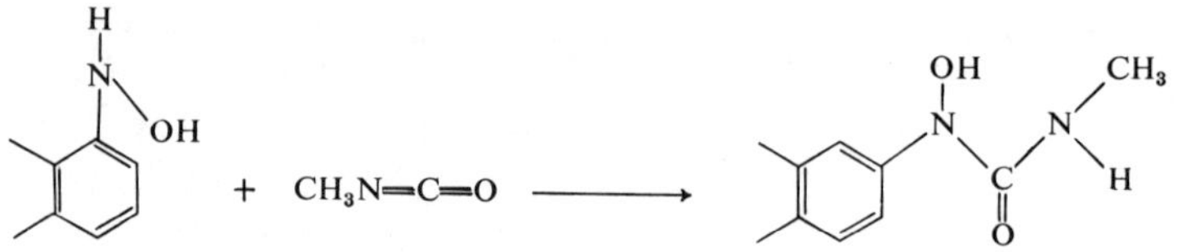

90   91

This procedure has been applied to monitoring micromolar levels of arylhydroxylamines in liver microsomal suspensions. Because of the low levels of hydroxylamine present in metabolic systems, an isocyanate with similar reactivity to methylisocyanate was sought which would yield a hydroxyurea detectable at much lower levels. As previously discussed, p-dimethylaminophenylisocyanate (**7**) was designed and synthesized to react with arylhydroxylamines to form a product (**9**) which was a strong chromophore to allow uv detection at 10^{-7} M, and was electro-oxidizable to permit amperometric monitoring of hydroxylamines at 5×10^{-9} M levels[41] in liver homogenates.

2.6. Thiols

No preseparation derivatization techniques have been described for HPLC analysis of thiols. However, benzoyl chloride was shown[226] to react with volatile thiols to yield thioamides which were separated by GLC. The application of this method to HPLC seems feasible based on the chromophoric nature of the product. Furthermore, modification of this method by use of *p*-nitrobenzoyl chloride would appear to offer better spectrophotometric sensitivity for HPLC analysis.

N-Dansylaziridine, **92**, reacts selectively with thiols to form fluorescent products, **93**.[227] The reagent, synthesized by condensation of ethyleneimine

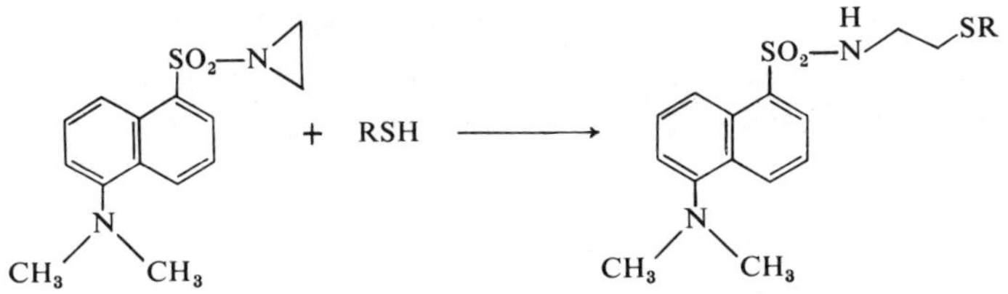

and dansyl chloride, has been used to label sulfhydryl portions of proteins. Although not yet applied to chromatographic analysis, this fluorogenic reagent should prove useful in the derivatization of thiols prior to their HPLC analysis with either spectrophotometric of fluorometric detection.

3. DETECTOR AMPLIFICATION BY PAIRED ION CHROMATOGRAPHY

3.1. Theoretical Considerations

Ion pair partition chromatography is a useful technique for isolation of ionizable organic compounds. High selectively and high sample capacity are possible, and by using chromophoric or fluorophoric counter ions high sensitivity can also be achieved. Thus ion pair chromatography can be considered a special form of derivatization in which the analyte and derivatizing agent combine primarily by coulombic rather than covalent forces. The theory and application of ion pairing has been methodically and elegantly described primarily by Schill and co-workers over the past 20 years.[231]

A cation ($R^\oplus$) may partition into a nonpolar organic phase by the addition of a counter ion ($X^\ominus$) that forms an ion pair (RX) with it:

$$R^\oplus_{aq} + X^\ominus_{aq} \rightleftharpoons RX_{org} \tag{43}$$

The charges are thus insulated in the ion pair, creating a hydrophobic species that is more compatible with less polar solvents. The ion pair extraction principle can be applied to partition chromatography both in reversed and normal phase systems; but amplification of detector response can only be obtained when the substrate migrates as an ion pair with an organic mobile phase[232] so that high concentrations of the chromophoric counter ions do not interfere with measurement of the ion pair. This is most often accomplished by incorporating the ion-pairing agent in an aqueous stationary phase, and operating in the straight phase mode.[233]

A quantitative expression for the distribution of an ion pair (RX) is given by the extraction constant, E_{RX}, which is defined as

$$E_{RX} = [RX]_{org} \cdot [R^+]_{aq}^{-1} \cdot [X^-]_{aq}^{-1} \qquad (44)$$

The distribution of $R^\oplus$ between the two phases is given by the distribution ratio, D_R:

$$D_R = [RX]_{org} \cdot [R^+]_{aq}^{-1} = E_{RX} \cdot [X^-]_{aq} \qquad (45)$$

In a system where X^- is present in the aqueous stationary phase, the capacity factor of the sample $R^\oplus$, k_R', is given by

$$k_R' = \frac{n_{R,s}}{n_{R,m}} = (E_{RX} \cdot [X]_{aq})^{-1} \cdot V_s \cdot V_m^{-1} \qquad (46)$$

where $(n_{R,s}/n_{R,m})$ is the mole ratio of total R species in stationary and mobile phases and $V_s \cdot V_m^{-1}$ is the phase volume ratio. The capacity factor can be varied by altering the nature and concentration of the counter ion, $X^\ominus$ used. The nature of X^- will have an influence on E_{RX}; exchange of the counter ion can produce changes in E_{RX} and D_{RX} of more than 10 log units.[233] Regulation of E_{RX} is mainly achieved by altering the composition of the organic phase. This may be most effectively accomplished using a binary organic phase that consists of one weakly and one strongly solvating agent, as this will provide a simple means for systematic variation of E_{RX} and k'.[232,234] Because all ion pairs have polar character, they are more efficiently extracted by slightly polar solvents (e.g., chloroform) than by nonpolar agents (e.g., aliphatic hydrocarbons). Strongly hydrogen-bonding solvents (e.g., lipophilic alcohols) have still greater extracting power, particularly for ion pairs with functionalities capable of forming hydrogen bonds. However, these agents offer much less selectivity than solvents with lower solvating capacity.[233,235] The concentration range of X^- which can be employed as counter ion is limited since very low concentrations yield unstable chromatographic systems while high concentrations often give rise to disturbing side reactions that can compromise chromatographic performance.[236,237]

The influence of side reactions (e.g., ion pair dissociation or dimerization in the organic phase; association processes or protolysis in the aqueous phase) on ion pair partitioning behavior can be considered by exchanging the stoichiometric extraction constant, E_{RX}, for a conditional extraction constant E_{RX}^*, where the α coefficients express the influence of side reactions within the phases:

$$E_{RX}^* = E_{RX}\alpha_{RX} \cdot \alpha_R^{-1} \cdot \alpha_X^{-1} \qquad (47)$$

The α coefficients have a value of 1 when side reactions are negligible and increase with rising influence of the side reactions.[238] All processes that affect E_{RX}^* will influence the chromatographic behavior of the ion-pairing system.

Association and dissociation processes involving ion pairs are particularly serious, since these effects will vary with concentration of the migrating species, producing asymmetric chromatographic bands. Dissociation of the ion pair in the organic phase is particularly pronounced in polar solvents (e.g., in 1-pentanol dissociation constants of $\sim 10^{-3}\ M$ are observed.[239]). The conditional extraction constant is given by

$$E_{RX}^* = E_{RX}\alpha_{RX} = E_{RX}(1 + K_{d,RX} \cdot [X^-]_{org}^{-1}) \qquad (48)$$

where $K_{d,RX}$ is defined as $[R^+]_{org} \cdot [X^-]_{org} \cdot [RX]_{org}^{-1}$. Dissociation of the ion pair in the organic phase will increase with decreasing sample concentration; thus $[R^+]_{org}$ and the capacity factor (k') change [equation (49)] with concentration. This decrease in k' with decreasing sample concentration leads to tailing peaks in the chromatogram:

$$k' = V_s(V_m E_{RX}^* C_X)^{-1} \qquad (49)$$

If $[X^-]_{org}$ can be maintained at a constant level [reaction (48)], the conditional constant will be independent of sample concentration. This situation can be achieved by introducing a second ion, $Y^{\oplus}$, into the aqueous phase which forms an ion pair with $X^{\ominus}$ (YX). If $C_{YX,org} \gg C_{RX,org}$ and both ion pairs have dissociation constants of similar magnitude, $[X^-]_{org}$ will be controlled by $C_{YX,org}$ and will be independent of $C_{RX,org}$.[238] The equation describing the conditional extraction constant:

$$E_{RX}^* = E_{RX}(1 + K_{d,RX} \cdot [K_{d,YX} \cdot E_{YX} \cdot (X^-)(Y^+)]) \qquad (50)$$

shows that if the aqueous concentration of the counter ion, $X^\ominus$, and foreign cation, Y^+, are kept constant, E^*_{RX} will be independent of the concentration of the sample, R^+. Such linear isotherms are a prerequisite for obtaining symmetrical peaks.

Dimerization of the ion pair in the organic phase and association of ion pair components in the aqueous phase will increase with increasing concentration and again produce asymmetric (tailing) peaks. Formation of the ion pair or other associates in the aqueous phase produces an increase in k' with increasing sample concentration and may result in leading peaks. The ion pair formation constants for a series of quaternary ammonium compounds with picrate ion in aqueous solution were determined to be $\sim 10^{1.2}$,[233] indicating that changes in the concentration of $R^\oplus$ or $X^\ominus$ smaller than $10^{-2.5}$ M should not influence k'.

3.2. Applications

Paired ion chromatography in which the counter ion is also used to amplify detector response has been applied to both cationic and anionic species. Acetylcholine has been determined in rat brain[240] by extraction into dichloromethane as an ion pair with hexanitrodiphenylamine, followed by partition chromatography on a cellulose column containing 0.06 M picrate in phosphate buffer (pH 6.5) as stationary phase and chloroform:n-pentanol (19:1) as mobile phase. The drug is thus detected spectrophotometrically as its picrate ion pair.

Alkylammonium ions have been chromatographed on cellulose and diatomaceous earth supports as ion pairs either directly with β-naphthalene sulfonate[241] or initially with chloride as counter ion followed by exchange of the chloride with β-naphthalene sulfonate[242] immediately prior to entering the detector, permitting low-level photometric monitoring. Chloroform–n-pentanol mobile phases were used to elute components. This system has also been extended to a wider variety of alkylamines, amino acids, and peptides using an aqueous stationary phase of β-naphthalene sulfonate on LiChrospher SI-100 with chloroform:n-pentanol mobile phase.[232] Alternatively, a series of nonabsorbing quaternary ammonium salts have been chromatographed as ion pairs with a stationary phase consisting of an aqueous 0.06 M picrate solution on an ethanolyzed cellulose support.[233]

The paired-ion approach to detector amplification has also been applied to the separation and quantitation of drug mixtures in tablet formulations. Mixtures of phenobarbital, hyoscyamine, and ergotamine were analyzed by separation on an SI-60 column coated with an aqueous buffered (pH 6) solution of picrate.[243] Components were eluted with chloroform saturated with stationary phase, and detected spectrophotometrically by monitoring the picrate ion pair. A similar approach[244] was applied to the quality control analysis of Cafergot-PB®, a medication used to treat migraine, where small amounts of ergotamine and hyoscyamine must be determined in the presence of a 1000-fold excess of caffeine and butalbital. In this case, advantage was taken of the ability of the two basic compounds, hyoscyamine and ergotamine, to form ion pairs with picrate, to develop an HPLC separation of components. Monitoring at λ_{max} for picrate (254 nm) allows partial resolution from the other two components (Figure 16). Complete supression of caffeine and butabital was achieved by detection at the second λ_{max} for picrate (345 nm) where these components show no absorbance (Figure 17).

Although most applications of paired ion chromatography to enhance detector response have involved anionic counter ions, carboxylic acids have also been analyzed by this technique using chromophoric cationic pairing agents. Benzilic, phenylbutanoic, and salicylic acid have been separated on a cellulose column with an aqueous stationary phase containing ion-pairing agent and a mobile phase of n-hexane:dichloromethane:n-pentanol. Two strongly uv absorbing quaternary ammonium salts, N,N-dimethylprotriptyline, **94**,[245] and N-methylimipramine, **95**,[246] have been used as chromophoric counter ion to permit low-level analysis of these acids. At 254 nm, ion

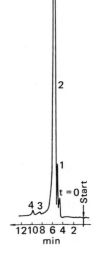

Fig. 16. Separation of components in cafergot PB®. Column SI-100 (5 μm), 15 cm × 3 mm i.d., Mobile Phase, chloroform saturated with stationary phase, 0.06 M picric acid at pH 6. Flow rate, 0.2 ml/min⁻¹. Detection at 254 nm (1.0 a.u.f.s.). Peaks: (1) butalbital; (2) caffeine; (3) hyoscyamine; (4) ergotamine (components 3 and 4 separated as ion pairs).[244] Reproduced from the *Journal of Chromatography* by permission of Elsevier Publishing Co.

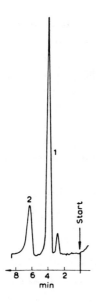

Fig. 17. Separation of Cafergot PB® components as ion pairs. System as described in Figure 16, except detection at 345 nm (0.05 a.u.f.s.). Peaks: (1) hyoscyamine; (2) ergotamine.[244] Reproduced from the *Journal of Chromatography* by permission of Elsevier Publishing Co.

pairs of **94** have a molar absorptivity of 2.9×10^3 with nonabsorbing anions, and ion pairs of **95** have an $a_m = 8.5 \times 10^3$. Thus, **95** offers three times greater sensitivity than **94** at 254 nm. At 293 nm (the λ_{max} for **94**), however, **94** has an a_m of 14.1×10^3, giving twice the sensitivity as **95** at the absorbance maximum for the respective quaternary ammonium compound.

$(CH_3)_3\overset{+}{N}$ $(CH_3)_3\overset{+}{N}$

94   **95**

Recently, a continuous ion pair extraction principle has been applied by Frei and co-workers[247] to the detection of ionizable solutes eluting from a reversed phase column after HPLC separation. Solutes present in aqueous mobile phases eluting from the column were continuously extracted as ion pairs into organic solvents and the organic phases then passed into the detector. Using the fluorescent anion, 9,10-dimethoxyanthracene-2-sulfonate as counter ion, a series of weakly absorbing tertiary amines have been monitored fluorimetrically.

4. CONCLUSIONS

The use of chemical derivatization prior to chromatographic separation in HPLC to enhance detectability, improve chromatographic properties, or stabilize analytes is rapidly increasing. Continued design of reagents, particularly those capable of proviging detector amplification and thereby broadening the utility of this separation technique, is anticipated. Particular growth is needed (1) in the design of reagents that are transparent to detectors but that react with specific functional groups to yield products detectable at trace levels (<1 ng/ml of test solution) and (2) in improving selectivity in the chemical transformation of individual functionalities in the presence of other groups. Development in these areas will yield an improvement in overall specificity and sensitivity of HPLC analysis. These capabilities will allow scientists to more specifically monitor drug species (or other chemical entities) for longer periods of time and to measure trace amounts of these materials which would otherwise go undetected in various biological and chemical matrices. In this way, the disposition of the drug can be better defined, and its safe usage often enhanced.

However, control of reactivity and rational design of reagents such as those just described demands a deeper understanding of the chemical mechanism associated with individual derivatization reactions. The analyst must therefore utilize the basic tools of physical organic chemistry, i.e., kinetics, thermodynamics, and extrathermodynamic (e.g., linear free energy relationships) principles to characterize and more effectively control and utilize the chemistry at his or her disposal.

REFERENCES

1. J. F. Lawrence and R. W. Frei, Chemical derivatization in liquid chromatography, in *Journal of Chromatography Library* 7, Elsevier, Amsterdam (1976).
2. T. H. Jupille, *Am. Lab.*, 85–92 (1976).
3. M. S. F. Ross, *J. Chromatogr.* 141, 107–119 (1977).
4. P. T. Kissinger, K. Bratin, G. C. Davis, and L. A. Pachla, *J. Chromatogr. Sci.* 17, 137–146 (1979).
5. J. F. Lawrence, *J. Chromatogr. Sci.* 17, 147–151 (1979).
6. T. Jupille, *J. Chromatogr. Sci.* 17, 160–167 (1979).
7. S. Ahuja, *J. Chromatogr. Sci.* 17, 168–172 (1979).
8. K. Blau and G. S. King, *Handbook of Derivatives for Chromatography*, Heyden and Son, London (1977).
9. D. R. Knapp, *Handbook of Analytical Derivatization Reactions*, Wiley Interscience, New York (1979).

10. D. J. Munger, L. A. Sternson, and A. J. Repta, *J. Chromatogr.* **143**, 375–382 (1977).
11. P. H. Boyle, *Chem. Soc. (London) Q. Rev.* **25**, 323–341 (1972).
12. I. S. Krull, in *Advances in Chromatography*, Vol. 16, J. C. Giddings, E. Grushka, J. Cases, and P. R. Brown, eds., Marcel Dekker, New York (1978), pp. 175–210.
13. J. Merzhauser, E. Roder, and C. Hesse, *Klin. Wochenschr.* **51**, 883–888 (1973).
14. K. Hostettmann and M. McNair, *J. Chromatogr.* **116**, 201–206 (1976).
15. G. Guilbault, *Practical Fluorescence*, Marcel Dekker, New York (1973).
16. I. R. Politzer, G. W. Griffen, B. J. Dowty, and L. J. Laseter, *Anal. Lett.* **6**, 539–546 (1973).
17. D. R. Knapp and S. Krueger, *Anal. Lett.* **8**, 603–610 (1975).
18. F. A. Fitzpatrick, M. A. Wynalda, and D. G. Kaiser, *Anal. Chem.* **49**, 1032–1035 (1977).
19. M. A. Carey and H. W. Persinger, *J. Chromatogr. Sci.* **10**, 537–543 (1972).
20. R. A. Henry, J. A. Schmit, and J. F. Dieckman, *J. Chromatogr. Sci.* **9**, 513–520 (1971).
21. L. J. Papa and L. P. Turner, *J. Chromatogr. Sci.* **10**, 747–750 (1972).
22. G. B. Cox, *J. Chromatogr.* **83**, 471–481 (1973).
23. J. W. Higgins, *J. Chromatogr.* **121**, 329–334 (1976).
24. F. A. Fitzpatrick and S. Siggia, *Anal. Chem.* **45**, 2310–2314 (1973).
25. C. R. Clark and M. M. Wells, *J. Chromatogr. Sci.* **16**, 332–339 (1978).
26. L. L. Needham and M. M. Kochhar, *J. Chromatogr.* **114**, 220–222 (1975).
27. P. J. Porcard and P. Shubiak, *Anal. Chem.* **44**, 1865–1867 (1972).
28. L. D. Bighley, D. E. Wurster, D. Cruden-Loeb, and R. V. Smith, *J. Chromatogr.* **110**, 375–380 (1975).
29. R. F. Borch, *Anal. Chem.* **47**, 2437–2439 (1975).
30. H. D. Durst, M. Milano, E. J. Kikta, S. A. Connelly, and E. Grushka, *Anal. Chem.* **47**, 1797–1801 (1975).
31. E. Grushka, H. D. Durst, and E. J. Kikta, *J. Chromatogr.* **112**, 673–678 (1975).
32. W. Morozowich and S. L. Douglas, *Prostaglandins* **10**, 19–40 (1975).
33. M. J. Cooper and M. W. Anders, *Anal. Chem.* **46**, 1849–1852 (1974).
34. T. Sugiura, T. Hayashi, S. Kawai, and T. Ohno, *J. Chromatogr.* **110**, 385–388 (1975).
35. R. W. Roos, *J. Chromatogr. Sci.* **14**, 505–512 (1976).
36. P. T. Kissinger, C. J. Refshauge, R. Dreiling, L. Blank, R. Freeman, and R. N. Adams, *Anal. Lett.* **6**, 465–477 (1973).
37. R. J. Davenport and D. C. Johnson, *Anal. Chem.* **46**, 1971–1978 (1974).
38. V. R. Taden, J. Landelma, H. Pope, and G. Muusze, *J. Chromatogr.* **125**, 275–286 (1976).
39. P. T. Kissinger, *Anal. Chem.* **49**, 447A–456A (1977).
40. R. N. Adams, *Electrochemistry at Solid Electrodes*, Marcel Dekker, New York (1969), pp. 305–383.
41. D. Musson and L. A. Sternson, *J. Chromatogr.* **188**, 159–167 (1980).
42. L. J. Felice and P. T. Kissinger, *Clin. Chim. Acta* **76**, 317–320 (1977).
43. D. Sawyer and J. L. Roberts, *Experimental Electrochemistry for Chemists*, John Wiley and Sons, New York (1974), pp. 378–382.
44. H. B. Hanekamp, P. Bos, U. A. T. Brinkman, and R. W. Frei, *Z. Anal. Chem.* **297**, 404–410 (1979).
45. R. W. Frei, personal communication.
46. I. Jane and J. F. Taylor, *J. Chromatogr.* **109**, 37–42 (1975).

47. Y. Golander and L. A. Sternson, *J. Chromatogr. Biomed. Appl.* **181**, 41–49 (1980).
48. D. Mendenhall, H. Kobayashi, F. M. L. Shih, L. A. Sternson, T. Higuchi, and C. J. Fabian, *Clin. Chem.* **24**, 1518–1524 (1978).
49. F. Panettiere and C. A. Coltman, Jr., *Cancer* **27**, 835–841 (1971).
50. S. J. Bannister, J. Stevens, D. Musson, and L. A. Sternson, *J. Chromatogr.* **176**, 381–390 (1979).
51. K. Connors, *Reaction Mechanisms in Organic Analytical Chemistry*, Wiley Interscience, New York (1973), pp. 516–601.
52. W. P. Jencks, *Catalysis in Chemistry and Enzymology*, McGraw-Hill, New York (1969), Chap. 10.
53. M. L. Bender, Chem. Rev. **60**, 53–113 (1960).
54. M. L. Bender and R. J. Thomas, *J. Am. Chem. Soc.* **83**, 4189–4193 (1961).
55. L. P. Hammett, *Physical Organic Chemistry*, McGraw-Hill, New York (1960), Chap. 7.
56. H. H. Jaffe, *Chem. Rev.* **53**, 191–261 (1953).
57. D. H. McDaniel and H. C. Brown, *J. Org. Chem.* **23**, 420–427 (1958).
58. J. F. Kirsch and W. P. Jencks, *J. Am. Chem. Soc.* **86**, 837–846 (1964).
59. M. L. Bender and B. W. Tornquest, *J. Am. Chem. Soc.* **79**, 1656–1662 (1957).
60. W. P. Jencks and J. Carriuolo, *J. Am. Chem. Soc.* **83**, 1743–1750 (1961).
61. D. G. Oakenhall, T. Riley, and V. Gold, *J. Chem. Soc. Chem. Commun.*, 385–386 (1966).
62. W. P. Jencks, *Catalysis in Chemistry and Enzymology*, McGraw-Hill, New York (1969), pp. 68–70.
63. T. C. Bruice and R. Lapinski, *J. Am. Chem. Soc.* **80**, 2265–2267 (1958).
64. A. R. Fersht and W. P. Jencks, *J. Am. Chem. Soc.* **91**, 2125–2126 (1969).
65. W. Steglich and G. Hofle, *Angew. Chem. Int. Ed. Engl.* **8**, 981 (1969).
66. W. P. Jencks and J. Carriulo, *J. Biol. Chem.* **234**, 1272–1279, 1280–1285 (1959).
67. K. A. Connors and N. K. Panoit, *Anal. Chem.* **50**, 1542–1545 (1978).
68. E. L. Rowe and S. M. Machkovech, *J. Pharm. Sci.* **66**, 273–275 (1977).
69. R. Wolfenden and W. P. Jencks, *J. Am. Chem. Soc.* **83**, 4390–4393 (1961).
70. J. F. Kirsch and W. P. Jencks, *J. Am. Chem. Soc.* **86**, 833–837 (1964).
71. S. Lapshin, V. A. Dadall, Y. S. Simanenko, and L. M. Litrinenko, *Zh. Org. Chim.* **13**, 586–594 (1977).
72. J. Vessman, P. Hartvig, and M. Molander, *Anal. Lett.* **6**, 699–707 (1973).
73. P. Hartvig and J. Vessman, *Anal. Lett.* **7**, 223–231 (1974).
74. P. Hartvig and J. Vessman, *Acta Pharm. Suecica* **11**, 115–124 (1974).
75. P. Hartvig, W. Handl, J. Vessman, and C. M. Svahn, *Anal. Chem.* **48**, 390–393 (1976).
76. L. A. Sternson and A. D. Cooper, *J. Chromatogr.* **150**, 257–258 (1978).
77. B. A. Davis, *Biomed. Mass. Spec.* **6**, 149–156 (1979).
78. N. Seiler and M. Wiechmann, in *Progress in Thin-Layer Chromatography and Related Methods*, Vol. 1, A. Niederweiser and G. Pataki, eds., Ann Arbor-Humphrey Science Publishers, Ann Arbor, Michigan (1970), p. 95.
79. B. Davis, *J. Chromatogr.* **151**, 252–255 (1978).
80. R. W. Frei, W. Santi, and M. Thomas, *J. Chromatogr.* **116**, 365–377 (1976).
81. W. Dunges, G. Naundorf, and N. Seiler, *J. Chromatogr. Sci.* **12**, 655–657 (1974).
82. P. J. Meffin, S. R. Harapat, and D. C. Harrison, *J. Pharm. Sci.* **66**, 583–586 (1977).
83. F. L. Vandemark, G. J. Schmidt, and W. Slavin, *J. Chromatogr. Sci.* **16**, 465–469 (1978).

84. N. D. Brown, R. B. Sweet, J. A. Kintzios, H. D. Cox, and B. P. Doctor, *J. Chromatogr.* **164**, 35–40 (1979).
85. M. M. Abdell Monem and K. Ohno, *J. Chromatogr.* **107**, 416–419 (1975).
86. A. Chimiak and T. Polonski, *J. Chromatogr.* **115**, 215–217 (1975).
87. A. Lang, H. E. Geissler, and E. Mutschler, *Arzneih. Forsch.* **28**, 575–577 (1978).
88. G. Powis and M. M. Ames, *J. Chromatogr.* **170**, 195–201 (1979).
89. R. Bongiovanni and W. Dutton, *J. Liq. Chromatogr.* **1**, 613–630 (1978).
90. R. W. Frei, M. Thomas, and I. Frei, *J. Liq. Chromatogr.* **1**, 443–445 (1978).
91. R. N. Kaul, M. W. Conway, M. L. Clarke, and J. Huffine, *J. Pharm. Sci.* **59**, 1745 (1970).
92. R. N. Kaul, M. W. Conway, and M. L. Clarke, *Nature* **226**, 372–373 (1970).
93. I. S. Forrest, D. E. Green, S. D. Rose, G. C. Skinner, and D. M. Torres, *Res. Commun. Chem. Pathol. Pharm.* **2**, 787–792 (1971).
94. J. N. Le Page, W. Lindner, G. Davies, D. E. Seitz, and B. L. Karger, *Anal. Chem.* **51**, 433–435 (1979).
95. J. E. Benson and P. E. Hare, *Proc. Natl. Acad. Sci. USA* **72**, 619–622 (1975).
96. S. Udenfriend, S. Stein, P. Bohlen, W. Dairman, W. Leihgruber, and M. Weigle, *Science* **178**, 871–872 (1972).
97. M. Roth, *Anal. Chem.* **43**, 880–882 (1971).
98. Y. Chang, A. J. Repta, and L. A. Sternson, unpublished results.
99. R. W. Frei, L. Michel, and W. Santi, *J. Chromatogr.* **126**, 665–677 (1976).
100. R. W. Frei, L. Michel, and W. Santi, *J. Chromatogr.* **142**, 271–281 (1977).
101. K. Imai, M. Tsukamoto, and Z. Tamura, *J. Chromatogr.* **137**, 357–362 (1977).
102. W. McHugh, R. A. Sandmann, W. G. Haney, S. P. Sood, and D. P. Wittmer, *J. Chromatogr.* **124**, 376–380 (1976).
103. J. T. Stewart and J. L. Williamson, *Anal. Chem.* **48**, 1182–1185 (1976).
104. J. Hodgin, *J. Liq. Chromatogr.* **2**, 1049–1059 (1979).
105. P. Lindroth and K. Mopper, *Anal. Chem.* **51**, 1667–1674 (1979).
106. L. D. Mell, A. R. Dasler, and A. B. Gustafson, *J. Liq. Chromatogr.* **1**, 261–277 (1978).
107. T. P. Davis, C. W. Gehrke, C. W. Gehrke, Jr., T. D. Cunningham, K. C. Kuo, K. O. Gerhardt, H. D. Johnson, and C. H. Williams, *Clin. Chem.* **24**, 1317–1324 (1978).
108. R. E. Suboen, R. G. Brown, and A. C. Noble, *J. Chromatogr.* **166**, 310–312 (1978).
109. S. Maitra, T. T. Yoshikawa, J. L. Hausen, I. Nilsson-Ehle, W. J. Palin, M. C. Schotz, and L. B. Guze, *Clin. Chem.* **23**, 2275–2278 (1977).
110. C. R. Clark and M. M. Wells, *J. Chromatogr. Sci.* **16**, 332–339 (1978).
111. L. L. Needham and M. A. Kochhar, *J. Chromatogr.* **114**, 220–222 (1975).
112. T. Suguira, T. Hayashi, S. Kawai, and T. Ohno, *J. Chromatogr.* **110**, 385–388 (1975).
113. S. L. Wellows and M. A. Carey, *J. Chromatogr.* **154**, 219–225 (1978).
114. C. R. Clark, J. D. Teague, M. M. Wells, and J. H. Ellis, *Anal. Chem.* **49**, 912–915 (1977).
115. Y. Y. Chang and E. H. Creaser, *J. Chromatogr.* **116**, 215–217 (1976).
116. Regis Lab Notes, No. 17, Regis Chemical Co., Morton Grove, Illinois, November 1974.
117. H. Furukawa, Y. Mori, Y. Takenchi, and K. Ito, *J. Chromatogr.* **136**, 428–431 (1977).
118. W. H. Pirkle and D. W. House, *J. Org. Chem.* **44**, 1957–1960 (1979).
119. S. D. Ross, *Progr. Phys. Org. Chem.* **1**, 31–74 (1963).
120. J. F. Bunnett, E. W. Garbisch, and K. M. Pruitt, *J. Am. Chem. Soc.* **79**, 385–391 (1957).

121. H. Beyer and U. Schenk, *J. Chromatogr.* **39**, 482–490 (1969).
122. H. Beyer and U. Schenk, *J. Chromatogr.* **39**, 491–495 (1969).
123. C. Martel and D. J. Phelps, *J. Chromatogr.* **115**, 633–634 (1975).
124. J. Franc and V. Koudelkova, *J. Chromatogr.* **170**, 89–97 (1979).
125. J. V. Janovsky and L. Erb, *Chem. Ber.* **19**, 2155–2158 (1886).
126. F. van Hoof and A. Heyndrickx, *Anal. Chem.* **46**, 286–288 (1974).
127. J. Monforte, R. J. Bath, and I. Sunshine, *Clin. Chem.* **18**, 1329–1333 (1972).
128. H. J. Klimisch and D. Ambrosius, *J. Chromatogr.* **121**, 93–95 (1976).
129. G. J. Krol, J. M. Banovsky, C. A. Mannan, R. E. Pickering, and B. T. Kuo, *J. Chromatogr.* **163**, 383–389 (1979).
130. H. J. Klimisch and L. Stadler, *J. Chromatogr.* **90**, 141–148 (1974).
131. G. L. Dadisch and P. Wolschann, *Z. Anal. Chem.* **292**, 219–227 (1978).
132. A. K. Mitra, C. L. Baustian, and T. J. Mikkelson, *J. Pharm. Sci.* **69**, 257–261 (1980).
133. F. A. Fitzpatrick and S. Siggia, *Anal. Chem.* **45**, 2310–2314 (1973).
134. J. Lehrfeld, *J. Chromatogr.* **120**, 141–147 (1976).
135. F. Nachtman, H. Spitzy, and R. Frei, *Anal. Chem.* **48**, 1576–1579 (1976).
136. F. Nachtman, H. Spitzy, and R. W. Frei, *J. Chromatogr.* **122**, 293–303 (1976).
137. F. Nachtman, *Z. Anal. Chem.* **282**, 209–213 (1976).
138. R. Schwarzenbach, *J. Chromatogr.* **140**, 304–309 (1977).
139. F. Nachtman and K. W. Budna, *J. Chromatogr.* **136**, 279–287 (1977).
140. M. A. Carey and H. E. Persinger, *J. Chromatogr. Sci.* **10**, 537–543 (1972).
141. P. J. Porcard and P. Shubiak, *Anal. Chem.* **44**, 1865–1867 (1972).
142. L. D. Bighley, D. E. Wurster, D. Crugen-Loeb, and R. V. Smith, *J. Chromatogr.* **110**, 375–380 (1975).
143. L. Penzes and G. W. Oertel, *J. Chromatogr.* **51**, 322–324 (1970).
144. R. W. Roos, *J. Pharm. Sci.* **67**, 1735–1739 (1978).
145. L. P. Penzes and G. W. Oertel, *J. Chromatogr.* **74**, 359–365 (1972).
146. L. P. Penzes and G. W. Oertel, *J. Chromatogr.* **51**, 325–327 (1970).
147. R. Duire, R. Chayen, *J. Chromatogr.* **45**, 76–81 (1969).
148. K. L. Oehrle and K. Vogt, *Acta Endocrinol.* **82**, Suppl. 202, 61–63 (1976).
149. M. Frei-Hausler, R. W. Frei, and O. Hutzinger, *J. Chromatogr.* **84**, 214–217 (1973).
150. O. Hotzinger, R. A. Heacock, and S. Safe, *J. Chromatogr.* **97**, 233–247 (1974).
151. R. M. Cassidy, D. S. LeGay, and R. W. Frei, *J. Chromatogr. Sci.* **12**, 85–89 (1974).
152. J. F. Fiechtl, G. Spiteller, W. W. Just, G. Werner, and M. Wichmann, *Naturwissenschaften* **60**, 207–208 (1973).
153. R. Chayen, S. Gould, A. Harell, and C. V. Stead, *Anal. Biochem.* **39**, 533–535 (1971).
154. N. E. Hoffman and J. C. Liao, *Anal. Chem.* **48**, 1104–1106 (1976).
155. C. G. Scott, M. J. Petrine, and T. McCorklie, *J. Chromatogr.* **125**, 157–161 (1976).
156. C. G. Swain and C. B. Scott, *J. Am. Chem. Soc.* **75**, 141–147 (1953).
157. J. D. Edwards and R. G. Pearson, *J. Am. Chem. Soc.* **84**, 16–24 (1962).
158. A. J. Parker, *Adv. Phys. Org. Chem.* **5**, 173–178 (1967).
159. H. D. Durst, *Tetrahedron Lett.* **28**, 2421–2424 (1974).
160. H. D. Durst, M. Milano, E. J. Kikta, Jr., S. A. Connelly, and E. Grushka, *Anal. Chem.* **47**, 1797–1801 (1975).
161. I. R. Politzer, G. W. Griffin, B. J. Dowty, and J. L. Laseter, *Anal. Lett.* **6**, 539–546 (1973).
162. Regis Chemical Co., Morton Grove, Illinois (Regis Lab Notes No. 16).
163. F. A. Schmid and D. J. Hutchinson, *Cancer Res.* **34**, 1671–1675 (1974).

164. D. R. Knapp and S. Krueger, *Anal. Lett.* **8**, 603–610 (1975).
165. E. Vowinker, *Chem. Ber.* **100**, 16–22 (1967).
166. B. Shaiki, N. J. Pontzer, J. E. Molina, and M. I. Kelsey, *Anal. Biochem.* **85**, 47–55 (1978).
167. P. D. Bartlett, in *Gilman's Organic Chemistry*, Vol. III, Wiley, New York (1953), p. 35.
168. R. F. Borch, *Anal. Chem.* **47**, 2437–2439 (1975).
169. M. D'Amboise and M. Gendreau, *Anal. Lett.* **12**, 381–395 (1979).
170. H. C. Jordi, *J. Liq. Chromatogr.* **1**, 215–230 (1978).
171. M. J. Cooper and M. W. Anders, *Anal. Chem.* **46**, 1849–1852 (1974).
172. E. Grushka, H. D. Durst, and E. J. Kheta, *J. Chromatogr.* **112**, 673–678 (1975).
173. W. Morozowich and S. L. Douglas, *Prostaglandins* **10**, 19–40 (1975).
174. T. J. Roseman, S. S. Butler, and S. L. Douglas, *J. Pharm. Sci.* **65**, 673–676 (1976).
175. F. A. Fitzpatrick, *J. Pharm. Sci.* **65**, 1609–1613 (1976).
176. S. Lam and E. Grushka, *J. Liq. Chromatogr.* **1**, 33–42 (1978).
177. A. Hulshoff, H. Roseboom, and J. Renema, *Adv. Chromatogr.*, A. Zlatkis, ed., Chromatography Symposium Publication, Houston, pp. 563–569 (1979).
178. M. Mueller, R. Pietschmann, C. Plachetta, R. Sehr, and H. Tuss, *Z. Anal. Chem.* **288**, 361–368 (1977).
179. W. Duenges, *Anal. Chem.* **49**, 442–445 (1977).
180. S. G. Zelenski and J. W. Huber, *Chromatographia* **11**, 645–646 (1978).
181. S. Lam and E. Grushka, *J. Chromatogr.* **158**, 207–214 (1978).
182. W. Duenges, *Chromatographia* **9**, 624–626 (1976).
183. Regis Lab Notes, Morton Grove, Illinois, No. 21 (1977).
184. W. Lindner and W. Santi, *J. Chromatogr.* **176**, 55–64 (1979).
185. G. Gübitz and W. Wendelin, *Anal. Chem.* **51**, 1690–1693 (1979).
186. W. P. Jencks, *Progr. Phys. Org. Chem.* **2**, 63–128 (1964).
187. W. P. Jencks, *J. Am. Chem. Soc.* **81**, 475–481 (1959).
188. E. H. Cordes and W. P. Jencks, *J. Am. Chem. Soc.* **84**, 4319–4328 (1962).
189. J. M. Sayer and W. P. Jencks, *J. Am. Chem. Soc.* **91**, 6353–6361 (1969).
190. A. Williams and M. L. Bender, *J. Am. Chem. Soc.* **88**, 2508–2513 (1966).
191. L. J. Papal and L. P. Turner, *J. Chromatogr. Sci.* **10**, 747–750 (1972).
192. S. Selim, *J. Chromatogr.* **136**, 271–277 (1977).
193. R. W. Frei and J. F. Lawrence, *J. Chromatogr.* **83**, 321–330 (1973).
194. C. T. Mansfield, B. T. Hodge, R. B. Hege, and W. C. Hamlin, *J. Chromatogr. Sci.* **15**, 301–302 (1977).
195. H. Terada, T. Hayashi, S. Kawal, and T. Ohno, *J. Chromatogr.* **130**, 281–286 (1977).
196. B. Rittich and M. Krska, *J. Chromatogr.* **130**, 189–194 (1977).
197. R. A. Henry, J. A. Schmit, and J. E. Dieckman, *J. Chromatogr. Sci.* **9**, 513–520 (1971).
198. P. Knapstein and J. C. Touchstone, *J. Chromatogr.* **37**, 83–88 (1968).
199. J. Haki, *J. Chromatogr.* **61**, 183–186 (1971).
200. F. R. Fitzpatrick, S. Siggia, and J. Dingman, *Anal. Chem.* **44**, 2211–2216 (1972).
201. R. W. Roos, *J. Chromatogr. Sci.* **14**, 505–512 (1976).
202. N. W. Jacobsen and R. G. Dickinson, *Anal. Chem.* **46**, 298–299 (1974).
203. F. A. Fitzpatrick, M. A. Wyalda, and D. G. Kaiser, *Anal. Chem.* **49**, 1032–1035 (1977).
204. T. Kawasaki, M. Maeda, and A. Tsuji, *J. Chromatogr.* **163**, 143–150 (1979).

205. R. Chayen, R. Dvir, S. Gould, and A. Harrell, *Anal. Biochem.* **42**, 282–286 (1971).
206. G. Avigao, *J. Chromatogr.* **139**, 343–347 (1977).
207. J. C. Liao, N. E. Hoffman, J. J. Barboriak, and D. A. Roth, *Clin. Chem.* **23**, 802–805 (1977).
208. T. Hayashi, T. Sugiura, H. Tereda, and S. Kawai, *J. Chromatogr.* **118**, 403–408 (1976).
209. H. Nakamura and Z. Tamura, *Anal. Chem.* **50**, 2047–2051 (1978).
210. H. Nakamura and Z. Tamura, *J. Chromatogr.* **168**, 481–487 (1979).
211. H. Nakamura and Z. Tamura, *Anal. Chem.* **51**, 1679–1682 (1979).
212. D. Bucher, *Chromatographia* **10**, 723–725 (1977).
213. J. X. de Vries, R. Frank, and Chr. Birn, *FEBS Lett.* **55**, 65–67 (1975).
214. A. P. Graffeo, A. Haag, and B. L. Karger, *Anal. Lett.* **6**, 505–511 (1973).
215. P. Erdtmansky and T. J. Goehl, *Anal. Chem.* **47**, 750–752 (1975).
216. S. L. Macolm and T. R. Marten, *Anal. Chem.* **48**, 807–809 (1976).
217. T. Kawabata, H. Ohshima, T. Ishibashi, M. Matsui, and T. Kitsuwa, *J. Chromatogr.* **140**, 47–56 (1977).
218. S. S. Mirvish, *J. Natl. Cancer Inst.* **46**, 1183–1193 (1971).
219. M. S. F. Ross, *J. Chromatogr.* **133**, 408–411 (1977).
220. G. Neurath and W. Lüttich, *J. Chromatogr.* **34**, 257–258 (1968).
221. E. C. Miller and J. A. Miller, *Pharmacol. Rev.* **18**, 805–838 (1966).
222. L. A. Sternson and W. J. DeWitte, *J. Chromatogr.* **137**, 305–314 (1977).
223. A. R. Becker and L. A. Sternson, *J. Org. Chem.*, **45**, 1708–1710 (1980).
224. T. Kalhorn, A. R. Becker, and L. A. Sternson, *Bioorg. Chem.*, in press.
225. L. A. Sternson, W. J. DeWitte, and J. G. Stevens, *J. Chromatogr.* **153**, 481–488 (1978).
226. L. Gasco and R. Barrera, *Anal. Chim. Acta* **61**, 253–264 (1972).
227. W. H. Scouten, R. Lubcher, and W. Baughman, *Biochim. Biophys. Acta* **336**, 421–426 (1974).
228. H. B. Hanekamp, P. Bos, and R. W. Frei, *Adv. Chromatogr.*, A. Zlatkis, ed., Chromatography Symposium Publication, Houston, pp. 517–524 (1979).
229. E. Berlinger and L. H. Liu, *J. Am. Chem. Soc.* **75**, 2417–2420 (1953).
230. T. J. Goehl, G. M. Sundaresan, and V. K. Prasad, *J. Pharm. Sci.* **68**, 1374–1376 (1979).
231. G. Schill, *Separation Methods for Drugs and Related Organic Compounsd*, Apotekarsocieteten, Stockholm (1978).
232. J. Crommen, B. Fransson, and G. Schill, *J. Chromatogr.* **142**, 283–297 (1977).
233. S. Eksborg and G. Schill, *Anal. Chem.*, 2092–2100 (1973).
234. R. Modin and G. Schill, *Talanta* **22**, 1017–1022 (1975).
235. B. A. Persson, *Acta Pharm. Suecica* **8**, 193–204 (1971).
236. K. G. Wahlund, *J. Chromatogr.* **115**, 411–422 (1975).
237. B. Fransson, K. G. Wahlund, I. M. Johansson, and G. Schill, *J. Chromatogr.* **125**, 327–344 (1976).
238. K. G. Wahlund, *J. Chromatogr.* **115**, 411–422 (1975).
239. R. Modin and S. Bäck, *Acta Pharm. Suecica* **8**, 585–590 (1971).
240. S. Eksborg and B. A. Persson, *Acta Pharm. Suecica* **8**, 205–216 (1971).
241. S. Eksborg, *Acta Pharm. Suecica* **12**, 19–36 (1975).
242. S. Eksborg, *Acta Pharm. Suecica* **12**, 243–252 (1975).
243. W. Santi, J. M. Huen, and R. W. Frei, *J. Chromatogr.* **115**, 423–436 (1975).

244. J. M. Huen, R. W. Frei, W. Santi, and J. P. Thevenin, *J. Chromatogr.* **149**, 359–376 (1978).
245. P. O. Lagerström, *Acta Pharm. Suecica* **12**, 215–234 (1975).
246. P. O. Lagerström and A. Theodorsen, *Acta Pharm. Suecica* **12**, 429–434 (1975).
247. R. W. Frei, J. F. Lawrence, U. A. Th. Brinkman, and I. L. Honigberg, *J. High Res. Chromatogr.* **2**, 11–14 (1979).

Chapter 4

Reaction Detectors
in Liquid Chromatography

R. W. Frei

1. INTRODUCTION

In the field of modern column liquid chromatography (LC) the use of chemical derivatization techniques has increased enormously in the past years. Evidence for this is the steadily growing number of research publications in this field and the appearance of several reviews and books dealing fully or partly with this topic.[1-7] Of the two major groups in this field, prechromatographic and postchromatographic techniques, the former has been a lot more popular in the recent past and it can well be said that prechromatographic techniques have already gained a status similar to the use of derivatization methodology in gas chromatography.[2,8] The reasons and advantages for using this approach have been discussed several times before.[1,7] The disadvantages, however, such as danger of artifact formation and the restriction to certain types of reactions have always prevented a very broad adaptation of this technique. This fact and the recent improvement of hardware and know-how in liquid chromatography have spurred a flurry of activity in the past two years in the field of postcolumn reaction detector development.[9] The lack of suitable detectors in LC for trace and

R. W. Frei • Department of Analytical Chemistry, Free University, De Boelelaan 1083, 1081 HV Amsterdam, The Netherlands.

ultratrace analysis in complex matrices has further catalyzed this trend. The uv detector, although being the most versatile and reliable detector, often lacks sensitivity and selectivity. As a result, sample preparation and cleanup can be excessively tedious. Fluorescence detectors have a better potential in this regard but only a few groups of compounds possess natural fluorescence. Similar arguments can be used for other detection modes such as electrochemical detection. In all these situations chemical derivatization in general and the use of postcolumn reaction detectors in particular can bring about a solution by enhancing the detection properties for the compounds in the matrix of interest.

The use of postcolumn reactions in liquid chromatography is by no means new. Reaction detectors have been used for many years in combination with classical column chromatography. The best-known techniques have been developed in the bioanalytical domain particularly for amino acids, where since the original work by Spackman, Stein, and Moore[10] considerable progress has been made. While 21 hr were necessary at that time to separate and analyze 20 amino acids, the time has been reduced to less than 1 hr within about ten years,[11,12] still using relatively classical designs.

The major advantages with the use of postcolumn reaction detectors are the absence of artifact formation and that it is not necessary that the reaction go to completion or be well defined. Good reproducibility is the only requirement. The coupling of different detection principles, i.e., a uv detector followed by a fluorescence reaction detector is quite a common arrangement. Coupling with other detection modes such as electrochemical detection should also be feasible. The disadvantages are also quite obvious. One is the influence of the mobile phase on the reaction medium. Very seldom will the optimal chromatographic eluent also be the best reaction medium and often severe compromises have to be made. This is in fact still the major drawback in our opinion and may be one reason why reaction detector development has been delayed. The same interdependence between reaction and eluent will also restrict the use of gradient elution with this detection mode.

The kinetics of a reaction will strongly influence the choice of a particular reaction detector type since the primary aim is to preserve the original chromatographic resolution as much as possible. Construction of proper reaction detectors hence reflects the constant struggle against band broadening.

Another requirement is that the reagent should not interfere with the detection signal of the derivative. Various reactor designs have been proposed in the recent past to cope with these drawbacks to various degrees and shall be discussed here, both from a theoretical and practical point of view. An up-to-date list of applications shall illuminate the current possibilities and state of the art in this field. Also discussed will be some reactions which have not been used in postcolumn derivatization but which at the current state of reactor development show good potential for adaptation to this end. Finally some special applications and trends in reaction detectors are discussed with the hope of stimulating some ideas in other working groups.

2. THEORETICAL ASPECTS

2.1. Tubular Reactors

The simplest type of reactor consists of straight or helically coiled tubes made of different materials such as glass, Teflon, steel, and of widely differing geometries.

The solvent streams are nonsegmented and laminar flow patterns prevail. The treatment of such systems in terms of flow patterns and corresponding zone broadening is relatively straightforward and based on classical flow dynamics. The typical band shapes which can be expected in a linear reaction tube usually start from a plug injection with a square shape. Then a peak profile is obtained whose deviation from a Gaussian profile is an indication of the (geometrical) quality of the reactor design. Two phenomena are primarily responsible for the concentration profiles. One is the axial dispersion due to velocity gradients in the laminar flow pattern. The velocity of the fluid stream in the center is twice the mean velocity of the stream and at the layer closest to the tube wall it would essentially be zero. If this were the only phenomenon, one could expect an indefinite dispersion pattern.

The second phenomenon is a diffusion-controlled process. With reasonably high flow rates, axial diffusion is negligible for the axial dispersion in comparison to the dispersion caused by the velocity gradient. However, molecular diffusion from the wall to the center will tend to effectively counteract the formation of the velocity profile and in narrow tubes will result in a decrease of the dispersion when going to lower flow rates. The extreme situation would be a flow rate at which dispersion becomes totally diffusion controlled. This would correspond to the so-called Taylor minimum[1] with an approximate Reynolds number of $R_e = 0.1$ (see Figure 1). The flow velocity corresponding to this point is clearly too low for practical purposes (sample throughput) and hence in practical use we would work in a region above this point but likely still in the lower portion of the raising laminar flow curve of Figure 1 ($R_e \approx 1$).

Under such conditions and for a straight open tube design, the combination of parabolic velocity profile and molecular diffusion leads to a reasonably defined residence time dispersion. The variance of the residence time dispersion function in the reaction tube, $\Delta\sigma_{tr}^2$, can then be defined as

$$\Delta\sigma_{tr}^2 = \frac{2D_m t_v^3}{L^2} + \frac{d_t^2 t_v}{96D_m} \tag{1}$$

assuming an inert noninteracting compound.[14] Here L is the length of the tube, D_m the molecular diffusion coefficient, t_v the mean residence time (related to flow velocity), and d_t the inner diameter of the tube.

For the flow rates used conventionally in reaction detectors (>0.1 ml/min) equation (1) can be rewritten as

$$\Delta\sigma_{tr}^2 = \frac{d_t^2 t_v}{96D_m} \tag{2}$$

or

$$\frac{\Delta\sigma_{tr}^2}{t_v} = \frac{d_t^2}{96D_m} \tag{3}$$

From Figure 1 it is apparent that on going to higher, more practical flow-rates in a straight reaction detector we are obtaining more dispersion; hence the price we pay is a loss in resolution of bands due to band broadening. The situation improves drastically when moving into the turbulent flow region (see Figure 1) but turbulent flow conditions cannot be achieved under practical reaction detector conditions (flow of a few ml/min and tube

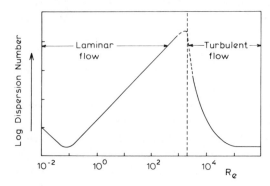

Fig. 1. The dispersion number as a function of the Reynolds number (R_e) for aqueous solutions in the laminar, transient, and turbulent regions of flow.

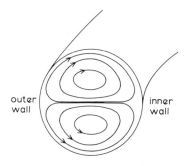

Fig. 2. Secondary flow pattern in the cross section of a coiled tube.

diameters of 0.2–2 mm). A means of improving this situation is to work with coiled reaction tubes. Owing to centrifugal forces acting on the flow pattern, secondary flows perpendicular to the main flow direction are produced (see Figure 2). As a result, better radial mixing and consequently a reduction of band broadening (flattening of the parabolic profile) occurs.

The variances in such a coiled tube can be expressed by modifying equation (3) to

$$\frac{\Delta\sigma_{tv}^2}{t_v} = \frac{k\,d_t^2}{96D_m} \tag{4}$$

The new factor k depends on the particular flow pattern and is a measure of the deviation from the parabolic profile.[15]

Studies on the flow in helically coiled tubes have been made by Dean,[16] who proposed the use of a dimensionless number (Dean number D_n) derived from the Reynolds number $R_e = 4\delta\Phi/\pi\,d_t\eta$. Φ is the volumetric flow rate, d_t the inner diameter of the tube, η the viscosity, and δ the solvent density. The expression for the Dean number is $D_n = R_e(d_t/d_c)^{1/2}$, where d_c gives the diameter of the coil.

k can be estimated from graphical plots[17] made against the dimensionless parameters $D_n \cdot Sc^{1/2}$. Sc is the so-called Schmidt number $Sc = \eta\delta/D_m$.

The pressure drops ΔP of such a tubular reaction system can be computed from the Poiseuille equation valid for laminar flow conditions, as follows:

$$\frac{\Delta P}{t_v} = \frac{512\eta\Phi^2}{\pi^2\,d_t^6} \tag{5}$$

A plot based on equation (4) that permits computation of the expected band broadening in a coiled tubular reactor is shown in Figure 3. A limiting

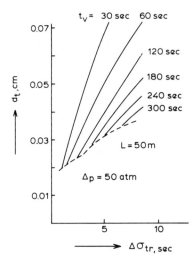

Fig. 3. Band broadening in coiled tubular reactors as a function of the tube diameter, d_t, for various reaction times, t_v.[18,19]

pressure drop of 50 atm and a maximum capillary length of 50 m were assumed. From Figure 3 it can be seen, for example, that for a residence time of 120 sec and with an inner diameter d_t of the reactor tube of 0.025 cm, a band broadening of $\simeq 3$ sec can be expected.

A more detailed discussion on band broadening phenomena has recently been given by Deelder *et al.*,[18] van den Berg,[19] and Tijssen.[20,21] Ruzicka and co-workers[22] have also discussed band broadening in tubular reactors in connection with continuous flow analysis in nonsegmented systems (flow injection analysis). Obviously zone spreading is not as critical under such autoanalyzer conditions since no chromatographic resolution is lost. The only disadvantage of excessive dispersion of injected sample plugs in such systems is dilution and lower sample throughput.

2.2. Bed Reactors

Tubular reactors are suitable and advantageous for reactions with fast kinetics and residence times below 30 sec. It has been argued by several authors[18,23−26] that bed reactors can be more favorable for reactions of intermediate speed and residence times of 30 sec and up to several minutes.

The theoretical aspects are just as well understood as for tubular systems and are relatively simple. The packed bed reactor can actually be viewed as a regular chromatographic column operated under nonretention (t_0) conditions. The band-broadening phenomena can be attributed to factors such as axial molecular diffusion and convective mixing. Naturally the geometry of the packed bed would be of prime importance, suggesting the use of good packing techniques analogous to HPLC (high-performance liquid chromatography). The following empirical equation proposed by Hilby[27] and later used by Deelder et al.[18] describes the situation that we can expect in such a system and permits us to recognize the parameters which have to be considered for minimizing band broadening:

$$H = \frac{L \, \Delta\sigma_{tr}^2}{t_v^2} = \frac{2\gamma D_m}{u} + \frac{\lambda_I \, d_p}{1 + \lambda_2 (D_m/u \, d_p)^{1/2}} \tag{6}$$

Here γ is the tortuosity factor, λ_1 and λ_2 are bed geometry constants, u is the interstitial fluid velocity, d_p the mean particle size, D_m the diffusion coefficient, and t_v the reaction time.

The pressure drop which would be lower for the bed reactor than for a tubular reactor for equivalent residence times can be calculated by equation (7):

$$\Delta p = \frac{u\eta L}{K_0 \, d_p^2} \tag{7}$$

where K_0 is the permeability constant, L the bed length, and η the viscosity coefficient.

From these relationships it can be concluded that within practical Δp limits, the use of small d_p will be advantageous to the construction of bed reactors. The band broadening and pressure drop that can be expected as a function of reaction time and d_p for bed reactors of 30-cm length and 0.46-cm i.d. at a flow rate of 1 ml/min ($\eta = 0.5$ cP, $D_m = 2 \times 10^{-5}$) are shown in Figure 4.[12] It can be seen that for a 1-min reaction using 10-μm particles a $\Delta\sigma_{tr}$ of 0.5 sec and a Δp of 50 atm could be expected.

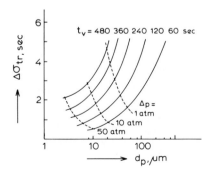

Fig. 4. Band broadening in packed bed reactors as a function of the particle diameter, d_p, for various reaction times, t_v, and for a reactor length of 30 cm.[18,19]

2.3. Segmented Stream Reactors

For reactions with relatively slow kinetics which require actual residence times of, say, above 4 min, band broadening in nonsegmented stream reactors can become critical.

The resulting loss of chromatographic resolution can be detrimental to the total technique. The concept of segmentation of flowing liquid stream with air or other gas bubbles such as used in autoanalyzers and introduced originally by Skeggs[28] can effectively suppress dispersion of sample plugs during extended residence times.

The goal of segmentation is actually to reduce axial diffusion of the sample zones.

Some attempts have been made to describe an air-segmented reaction system.[29,30] The variance of the residence time distribution can be described as

$$\Delta\sigma_{\mathrm{td}}^2 = \frac{2\pi^2 \, dLlr^3}{Q^2} \tag{8}$$

where d represents the uniform thickness of the liquid layer on the tube wall, L is the length of the tube, l the length of a liquid segment, and Q the flow rate.

This equation tells us that small $\Delta\sigma_{\mathrm{td}}^2$ values can be obtained by using small liquid segments obtained with a high frequency of air bubbles. The use of short tubes, a high flow rate, and a small inner tube diameter are also recommended. In recent publications Snyder[31,32] proposed a more theoretical description of the phenomenon of dispersion in segmented flows. He gives an expression for the variance in time units [see equation (9)] for the dispersion in segmented flow through open tubes:

$$\Delta\sigma_{\mathrm{tr}}^2 = \left| \frac{0.2 \, d_t^{2/3}(\Phi + nV_G)^{5/3}\eta^{2/3}}{\sigma^{2/3}D_R\Phi} + \frac{1}{n} \right|$$

$$= \left| \frac{2.35(\Phi + nV_G)^{5/3}\eta^{2/3}t_v}{\sigma^{2/3}\Phi \, d_t^{4/3}} \right| \tag{9}$$

Here d_t is the internal diameter of the reactor tube, Φ the liquid flow through the reactor, n the segmentation frequency, V_G the volume of a gas bubble, η the viscosity of the reaction mixture, and σ its surface tension. D_R is a dispersion coefficient for the tracer substance in water. The optimization of these various parameters is described.[31,32] The most important terms for the experimentalist are the internal tube diameter d_t, the flow rate Φ, and the bubble frequency n.

As a practical lower limit for d_t, 0.05 cm is assumed. For such a value the optimal flow rate would be 0.04 ml/sec with $\sigma_t = 0.7$ sec and $n = 35/\text{sec}$.

Other parameters to be considered are the surface tension, viscosity, diffusion coefficient, and reaction times. In situations with high surface tension, surfactants are frequently added to the system. Of considerable importance is the material, the quality and geometry of connections, coils, mixing tees, debubblers, and phase separators which can contribute to band broadening to a major degree. One of the major aspects which becomes apparent is the phenomenon of leakage from one segment to another through wetting of the inner tube walls. As a result, small transfers of substance from one segment to another occur. Nevertheless suppression of band broadening is quite efficient with this approach.

In some publications it has been shown that the critical bubble-removal step prior to detection can be circumvented by adopting "electronic debubbling" techniques.[34] However, it appears that physical debubbling before detection can be marginally advantageous with flow cells of >20-μl volumes.[32]

Recently solvent segmentation has been proposed instead of air segmentation for the suppression of band broadening.[35-37] This principle is based on, for example, segmentation of the aqueous column effluent stream with organic nonmiscible solvent plugs such as chloroform[35] or the segmentation of an organic chromatographic eluent such as used in normal phase chromatography with aqueous solvent plugs.[36] The immiscible solvent cannot only act as spacer to suppress zone dispersion but also as reagent carrier and as extraction medium in so-called extraction detectors.

Theoretical aspects for solvent-segmented streams can be expected to be even more complex than for gas segmentation but investigations are underway to find out to what degree the relationships proposed by Snyder[31,32] can be adapted to this end. One aspect which again plays a major role in band broadening is the wetting factor, as has been shown by Lawrence et al.[36] for a model system of postcolumn fluorescent ion pair formation and extraction. The fluorescent partner was dimethoxyanthracene sulfonate (DAS), and tertiary amines of pharmaceutical and agricultural interest[35] were analyzed. Figure 5 illustrates the influence of glass as compared to Teflon coils (coil length 1 m) on band broadening at varying DAS flow rates and constant HPLC flow rate. With glass coils, band broadening is seen to be almost independent of the flow rate of the aqueous phase, whereas it strongly increases with decreasing flow rate in the case of Teflon coils. The major factor responsible for this is undoubtedly the wetting phenome-

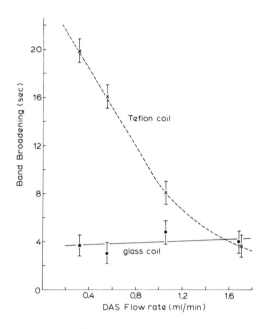

Fig. 5. Influence of DAS flow rate on band broadening for HA in teflon (- - - -) and glass (———) coils. Band broadening measured as the increase in peak width at 1/2 height.[36]

non. Since the ion pair to be detected is carried in the organic phase, band broadening can be expected to be more severe in Teflon tubes with strong wetting of the organic segments on the wall; this is born out in Figure 5 at a low DAS flow rate.

The decrease in band broadening with a higher flow rate of the aqueous DAS solution (Figure 5) can be partly attributed to the increased rate of segmentation, resulting in a more efficient suppression of band broadening.[31,32] It might also be attributed to some degree to the higher total flow rate which results in changed flow characteristics, but here a better understanding of the Teflon system would be needed for a definite answer.

In principle one can stipulate the following rule of thumb: when the compound (ion pair, derivative, etc.) to be detected is contained primarily in the organic solvent plug, one would chose a coil material which is non-wetting for the organic phase (i.e., glass or steel tubing) and viceversa.

The efficiency of solvent segmentation for suppression of band broadening, assuming optimal reactor geometry and material, has been demonstrated for actual chemical derivatization reactions which require residence times of up to 24 min.[37] Even under these conditions band broadening in the reaction coil was negligibly small in comparison to band broadening obtained from other sources such as mixing tees, connections, and phase separators. Since the sequence of solvent segments can be controlled just as accurately as in air segmentation, electronic desegmentation (analogous to electronic debubbling[34]) could be a feasible approach to further reduce band broadening in these detectors.

2.4. Mixing Units

According to the following equation,

$$\sigma_t^2(\text{reaction detector}) = \Delta\sigma_t^2(\text{connections}) + \Delta\sigma_t^2(\text{mixing units})$$
$$+ \Delta\sigma_t^2(\text{reactor unit}) + \Delta\sigma_t^2(\text{detector}) \qquad (10)$$

the total band broadening in the reaction detector is composed of the individual contributions from connections, mixing units, reactor unit, and detector cell plus, in segmented flow reactors, debubbler and/or phase separator.

From this it is understandable that particularly for fast reactions (short residence time) the contribution from reagent mixing units can become quite significant.

It is also important that the streams be homogeneously mixed before the zones enter the reaction unit, particularly when a bed reactor is used. Proper construction of the mixing units (mixing tees) is therefore a major condition for good functioning of a reactor detector. The different mixing geometries that have so far appeared in the literature are shown in Figure 6. To establish generally valid rules for the optimal construction is difficult, since mixing ratios and solution composition (viscosity, density, etc.) play an important role. Construction of such mixing units has been described by several authors[19,25,38–41] and so-called zero dead volume tees are also commercially available.

For a model system consisting of nonapeptides derivatized with Fluram® with a residence time of 10 sec, construction of a proper mixing unit was problematic since the reagent was added at only 1/10 of the eluent flow and in acetonitrile. The eluent composition was an aqueous buffer solution with quite different density and viscosity. Working with a mixing tee according to version I in Figure 6 resulted in layering effects and so-called "Schlieren formation" and consequently significant band broadening.[39] Version II was not much better owing to the large differences in flow rates. By adopting version III a significant reduction in band broadening was observed,[39] and version IV, suggested by Zech and Voelter,[38] was finally adopted as the optimal design for this derivatization.[39,40]

The important feature of versions III and IV is the increased turbulence

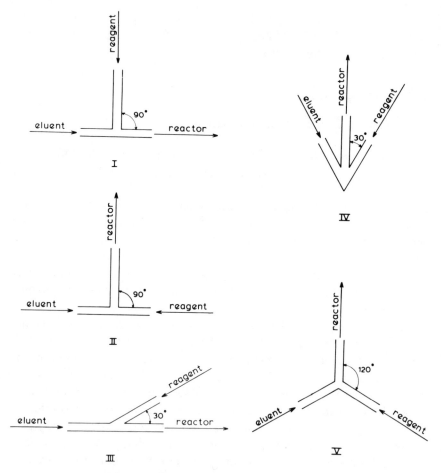

Fig. 6. Different geometries for mixing units.

which is produced in the mixing area, causing a more efficient, mixing of the streams with a decrease in band broadening.

When mixing streams with similar properties, i.e., two predominantly aqueous streams, and at flow rates which are not widely dissimilar, these aspects seem to be of less importance, as has also been shown by Oelrich and Theuerkauf[41] for the derivatization of gentamicin components with *o*-phthalaldehyde (OPA) to form fluorescent products.

In Figure 7 one sees the corresponding chromatograms obtained with the two different geometries. They are practically identical although with the 90° geometry (version I) a slight asymmetry (tailing) of peaks 2 and 3 can be observed on close examination.

Fig. 7. 90° T-piece and 30°/30° T-piece. Packing: LiChrosorb RP-8 (5 μm) 250 × 4 mm; eluent: 0.015 M 1-butanesulfonic acid–Na, 2% sodium sulfate, 0.1% (v/v) acetic acid in water; flow rate: 1.6 ml/min; pressure: 220 bar; detection (fluorescence): excitation: filter UG5; emission: filter KV 408 (Knauer uv-fluorescence-detector model 7200); reagent: titrate 24.7 g (0.4 mol) of boric acid dissolved in 900 ml of water with concentrated KOH solution to pH 10.4 and dilute with water to 1000 ml. Dissolve 60 mg of o-phthalaldehyde in 1 ml of methanol, add 0.2 ml of 2-mercaptoethanol and mix until decolorization occurs. Add then 100 ml of potassium borate buffer. Flow rate (reagent): 0.4 ml/min; reaction capillary: PTFE tube (2.5-m × 0.3-mm i.d.); temperature: RT.[41]

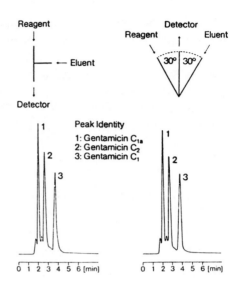

Peak Identity
1: Gentamicin C_{1a}
2: Gentamicin C_2
3: Gentamicin C_1

The production of reproducible mixing units still presents some problems and puts high demands on workshop facilities.

An interesting approach to efficient mixing of fluid streams in reaction detector technology has been proposed by Jonker et al.[25,26,42] This group used short, narrow-diameter tubes with relatively large-diameter glass beads for efficient radial mixing of reagents with column effluents. The theoretical basis has also been discussed by the same author(s).[42] Radial mixing of a component in a fluid stream is analogous to longitudinal mixing, as long as the radial distribution curve of the components does not reach the walls of the flow bed:

$$\sigma_{ri}^2 = H_{ri}z_i \tag{11}$$

where σ_{ri} is the standard deviation of the radial concentration profile of component i, generated by a point source in the center of the fluid stream, H_{ri} is the theoretical plate height for radial dispersion of component i, and z_i is the distance downstream from the point source. If the radial distribution curve reaches the boundary of the fluid, equation (11) is no longer valid. This is obvious if one realizes that the standard deviation cannot exceed that of a rectangular function, describing the uniform distribution of the test compound over the cross section of the fluid stream. However, it is reasonable to expect, in chemical flow reactors, that this latter situation is practically reached at a given distance $z_i = L_c$:

$$L_c = \alpha(r_m^2/H_{ri}) \tag{12}$$

where L_c is the critical length below which the mixing is not complete and which in a flow reactor defines a region in which the conversion is at least partly transport limited, α is the numerical factor depending on the criteria for total mixing and the initial distribution, and r_m is the radius of the flow stream. The theoretical plate height H_{ri} for radial dispersion depends strongly on the geometry of the flow system. For laminar flow in an open tube, in the absence of secondary flow caused by coiling, the expression for H_{ri} is simply

$$H_{ri} = 2D_{if}/v \qquad (13)$$

where D_{if} is the diffusion coefficient of component i in the stream, and v is the flow velocity. Combination with equation (12) results in the expression

$$L_c = \frac{\alpha}{2\pi} \frac{w}{D_{if}} \qquad (14)$$

in which $w = \pi r_m^2$ and v is the flow rate. If we take as an example $w = 30$ µl/sec, with $D_{if} = 10^{-5}$ cm²/sec and $\alpha = 1$, we obtain L_c 477 cm.

Clearly, a straight short tube does not permit us to achieve complete radial mixing. A narrow coiled tube, as we have seen earlier,[18-20] will give better radial mixing, because of secondary effects, but still it is not too efficient, and hence the requirement stated earlier, that mixing should be complete before the fluid stream enters a reactor, prevails.

For packed bed reactors an expression for the theoretical plate height can be given as follows:

$$H_{ri} = \lambda_r d_p + \frac{2D_{if}}{\tau_r v} \qquad (15)$$

in which λ_r is the numerical factor, characterizing convective radial dispersion, d_p the particle diameter, and τ_r the radial tortuosity factor. Values between 0.15 and 0.06 have been given[43,44] for λ_r. For most practical cases the first term in equation (15) dominates because of the flow velocity used. Taking this into account and combining equations (15) and (12) results in

$$L_c = \frac{\alpha r_m^2}{\lambda_r d_p} \qquad (16)$$

From equation (16) it can be seen that a packed mixing column should have a low ratio of column-to-particle diameter in order to give good radial mixing within a short length. The typical designs hence recommended[42] are made of 1.1-mm-internal-diameter tubing filled with 150-μm-diameter glass beads. Column length can then typically be a few centimeters ($\geq$20 cm) for complete mixing. The advantage of such mixing devices would be good reproducibility while still maintaining low pressure drop and keeping longitudinal dispersion (band broadening) at a minimum.

3. TECHNICAL ASPECTS AND APPLICATIONS

3.1. Tubular Reactors (Nonsegmented Streams)

3.1.1. Fluorescence Detection Techniques

From what has been said in the Introduction, it seems clear that an ideal reagent for postcolumn reactions would possess no or only little native fluorescence and would react rapidly. Fluorigenic reagents, which are coming close to fullfilling these prerequisites, are fluorescamine (Fluram) developed by Udenfriend et al.[45] (Hoffmann–La Roche Ltd., Basle, Switzerland) and o-phthalaldehyde (OPA) originally proposed by Roth.[46] Both reagents yield highly fluorescent derivatives and are selective for primary amines. The reactions go to completion within from a few seconds up to several minutes, depending on the reaction environment and on the specific compounds (steric factors, reactivity of the $-NH_2$ group) investigated. With Fluram, some substituted anilines (aminophenoles, chloroanilines), for example, have been found to react much more slowly,[47] the reactions reaching a plateau in some cases only after several minutes residence time in the reactor.

Fluram. The reaction of Fluram is given below:

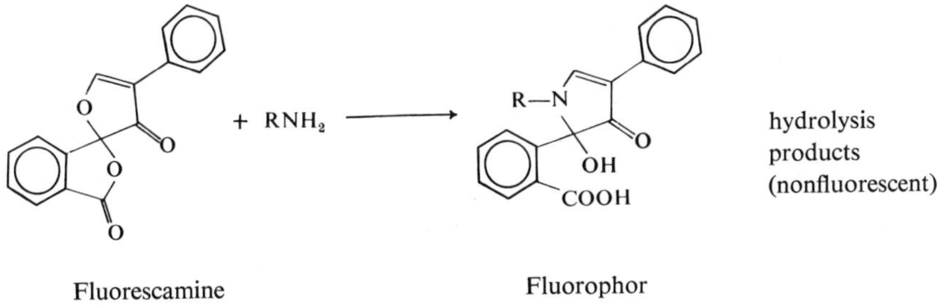

| Fluorescamine | Fluorophor | |

For reactions with slow kinetics the competition reaction becomes dominant and the derivatization problematic.[47] It is also possible to use Fluram for prechromatographic derivatization,[49] but from what has been said it is obvious that the main advantage of this system lies in postcolumn reaction detection.

A typical kinetics curve for the Fluram reaction is shown in Figure 8. The nonapeptide oxytocine served as model compound but similar curve shapes were obtained for other systems,[40] pointing to a pseudo-first-order reaction at least for shorter reaction times and with a large reagent excess being used. Here one can profit from an advantage of reaction detectors mentioned before, that a complete reaction is not necessary for analytical purposes as long as the reaction conditions are reproducible. For the reaction in Figure 8 this means that even though it takes up to 50 sec to come to a plateau for the oxytocine, the reaction can be stopped at, for example, 10 sec while still giving 85% of the total fluorescence yield. The advantage is obvious: shorter residence time means less reaction coil, hence lower back-pressure and less band broadening, and secondary reactions become less dominant. The same concept is of general importance in reaction detector design since with the (usually) large reagent excess used, pseudo-first-order reactions are encountered in most cases. Taking, for example, a 10-min reaction time, needed to reach completion of the derivatization, one can by

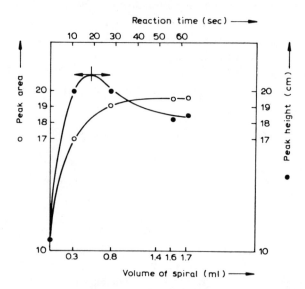

Fig. 8. Influence of length of spiral (reaction volume and time) on the fluorescence signal. Conditions: column, RP-8 (10 μm), 10-cm × 0.4-cm i.d.; eluent, acetonitrile–water (20:80), pH 7; reagent, Fluram (30 mg per 100 ml of acetonitrile). Eluent flow rate, 1.6 ml/min; reagent flow rate, 0.15 ml/min; injection of 100 μl of oxytocine (5 I.U./ml) via loop; fluorescence detection with Aminco fluoromonitor.[40]

the same token select a reaction (residence) time of 2–4 min, making such a reaction accessible to relatively simple nonsegmented reactor designs.

Another aspect of importance is the composition of the reaction and detection medium and the compromise that has to be made between reaction medium and chromatographic eluent. This demands a concise study of solvent and solution parameters for any new system to be studied. The choice of a proper organic solvent as a polarity modifier in the eluent and as a solvent for the Fluram has been shown to be quite critical. Acetonitrile exhibited optimal properties both from the separation and detection (fluorescence background and quenching) point of view for the nonapeptide systems studied.[39,40] DMSO has also been found to be a good fluorescence enhancer for Fluram derivatives and for o-phthalaldehyde derivatives by Fröhlich et al.[50,51]

The importance of proper designs of mixing units has been mentioned earlier.[40] Amino acids have been among the first systems for which Fluram has been used in reaction detectors.[38,52–54]

Felix and Terkelsen[52] have proposed a technique for simultaneous determination of primary and secondary amino acids. Secondary amino acids such as proline or N-methylamino acids were transformed into Fluram-reactive primary amines via oxidative decarboxylation with N-chlorosuccinimide. The method readily permits distinction between secondary and primary amino acids and is sensitive down to 100 pmole detection levels, for L-N-methylalanine. Four minipumps (Milton Roy Co.) were used to introduce appropriate reagents. All valves, mixing tees, connectors, columns, gauges, and tubing were manufactured by Chromatronix (Rainin Instrument Co.)

Pump 1 introduced the eluting buffers by means of a rotary valve which permitted the selection of the appropriate buffer (buffer 1: sodium citrate, pH 3.28, 0.2 M Na$^+$; buffer 2: sodium citrate, pH 4.24, 0.2 M Na$^+$; flow rate, 9.2 ml/hr). Pump 2 was used for introduction of 10^{-3} M N-chlorosuccinimide in 0.05 M HCl. This pump was kept on throughout the analysis, and the pH of the eluent was set at 2–2.4 by adjusting the flow rate from pump 2 (6.6 ml/hr). Pump 3 was used for the introduction of the borate buffer (pH 9.7, 0.10 M), and the pH of the eluent was set at 8.5–9.0 by adjusting the flow rate from the pump (25.2 ml/hr). Pump 4 was used for the introduction of fluorescamine (300 mg/l in acetone; flow rate, 19.6 ml/hr). The fluorescent mixtures were detected in an Aminco Fluoro-Microphotometer equipped with an 85-W mercury vapor lamp assembly, 2-mm-i.d. high-pressure flow cell, Corning No. 7-51 primary filter, and Wratten No. 4 secondary filter (American Instrument Co.).

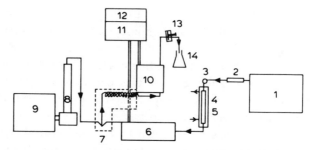

Fig. 9. HPLC unit for postcolumn derivatization (schematic). Components: (1) pump (Haskel Engineering Co.); (2) precolumn of Si-60, 30 μm, 5 cm long, to remove impurities in the solvent stream; (3) loop injection system, 20–150 μl (Rheodyne Co.); (4) analytical column; (5) mantle for thermostatting with water; (6) uv detector PE LC 55 (Perking-Elmer); (7) the mixing and reaction device (encircling interrupted line to emphasize its importance); (8) pump (Isco Ltd.); (9) electronic control for Isco pump; (10) fluorescence detector (American Instrument Co.); (11) two-pen recorder (W & W Electronics, Inc.); (12) integration system (Hewlett-Packard HP 3352); (13) metal clamp for backpressure; (14) reagent reservoir (waste).[39]

Voelter and Zech[38,54] analyzed amino acids and peptide hormone hydrolysates in the concentration ranges 100 pmol–5 nmol with a fluorescamine solution of 15–20 mg/100 ml also in acetone. Separation of the compounds was again carried out on ion exchange or on reversed phase materials.

A 4-m, 0.3-mm-i.d. steel capillary was used for the reaction, which at the flow rates used (0.75–1.25 ml/min) corresponds to a few seconds reaction time.

A typical and more universally usable system for nonsegmented stream postcolumn fluorescence reactions is shown in Figure 9. This has been used for the detection of nonapeptides and by-products[39,40] in pharmaceutical formulations. When relatively high acetonitrile concentrations had to be used a significantly enhanced fluorescence background resulted. Because of this background pump pulsations were amplified and a noisy base line was obtained. Pulseless pumps have therefore been used in this apparatus. However, well-damped piston pumps of the newer line can also be used.[47]

With this system (Figure 9) and a 10-sec residence time in the reactor a band broadening of 3.5% relative to the uv signal was obtained for the oxytocine peak. Detection limits (1–10 ng) were about two orders of magnitude better than for uv detection at 210 nm. The reproducibility of signals for repetitive injection was better than ±2% relative s.d.

A typical chromatogram as obtained with this apparatus for a duplicate injection of an ampoule solution is shown in Figure 10. A further simpli-

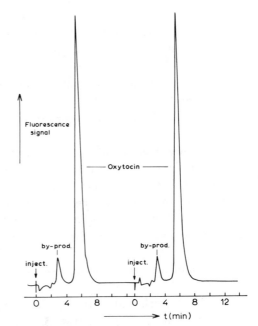

Fig. 10. Chromatogram of 100 μl of Syntocinon injection solution (5 I.U./ml). Stationary phase, RP-8 (Merck), 10 μm, column i.d. 0.3 mm, length 25 cm; mobile phase, phosphate buffer (pH 7)–acetonitrile (80:20, v/v), flow rate 1.47 ml/min, Δ_p = 132 kg/cm²; Fluram, 30 mg per 100 ml of acetonitrile, flow rate 0.16 ml/min.[39]

fication was proposed by Oelrich *et al.*[41,55] for the derivatization of gentamicin components with OPA. This group used a cassette principle as proposed by Kaiser[56] and as shown in Figure 11.

Figure 11 shows a possible cassette arrangement for postcolumn derivatization. The sample is chromatographed in the usual way; the reagent

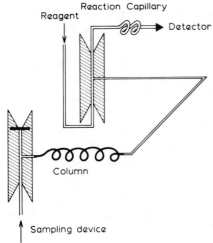

Fig. 11. Cassette arrangement for postcolumn derivatization.[41]

solution is then added via the column outlet T-piece, e.g., by means of a second pump. Eluate and reagent are mixed in the T-piece and passed through a reaction capillary into the detector.

A chromatogram run in such a way has been shown in Figure 7; it shows gentamicin after ion pair chromatography, derivatization with *o*-phthalaldehyde, and fluorescence detection. Gentamicin is an aminoglycoside antibiotic which is composed of three major and several minor components.

OPA (*o-phthalaldehyde*). OPA reacts with amino acids in alkaline medium in the presence of a reducing agent such as 2-mercaptoethanol to produce highly fluorescent derivatives which may be excited at 340 nm and which emit at 445 nm. The sensitivity of the technique extends to the low-nmol region. The reaction is carried out at room temperature and is applicable to the determination of amino acids in effluents from column chromatography.[57,61]

An automated autoanalyzer based on this principle has been described by Roth.[58] In his method the amino acids are separated on a column $(0.6 \times 25$ cm) which is filled with Aminex 6 ion exchange resin. For elution, three citrate buffers of pH 3.20, 4.25, and 6.40 are pumped successively at 25 ml/hr, the first change being made after elution of glutamic acid and the second change after elution of phenylalanine. The reagent solution, which consists of *o*-phthalaldehyde (0.8 g/l), ethanol (10 ml/l), and 2-mercaptoethanol (200 μl/l) in 0.1 M borate buffer (pH 10), is pumped at a rate of 30 ml/hr. The effluent and reagent are mixed in a reaction coil for 5 min at room temperature. The mixture then enters a fluorimeter which is equipped with a microflow cell for fluorescence measurements.

An analysis of amino acids in human plasma carried out by this technique is shown in Figure 12 and the relative fluorescences for the different acids are indicated in Table 1.

The general disadvantage of OPA and Fluram over the conventional ninhydrin reactors is the absence of a reaction with acids such as proline and hydroxyproline and a poor reaction with cysteine. Lund *et al.*[60] circumvented this problem by modifying a conventional Beckman model 120C amino acid analyzer to include the OPA reaction. They used the OPA section primarily to detect phenylthiohydantoin (PTH) amino acids after back hydrolysis for sequencing of nanomole amounts of protein. The modified apparatus is shown in Figure 13. A different approach was chosen by Roth[58,62] who chose the reagent 7-chloro-4-nitro-benzofurazan first described by Gosh and Whitehouse[63] for a fluorigenic reaction with proline and hydroxyproline and for prolyl peptides. This selective and sensitive

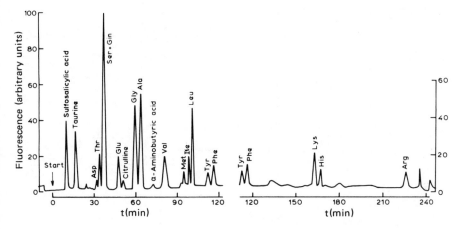

Fig. 12. Automated amino acid analysis with sensitive fluorescence detection. Analysis in normal blood plasma.[58]

TABLE 1. Relative Fluorescence from Amino Acids as Eluted with Program 2[a,b]

Substance	Relative fluorescence
Cysteic acid	81
Aspartic acid	97
Threonine	94
Serine	113
Glutamic acid	100
Glycine	101
Alanine	101
Valine	85
Methionine	85
Isoleucine	112
Leucine	111
β-2-Thienylalanine	103
Tyrosine	80
Phenylalanine	108
Lysine	105
Histidine	109
Arginine	89

[a] Reference 58.
[b] The values are the relative integrated peak areas with glutamic acid taken as 100. Sample: reference mixture with cysteic acid and β-2-thienylalanine added; 2 nmol of each compound.

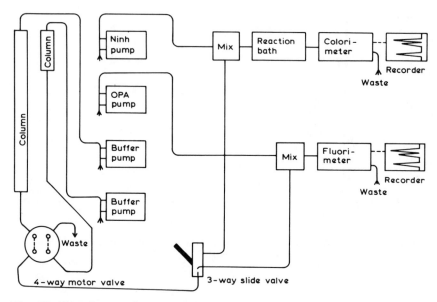

Fig. 13. Flow diagram of a conventional amino acid analyzer modified for fluorescence detection.[60] OPA = o-phthalaldehyde reagent; Ninh = ninhydrin reagent.

assay was used for blood plasma samples after ion exchange separation of the amino acids. The derivatives yield an intense fluorescence at relatively long wavelengths ($\lambda_{ex} = 465$ nm, $\lambda_{em} = 535$ nm), which further enhances its selectivity potential. The method was reported to be considerably more sensitive than the ninhydrin reaction.

The reagent OPA was also used for a number of other compounds. Creaser and Hughes[64] used the same reagent for detection of peptides and amino acids. Reaction times of 1 min were used for the peptides, which were quantitated at >5-nmol concentration levels. Polyamines can also lend themselves to analysis via OPA.[65] Spermine, spermidine, and putrescine, for example, were detected at the 3–15-p mol level after reaction times of only 15 sec. Work with gentamicin was reported earlier.[41,55]

Other reactions. Ueda et al.[66] reported the use of a trihydroxyindole reaction for the determination of catecholamines. The method is based on the oxidation of the catecholamines by ferricyanide followed by alkaline rearrangement of the adrenochrome to the fluorescent lutine[66] and has been discussed in an earlier reaction detector version by Mori.[67] Since the total reaction proceeds relatively fast (1 min), band broadening as a result of dispersion in the reaction coil for the use of an unsegmented flow reactor[66] seemed to be on the order of 15% compared to uv detection.

The selective analysis of phenothiazines by oxidation to fluorescent products has been described for the HPLC determination of these compounds.[68] The method involves the separation of the drugs prior to oxidation, followed by fluorometric determination of the products, and is one of the earlier reaction detectors for actual high-performance liquid chromatography (HPLC). In this technique a 25-μl volume sample (corresponding to 50 ng of thioridazine or of its metabolites) is injected into an HPLC system consisting of a silica (diameter 9 μm) stationary phase and 2,2,4-trimethyl-pentane–2-aminopropane–acetonitrile–ethanol (95.8:0.96:2.6:0.48) as the mobile phase at a flow rate of 1.14 ml/min. The column effluent is mixed with the oxidation reagents at a ratio of 1:1 by adding a potassium permanganate–acetic acid oxidizing solution (400 ng/liter of acetic acid). This solution and the column effluent are mixed in a reaction coil (volume 0.17 ml) in order to produce the fluorescent oxidation product. The time required for oxidation is 5 sec. The effluent stream is then mixed with hydrogen peroxide–ethanol in order to destroy the excess of permanganate which causes fluorescence quenching. This reaction is very fast, only 1–2 sec being required for completion. The effluent is then passed through the flow fluorimeter for detection at 365 nm (excitation) and 440 nm (emission).

The selectivity and sensitivity obtained makes this technique suitable for plasma level controls.

3.1.2. uv-Visible Detection

A colorimetric reaction for "true" serum creatinine detection was reported by Brown et al.[70] Following ion exchange separation the column effluent was mixed with saturated picric acid and NaOH and after a short reaction time at 37°C (total analysis 6 min/sample) the detection was carried out colorimetrically at 510 nm. A detection limit of 3 mg/l was reported. Sugars are also frequently detected by postcolumn reaction although most of the reported procedures involve quite long and complex reactions or use corrosive reagents and employ air-segmentation principles.

A noncorrosive procedure has been proposed by Mopper and Degens[71] which is based on the postcolumn reduction of 3,3'-(3,3'-dimethyloxy 1',1-biphenyl-4,4'diyl)bis(2.5-diphenyl)-2H-tetrazolium dichloride (Sigma Chemicals Inc.) by compounds such as reducing sugars. The reaction is also known under the name tetrazolium blue technique and has been further perfected and described.[72,73] The reactor was connected to the outlet of the chromatographic column, to an auxiliary pump, and to a detector through an Omnifit three-port valve, Unimetrics Co. model 10009. The first 3 m of the

4-m reactor were immersed in a water bath kept at 85°C; the last meter
was at room temperature. The reactor was collecting the effluent from the
analytical column and the tetrazolium blue solution was added by an auxiliary pump as a 2% solution prepared in 50% ethanol–water. The solution
was also 0.18 M in sodium hydroxide. The chromatograms were measured
at 530 nm, which is the absorption maximum for tetrazolium blue monoformazan.

   A chromatogram of polysaccharides obtained from a wood extract
with final detection by this technique is shown in Figure 14. Other reactions
have been described for carbohydrates using segmented systems.[73–76]

   Postcolumn reaction detectors can also be operated on the bases of
complexation, ion pair formation, and indicator techniques (see also the
later section on extraction detectors). In other words it does not necessarily
have to be a true chemical reaction. The formation of iodine–amine charge-
transfer complexes has been used by Clark et al.[77] for the postcolumn
detection of N,N-dimethylbenzylamine. The principle should be generally
applicable to amines. Since the complexation reaction takes place in a
matter of seconds this type of detection mode can be handled easily with
tubular reactors. The complex was measured at 254 nm and for the above
amine a 20-fold improvement in detection limit was observed. A disadvantage of this approach is the serious limitations in the choice of mobile phase
which strongly affects the complexation constant.

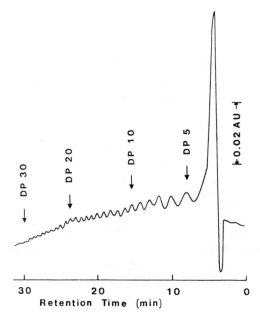

Fig. 14. Chromatogram of a
wood extract fractionated by
GPC using reversed phase chromatography. Column: 25-cm ×
× 0.43-cm i.d. filled with Chromosorb LC 9. Elution: linear
gradient from 70% ACN to 62.5%
ACN in water in 30 min. Flow
rate: 1.0 ml/min⁻¹. Detection:
visible detection of the tetrazolium blue monoformazan at
530 nm. Flow rate for tetrazolium blue: 1.5 ml/min⁻¹. Temperature of reactor: 85°C.[72]

Dichloromethane on a CN-bonded column was used in their study. The possibility of using indicator techniques for postcolumn reaction detection of acids has been mentioned quite early.[78,79] The principle involved the partition chromatographic separation or organic acids on silica gel and postcolumn addition of an indicator salt such as o-nitrophenol–Na salt.

The presence of acid in the eluent will produce the o-nitrophenol which may be measured by an increase in the absorption of the free phenol or by a decrease in the absorption of the free sodium salt at 432 nm.

Uv or fluorescence detectability can often be drastically enhanced by very simple modifications of original molecules, for example through postcolumn pH adjustment, simple hydrolysis, or redox reactions.

Clark and Chan,[80] for example, have shown that the uv detection properties of barbiturates can be about 20-fold enhanced by postcolumn ionization. This was achieved by simply adding a pH 10 borate buffer to the column effluent.

A recent example for an enzyme detector has been discussed by Schroeder et al.[69] for the lactate dehydrogenase isoenzyme system (LD) using nicotinamide adenine dinucleotide (NAD) as the substrate. The reaction is as follows:

$$\text{lactate} + \text{NAD} \xrightleftharpoons{\text{LD}} \text{pyruvate} + \text{NADH}$$

Either absorbance or fluorescence detectors can be used to detect the products. In this case, a tubular reactor system was used, as can be seen in Figure 15, which shows the flow diagram for a reaction detector using an enzymatic reaction and dual detection.

The high-pressure, gradient-elution pumping system shown in Figure 15 was obtained from Spectra-Physics (Santa Clara, California). The column packing was a 5–10-μm particle size ion exchanger (Corning Medical, Medfield, Massachusetts); DEAE-glycophase/CPG-250/columns were packed with a model 705 stirred-slurry apparatus (Micromeritics, Norcross, Georgia) using isopropanol as the solvent.

The primary eluent was a 20-mmol/liter tris(hydroxymethyl)aminomethane (Tris, "Trizma base," reagent grade; Sigma, St. Louis, Missouri) buffer adjusted to pH 7.80 with acetic acid. The secondary eluent was a 20-mmol/liter Tris buffer, containing also 150 mmol/liter of NaCl, adjusted to pH 7.80 with acetic acid after addition of NaCl.

The "standard" loading valve, situated at the end of the column and used to inject known volumes of preassayed sample for calibration of the reaction detector, was a model 204590 six-port, sample loop valve (Du Pont,

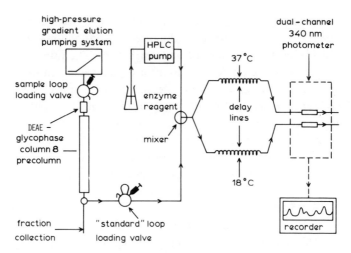

Fig. 15. Flow diagram of the HPLC system with a parallel stream reaction detector. This system affords a directly measurable signal, related to the amount of enzyme eluting from the column.[69]

Wilmington, Delaware). The mixer was a combination of a Swagelok union tee (Crawford Fitting, Cleveland, Ohio; Cat. No. ss-100-3) followed by a 3-cm tube (about 3/32-in. i.d.) filled with 3/32-in. ball bearings. The delay line(s) was a 60-ft length(s) of 0.020-in.-i.d., 1/16-in.-o.d. stainless steel tube.

The enzyme reagent was made by mixing 135 g of 2-amino-2-methyl-1-propanol, 22.5 ml of a 60% sodium lactate solution (Pfanstiel, Waukegan, Illinois), and 4.5 g of β-nicotinamide adenine dinucleotide (β-NAD; Sigma, Cat. No. N-7524), adjusting to pH 9.0 with HCl and diluting to 1.01. This reagent was delivered to the mixer under constant flow by a model 712-31 solvent delivery system (Laboratory Data Control, Riviera Beach, Florida).

A Waters model No. 440 dual-channel absorbance detector, fitted with two 340-nm conversion kits (Waters Assoc., Mildorf, Massachusetts), was used for monitoring the absorbance. A filter fluorometer was placed in series immediately after the second absorbance detector to enable comparisons of on-line fluorescence versus absorbance. This detector was an Aminco model No. J4-7461 fluorometric analyzer system (American Instrument, Silver Spring, Maryland) containing a General Electric No. F4TF/B1 ultraviolet (uv) source, a Corning 7-60 primary filter, a Wratten 2A secondary filter, and an 18-μl sample volume flow cell (Aminco, Cat. No. J4-7476).

The use of metal-complexation phenomena for postcolumn reaction detection of metals has been propagated by Fritz and co-workers[81,82] for classical LC. The same concept has been developed for HPLC by Elchuck and Cassidy[83] for the separation and detection of lanthanides.

The reagents used for complexation were alizarine red S, arsenazo I, and PAR. For all three colors, formation occurred instantaneously after mixing permitting construction of a reactor with very little band broadening ($\sim\sigma_t^2 = 2$ sec²). A typical separation obtained for a group of lanthanides and detected by this technique is shown in Figure 16.

The separation was usually carried out in an ion exchanger medium containing hydroxyisobutyric acid (HIBA) as a complexing agent. The reagent Arsenazo I (3-(2-arsonophenylazo)-4,5-dihydroxy-2,7-naphthalene-disulfonic acid trisodium salt) (Eastman Organic Chemicals, Rochester New York) was prepared in a 1.3×10^{-1} M solution containing 3 M ammonia. Total flow rates where 1–2 ml/min.

3.2. Bed Reactors

The bed reactor principle is the least used in reaction detector design although it has distinct merits for certain types of reactions, particularly of intermediate kinetics. Other advantages are the possibilities of operating them under high pressures and of having the reactor bed directly participate in the reaction.

Applications for the bed reactor principle include a colorimetric detection of hydroxyperoxides[23] which were separated on 5-μm silica gel

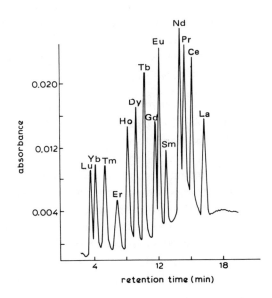

Fig. 16. Separation of the lanthanides on 5 μm Nucleosil SCX. Experimental conditions: 10-cm × 4-mm column; 10 μl of a solution containing ∼10 μg ml⁻¹ of each lanthanide; linear program from 0.018 mol l⁻¹ and pH = 4.6; detection at 600 nm after postcolumn reaction with Arsenazo I.[83]

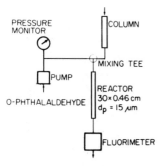

Fig. 17. Schematic diagram for the postcolumn reaction detection of amino acids.[18]

SI-60 with a 50% water-saturated mixture of 2,2,4-trimethylpentane–ethanol (95:5). In the reactor, iodine is formed by the peroxides acting on an acidic reagent solution of sodium iodide. Absorbance was measured at 362 nm after a reaction time of 1.5 min at 70°C. Detection limits in the low-nanogram range were reported. The band broadening (<1 sec) was distinctly below what one would expect for a tubular reactor using the same residence time. The same group has also used a bed reactor for the detection of amino acids with o-phthalaldehyde.[18,19]

Figure 17 gives a scheme of a reaction system for the fluorometric detection of amino acids which were separated by ion pair chromatography prior to the o-phthalaldehyde reaction. The linear eluent velocity in the chromatographic column corresponds to a volume flow rate of 1 cm³ min⁻¹. The flow rate for the reagent was set at the same value. Mixing of the column effluent and reagent was achieved by using a tee in which both streams were fed through 0.15-mm-i.d. channels. At 20°C the reaction was almost complete for $t_v = 60$ sec. The reaction was carried out in a 30 cm × 4.6 mm stainless steel reactor packed with 15-μm glass beads.[23] The fluorescence of the reaction mixture was continuously measured at 445 mm (filter GG 435) with a type FS 970 fluorometer (Schoeffel Instrument Corp., Westwood, New Jersey); the excitation wavelength was 340 nm.

The additional band broadening due to the reactor $\Delta\sigma_{\text{tr}}$ was measured to be about 1 sec. Using a tubular reactor for the same residence time would have resulted in a $\Delta\sigma_{\text{tr}}$ of up to five times as high and a considerably higher pressure drop. On the other hand it can be argued for this particular application that a 10–15-sec residence time would have been sufficient (see discussion about kinetics, Figure 8) since a plateau condition is not necessary, in which case a simple tubular reactor would have been just as feasible.

Bed reactor technology has also been proposed by Jonker *et al.*[25,26]

for adaptation to an improved ninhydrin reaction. Using a pressure cooker effect in the pressure-resistant reactor allowed for an increase in reaction temperature to above 100°C. The resulting reaction time for complete reaction was on the order of 1 min.

One of the prime areas for the use of bed reactors are enzyme reactions similar to the ones already discussed in the previous section.[69] This area had been pioneered by Regnier et al.[84–86] for the detection of enzymes of clinical importance. The underlying principle is again the catalysis of a given reaction by the enzyme, permitting a 10^4–10^5 amplification for the enzyme to be detected. The residence time of the enzyme reaction system in the bed reactor is ~3 min (ideal for a bed reactor design) and the reaction is maintained under zero-order reaction conditions with respect to the substrate concentration during the detection process. The apparatus used can be essentially of the same basic design as shown in Figure 15. Such a detection device for the detection of, e.g., isoenzymes, is unsurpassed with regard to selectivity and sensitivity.

In Figure 18 the chromatogram for LDH isoenzymes in serum can be seen. The detection mode and the underlying reaction were identical to the previously discussed system,[69] except that a bed reactor had been used. The resulting product NADH is detected by fluorescence and by absorbance and the resulting signals are compared in the figure.

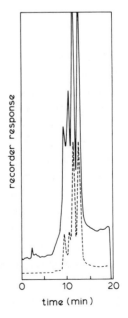

Fig. 18. The comparison of fluorescence with absorption detection in the high-speed separation of 4 LDH isoenzymes. The analytical column was packed with a 5–10-μm, 250-Å-pore diameter DEAE-glycophase/CPG; the postcolumn reaction vessel was a 600 × 4.1-mm precision bore tube and was packed with Whatman glass spheres. The total pressure at the head of the system varied between 1500 and 2000 p.s.i. Buffer pumps: A, 0.025 M Tris (pH 8.3); B, 0.025 M Tris (pH 8.3)–0.2 M NaCl, 10 min linear gradient, flow rate 1.5 ml/min. Substrate pump: 10-mm NAD, 100-mm lactate solution (pH 8.3) at a flow rate of 50 ml/hr. Detection: ———, fluorescence; - - - -, absorbance at 340 nm.[85]

3.3. Segmented Stream Reactors

3.3.1. Air Segmentation

3.3.1.1. uv-Visible Detection. The efficient suppression of band broadening in an air-segmented system such as that used for many years in autoanalyzers makes this technique suitable for reaction detector design with relatively slow reactions (i.e., ≥ 4 min and up to 1/2-hr residence times). A number of reviews and overviews have recently dealt with the possibility of coupling HPLC with autoanalyzers for detection purpose.[6-9,87-89]

Although actual combinations of the two concepts are currently not too numerous in the literature, one can safely state that a majority of the reactions used in autoanalyzers could in principle and with only minor modifications be adapted to HPLC detection. A survey, for example, of the available autoanalyzer application literature[90] will reveal many of these possibilities. Not surprisingly then, the majority of existing applications use the uv–visible spectrophotometric detection principle for the detection of reaction products, following the autoanalyzer tradition.[91]

The earliest applications have been reported for amino acid and sugar analyses, some still in connection with classical or low-pressure LC.

Amino acids. Postchromatographic detection of amino acids with ninhydrin is well known and routinely used for amino acid analysis.[92-94] Usually a modified Technicon AutoAnalyser system is used for the detection apparatus. The reaction is long ($\sim$13 min) and requires a temperature of 95°C. Under such conditions bubble segmentation must be used to keep band broadening to a minimum. The order of sensitivity is $\sim$10 nmol of amino acid. Although reliable, this method is inferior to the fluorescamine or the *o*-phthalaldehyde methods where reaction times are fast and sensitivity better by 2–3 orders of magnitude.

Ertinghausen *et al.*[92] in 1969 dealt quite extensively with aspects of band broadening in such a system. The procedures are usually based on ion exchange separation with automatic elution cycles, followed by development of the ninhydrin color reaction.

A typical system can be described as follows:

A Technicon AutoAnalyser L with the following modifications (or a similar system) may be used. The standard AutoAnalyser columns are replaced with microcolumns (0.45×80 cm) made from precision bore heavy-walled glass tubes or their equivalent. The ion exchange resin consists of Chromobeads B (17-μm spherical beads) which are packed to a height of 70 cm using a final operating pressure of 350–400 p.s.i. throughout. The column temperature is 61°C. The buffer used for development is pumped at a flow rate of 0.78 ml/min and the column effluent flows directly into a nitrogen bubble-segmented ninhydrin stream (1.69 ml/min) through a "T" junction. The mixture then passes through a color development coil which is heated to ∼95°C and then to the detector. The ninhydrin solution consists of 5 g of ninhydrin and 0.5 g of hydrindantin dissolved in 162 ml of methyl cellosolve and 87.5 ml of sodium acetate buffer at pH 5.5. This solution is diluted with 375 ml of methyl cellosolve and 375 ml of water. The buffer gradient for elution of acidic, neutral, and basic amino acids is generated automatically from three reservoirs (pH 2.58, 3.80, and 12.00). These are prepared from a stock solution consisting of 110.32 g of sodium citrate and 167.5 ml of 2 N sodium hydroxide in water which is heated to boiling for 30 min in order to remove trace amounts of ammonia; 75 ml of Brij 35 and 0.8 g of sodium azide are then added and the volume of the resulting solution is diluted to 7 liters. The buffer at pH 12.00 also contains 1.8 M sodium chloride. The developed colors are examined at 570 and at 440 nm.

A representative chromatogram obtained with such an automated analyzer system[92] is shown in Figure 19. The resolution obtained with these low-pressure column LC systems was already quite impressive ten years ago. Hatano et al.[95] have also used a standard amino acid analyzer system for the determination of mono- and diamines.

Carbohydrates. Several methods have appeared in the literature for carbohydrates which involve rigorous treatment of the chromatographic effluent. Because of the time required they are not suited to high-efficiency liquid chromatographic separations. Heating the effluent in an oil bath at 170–190°C produced uv-absorbing substances (260 nm) and enabled detection of many sugars at 1 μmol.[73] Concentrated sulfuric acid, when mixed and heated with the column effluent at 100°C, also was found to produce uv-absorbing chromophors.[74] This same method with the inclusion of phenol as an additional reagent has also been evaluated.[75,76] All of these methods destroy the carbohydrates, thus collection of the separated intact substances for further characterization is impossible. Also, as mentioned above, these methods are not suited to high-efficiency separations. However, as a routine method where sensitivity is not required, they appear satisfactory.

The principle involved for heat treatment[73] is heating of the column effluent at 170–190°C for 10–20 min in a delay coil and measuring the products by photometry. The carbohydrates are separated on a column (10 × 0.6-cm i.d.) consisting of Aminex A-14 at 60°C and are eluted with 0.3–0.8 M

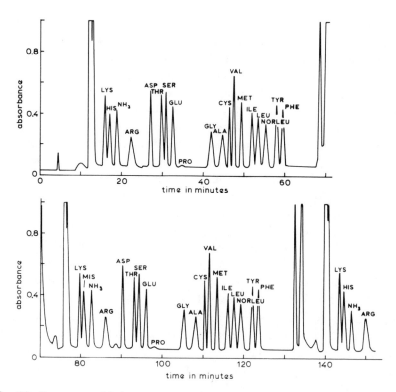

Fig. 19. Two sequential chromatograms of an amino acid standard run automatically. The sample contained 0.03 μmol of each amino acid. Absorbance read at 570 nm.[92]

borate buffer (pH 7.0–9.5) depending on the components to be separated. The column effluent is mixed via a secondary pumping system with more borate buffer of sufficient concentration and pH to create a final mixture consisting of 0.8 M borate (pH 7.0). The mixed effluent is then passed through a 0.5-mm PTFE mixing coil which is immersed in an oil bath maintained at 190°C. The time spent by the effluent in the bath is ~9.5 min. The effluent is measured at 260 nm (detection limit ≤1 μmol).

The reaction of carbohydrates with phenol in the presence of sulfuric acid[75,76] produces a uv-absorbing chromophore which permits detection of less than 1-μg amounts of carbohydrates separated on a strongly basic Aminex A-27 anion exchange column (Bio-Rad Labs.). The mobile phase consists of a concentration gradient of borate buffer (prepared by mixing 0.147 mol of sodium borate with 0.283 mol of boric acid per liter) and diluted to 50%, 30%, and 10% of its concentration for sequential elution of the column. The flow rate of the buffer is 11 ml/hr. The column effluent is mixed with the color-producing reagents in a mixer reactor. The phenol reagent (5% aqueous solution) is mixed with the column effluent at a flow rate of 3 ml/hr. Concentrated sulfuric acid is then introduced at a flow rate of 18 ml/hr. The reaction mixture is passed through 36 cm of capillary tubing which is heated at 100°C. The developed color is measured at 490 nm.

Noncorrosive techniques were also proposed by different groups[96-97] for carbohydrate analysis.

A 2,2'-bicinchoninate reagent proposed earlier by Mopper and Gindler[98] was adapted to an automated borate complex ion exchange chromatographic system by Sinner and Puls.[96] Ethanol–water mixtures were used as mobile phases. The reaction is based on a complex formation [copper(I)-2,2'-bicinchoninate (BCA) complex] in the presence of reducing sugars. The complex is detected at 560 nm. A first-order kinetics was observed for the reaction which comes to an optimum after ∼10 min with reaction temperatures close to the boiling point.

A similar principle was introduced by Simatupang and Dietrichs.[97] They proposed Neocuproin (2,9-dimethyl-1,10-phenanthrolinhydrochloride), a reagent which is being used for a long time for copper determination.[99] The reaction with reducing sugars is proposed as follows:

$$Cu^{2+} + (red)sugar + Neocuproin \longrightarrow Cu^{+}-Neocuproin\ complex + (ox)sugar$$

The Cu^+ complex exhibits an intense color measured at 460 nm. Reaction temperatures ∼97°C and reaction times on the order of 4 min were used (plateau condition, i.e., for xylose 10 min). In principle one could use bed reactors for this method with the possibility of working at still higher temperatures (nonsegmented condition) and shorter reaction times.

The modified Technicon analyzer coupled to the HPLC system is shown in Figure 20. A chromatogram of some sugars and the conditions used are depicted in Figure 21. Some details for the detection are given below: The optimal Na_2CO_3 solution at pH 9 (borate buffer) was 0.5 M; the Neocuproin solution 0.04% Neocuproin HCl and 0.02% Cu $SO_4 \cdot 5H_2O$ (Technicon, Tarrytown, New York).

Pesticides. Analysis of cholinesterase-inhibiting pesticides has been carried out by coupling an autoanalyzer to a liquid chromatograph.[100] This approach to postchromatographic reaction detection is extremely selective and often very sensitive. However, because enzymes are involved, chromatography conditions are somewhat limited. Since reaction time is usually 5–15 min total, air segmentation is required to keep band broadening to a minimum. Figure 22 clearly illustrates the selectivity of a system for the

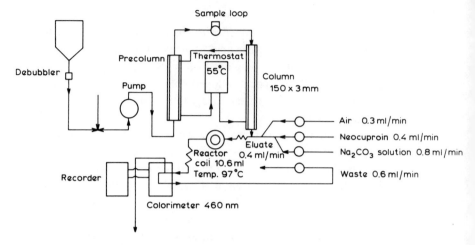

Fig. 20. Schematic of apparatus detection system: Technicon AutoAnalyser NC II (modified).[97]

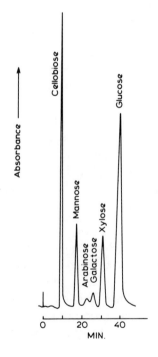

Fig. 21. Chromatogram of a standard mixture of cellobiose, mannose, arabinose, galactose, xylose, and glucose. Column 150 × 3 mm filled with Hamilton Ha-X 4.00, 7–10 μm. Borate buffer 0.4 *M*, pH 9.0. Flow rate 0.4 ml/min. Temperature 55°C. Detection at 460 nm with neocuproin Apparatus Technicon NC II (modified).[97]

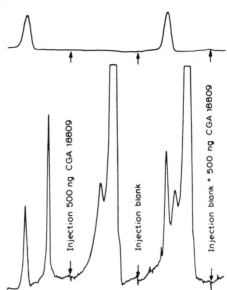

Fig. 22. Determination of CGA
18809 in plum leaf extract.[100]

organophosphate, CGA 18809, compared to uv detection at 297 mn. Only the cholinesterase-inhibiting pesticide shows a response with the autoanalyzer system. For strong inhibitors, such as diazoxon, low picogram quantities can be detected. A fluorescence detection system based on a similar approach will be discussed in the appropriate section. Ott[101] has proposed a colorimeter reaction detector for aromatic nitro-group-containing pesticides with similar potential claimed to the one described by Ramsteiner and Hörman.[100] The technique has been used for residue analysis of paraoxon and parathion in soil.

Cyclohexanone. Postcolumn derivatization has been applied to the determination of trace amounts of cyclohexanone in cyclohexanone oxime, and consists of the initial separation of the ketone from the oxime by liquid chromatography, followed by colorimetric reaction with 2,4-dinitrophenylhydrazine (DNPH).[102] The product is determined at 430 nm under basic conditions. A ternary liquid–liquid system was used for partition chromatography consisting of 2,2,4-trimethylpentane, ethanol, and water (34:5:1). The less polar upper layer served as the stationary phase. A diatomaceous material, Hyflow Super Cel (particle size, 7–11 μm), was used as the solid support. The columns (40 × 4-mm i.d.) were packed by the slurry technique, and the support material was coated *in situ* with the liquid stationary phase. A precolumn was inserted in order to maintain equilibrium between the mobile and stationary phases. Both columns were thermostatted at 25°C.

The flow rate of the mobile phase is 1.0 ml/min. The reagents were mixed with the effluent stream by a proportionating pump. DNPH in ethanol was added first at a flow rate of 0.7 ml/min and an ethanolic KOH solution at 2 ml/min. The first reaction coil was used to form the hydrazone, while the second coil is used to change the environment to basic conditions just before absorption measurements were made at 430 nm. Both reactions take place at room temperature. Air segmentation was used in the postcolumn system in order to keep band spreading to a minimum. The limit of detection of cyclohexanone was given as $\sim$100 ng at a signal-to-noise ratio of 10:1.

Deelder and Hendricks[102] also took a look at band broadening. They attributed a major portion of the broadening to debubbler and mixing tees. With two reaction coils used (2 and 1 m) of i.d. 2.4 mm a delay or residence time in the reactor of 3 min was reported. The loss of resolution due to the reaction detector was given as less than 20% ($k' \simeq 0.5$).

Phosphates. A more complicated system was also described[18] for the determination of condensed phosphates used as stabilizers in liquid fertilizers. The condensed phosphates were hydrolyzed by adding 1 ml/min 5 N H_2SO_4 to 0.33 ml/min phosphate solution. The hydrolysis took place in a 2.2-mm-i.d. glass tube at 90°C. Residence time was 8 min. The resulting ortho phosphate was determined with the phosphovanadomolybdate colorimetric method at pH = 1 in 2 min. The absorbance was measured at 420 nm. The variances for the various parts in the reactor were given as follows: hydrolysis and cooling section, $\sigma_{tr}^2 = 39$ sec^2; color reaction, $\sigma_{tr}^2 = 2$ sec^2; mixing tees, $\sigma_{tr}^2 = 8$ sec^2. The apparatus used for this work is shown in Figure 23 and a chromatogram of a mixture of phosphates is presented in Figure 24.

Metals. Automatic ion exchange chromatography of zinc(II) and other metals by postcolumn reaction has much potential for analysis of trace amounts in water, soil, and biological matrices. For zinc(II), Zincon (2-carboxy-2'-hydroxy-5'-sulfoformazyl-benzene) was used to produce the absorbing species in a reaction coil prior to absorption measurements using a conventional dual-coil autoanalyzer setup. Eluted zinc was recorded as Zincon complex. The limit of detection for zinc was $\sim$2 μg per injection. Other metals which react with Zincon are aluminium(III), beryllium(II), cadmium(II), cobalt(II), chromium(III), copper(II), iron(III), manganese(II), molybdenum(VI), nickel(II), and titanium(IV). They could conceivably be detected by the same procedure. The potential of postcolumn complexation reactions has been discussed earlier.[81-83] Since these reactions are quite rapid nonsegmented techniques can often be applied.

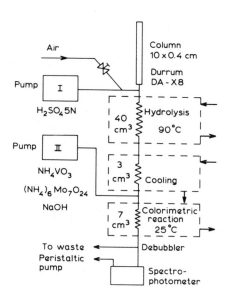

Fig. 23. Schematic drawing of apparatus used for postcolumn reaction of condensed phosphates.[18]

Penicillins. The analysis of penicillins in biological fluids is traditionally performed using mainly microbiological methods. Such assays often have an adequate sensitivity, but a low precision (typically an experimental error of ±15%) and a doubtful specificity since antibacterially active metabolites may also interfere. Finally, such assays are very slow. Consequently there is a great demand for alternative methods; one method recently developed for use in biological samples involves separation by reversed phase chromatography, and postcolumn derivatization in an air-segmented flow.[103]

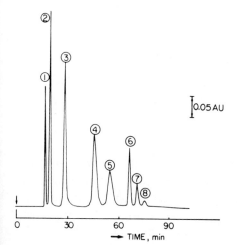

Fig. 24. Chromatogram for a mixture of phosphates. Column 10-cm × ×4-mm packed with Durrum DA-X 8 anion exchanger. Eluent sodium chloride (gradient) in acetate buffer pH 4.5, flow rate 0.33 cm³ min⁻¹. (1) Orthophosphate; (2) pyrophosphate; (3) tripolyphosphate; (4) tetrapolyphosphate; (5) pentapolyphosphate; (6) hexapolyphosphate; (7) heptapolyphosphate; (8) octapolyphosphate.[18]

Studies on the optimization of the different steps in the analytical procedure have been performed by the same authors. The reversed phase system used for separation allows injecting biological samples onto chromatographic system after a minimum of pretreatment. Many penicillins are furthermore rather polar and difficult to extract into an organic phase (e.g., ampicillin, carbenicillin, epicillin, and cyclacillin), and thus it is more suitable to perform the analysis in an aqueous environment.

The simple procedure for plasma sample preparation, for example, was as follows: (1) 1.0 ml of plasma is taken and the proteins precipitated by addition of 100 µl of trichloroacetic acid (70%) and mixing on a Whirlimixer for about 30 sec. (2) After centrifugation at 4000 rpm for 3 min, the supernatant is filtered through a piece of cotton inserted near the tip of a Pasteur pipette, and 100–500 µl of the clear situation is injected without delay onto the chromatographic column.

The inherent light absorbance properties of the penicillins are difficult to utilize for bioanalytical purposes since these compounds have a sufficiently high molar absorptivity only in the wavelength range 220–240 nm where many endogenous compounds also absorb.

A derivatization procedure was therefore developed that gives a product with high molar absorbance at about 310 nm where the endogenous disturbances are much less. The derivatization is performed in the postcolumn mode with air segmentation to minimize band broadening.

The derivatization method is based on a procedure developed by Bundgaard and Ilver[104] and proceeds via a selfcatalyzed attack of imidazole on the β-lactam bond followed by the formation of the mercuric mercaptide of the penicillinic acid:

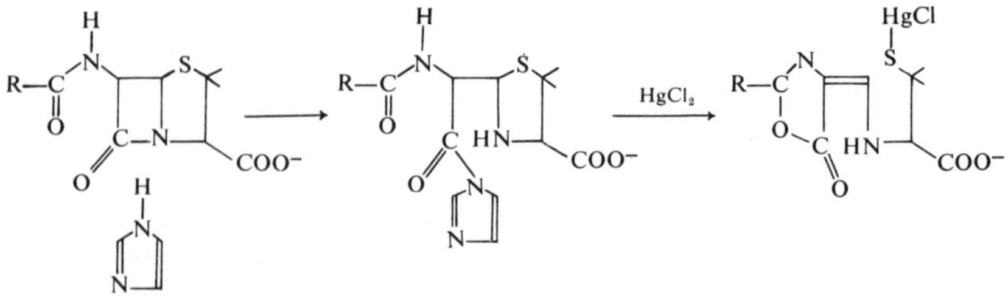

The product has a molecular absorbance of about 20,000 in the wavelength range 308–345 nm depending on the actual penicillin. A procedure which utilizes this reaction for the bioanalysis has been developed for an automatic air-segmented analytical system analogous to the detection system described here. The reaction is specific for penicillins with an intact β-lactam ring,

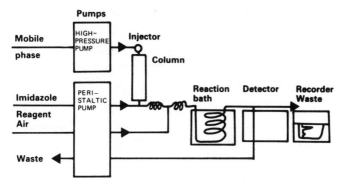

Fig. 25. Schematic of reaction detector for penicillins.[103]

which corresponds to the requirement for bactericidal effect on micro-organisms. Only the intact penicillin and possible active metabolites will consequently give a response in the method, but penicilloic acids of penicillins with a free amino group, such as ampicillin and amoxicillin, also react to some extent giving a chromatographic peak corresponding to 1–2% of the parent penicillin peak height.

Ampicillin differs from other penicillins by giving an unstable reaction product. However, when the reaction is performed in a flow system and the time between derivative formation and measurement is constant for all samples, it is neither necessary for the product to be stable, nor for the reaction to come to completion. The following procedure was recommended for the postcolumn reaction with the apparatus shown in Figure 25.

Aqueous solutions of imidazole (33%) and mercuric chloride (0.11%) adjusted to pH 7.2 by hydrochloric acid and containing Brij 35 (0.12%) were used as reagents. The flow rate was 0.3 ml/min and the air bubble rate: 1 sec^{-1}. The absorbance was measured at 310 nm. It was found necessary to include an unsegmented mixing coil for the eluate and the reagent stream directly after the column and before the air segmentation in order to decrease the base-line noise, but at the price of an increase in the total dispersion. The total dispersion in the system may be described by

$$\sigma_v{}^2 = \sigma_{vc}^2 + \sigma_{vd}^2 \tag{17}$$

where σ_{vc}^2 is the dispersion in the column and σ_{vd}^2 represents the dispersion in the reactor, while the dispersion in other parts of the chromatographic system, i.e., injection valve, connecting tubings, and the detector, is assumed to be insignificant. The quotient between the actual resolution, R_{BA}, and the maximal resolution, $R_{BA,\max}$, between the two components A and B,

TABLE 2. Dispersion in Column and Reactor[a][(103)]

Compound	σ_{vc}	σ_{vd}	σ_v	$R_{BA}/R_{BA,\mathrm{max}}$
Ampicillin	85	171	191	0.45
Mecillinam	175	217	278	0.63

[a] The symbols are defined in equations (17) and (18). The dispersion is given in volume units, μl.

is given by the following equation, assuming that $\sigma_A{}^2 = \sigma_B{}^2$ and that $R_{BA,\mathrm{max}}$ is obtained if $\sigma_{vd}^2 = 0$:

$$R_{BA}/R_{BA,\mathrm{max}} = [1 + (\sigma_{vd}^2/\sigma_{vc}^2)]^{-1/2} \tag{18}$$

Representative data for the dispersion of ampicillin and mecillinam in the column and reactor together with quotients according to equation (18) are given in Table 2. The dispersion in the reactor part is larger than on the column for both compounds, which indicates that further refinements of the reactor are possible involving the tubing inner diameter, mixing tees, and debubbler.

Some results from quantitative determinations in plasma are given in Table 3. The precision ($s_{\mathrm{rel}\%}$) is of the order of 2% from levels above 1 μg/ml and 5–8% down to 100 ng/ml provided 100 μl of the supernatant is injected.

By utilizing a preconcentration phenomenon as described earlier[(105)] larger injection volumes could be used and the detection limit accordingly

TABLE 3. Quantitative Determination of Ampicillin and Mecillinam in Plasma[a][(103)]

Compound	Amount added, ng/ml	Amount found, %	s_{rel}, %	n
Ampicillin	95	97.4	8.3	6
Ampicillin	284	101.1	5.6	5
Mecillinam	252	100.1	6.4	6
Mecillinam	503	101.0	5.3	5

[a] Standard curve: 5 standards in the ranges 48–665 ng/ml (ampicillin) and 192–962 ng/ml (mecillinam).

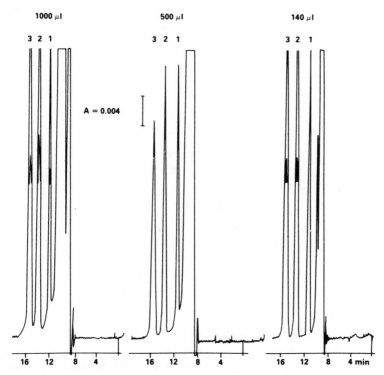

Fig. 26. Bioanalysis of penicillins—injection of large plasma volumes.[103] (1) Ampicillin, 710 ng; (2) cyclacillin, 780–980 ng; (3) mecillinam, 1400–1665 ng. Support: LiChrosorb RP-8, 5 μm (100 × 4 mm). Mobile phase: phosphate buffer pH 8 (I = 0.1) + methanol (7 + 3 v/v).

improved without seriously reducing the quality of chromatograms. The highest efficiencies for ampicillins were obtained when acidic solutions that give the highest capacity ratios were injected, while intermediate pH values corresponding to the lowest retentions gave much higher H values.

A much stronger gradient effect is obtained by injecting the samples dissolved in pure aqueous solutions, as is the case in the bioanalytical method. It was found that the capacity ratios and total dispersions for ampicillin, cyclacillin, and mecillinam are almost unaffected by the injection of up to 2 ml of the acid plasma supernatant. The resolution between the penicillins is then also independent of the injected volume and the limiting factor is the disturbance from eluted endogenous compounds as illustrated in Figure 26. For quantitation of ampicillin, it is not possible to inject more than about 1 ml, while the limit for mecillinam is not reached and seems to be considerably above 2 ml. Some quantitative data are given in Table 4. The

TABLE 4. Quantitative Determination of Ampicillin and Mecillinam in Urine[a] [(103)]

Compound	Amount added, μg/ml	Amount found, %	s_{rel}, %	n
Ampicillin	1.00	101.8	3.00	5
Ampicillin	50.2	100.9	1.35	5
Mecillinam	1.00	100.8	10.1	5
Mecillinam	49.9	102.0	0.45	5

[a] Injected volume: 10 μl; standard curve: Five standards in the range 1–50 μg/ml.

lowest level (1 μg/ml) corresponds to the injection of 10 ng and is for mecillinam near the limit of detection, but lower concentrations can probably be measured by the injection of larger volumes.

The method may also be applicable to whole blood and to urine and lymph samples.

3.3.1.2. Fluorescence Detection. Biogenic amines. A considerable amount of work has been carried out with groups of compounds containing primary amino groups and hence reacting in an HPLC postcolumn reactor with the *o*-phthalaldehyde reagent as shown below and as described earlier[46]:

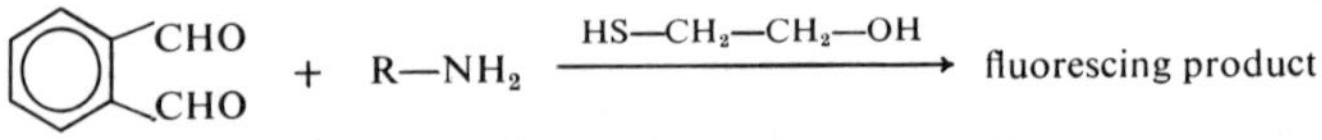

$$\underset{\text{CHO}}{\overset{\text{CHO}}{\bigcirc}} + \text{R—NH}_2 \xrightarrow{\text{HS—CH}_2\text{—CH}_2\text{—OH}} \text{fluorescing product}$$

The groups of compounds derivatized by this approach comprise different biogenic amines,[106] peptides,[107] 5-hydroxyindoleacetic acid and derivatives,[108–110] and finally amino acids,[53,58] the latter reacting sufficiently rapidly that nonsegmented stream technology has been applied (see earlier section).

One of the earlier applications of autoanalyzer technology to this type of fluorigenic reaction is the analysis of 5-hydroxyindoleacetic acid by reaction with *o*-phthalaldehyde in the presence of hydrochloric acid. The total system involves classical ion exchange chromatography with an amino acid analyzer[109,110] The column effluent is mixed with the reagents in order to produce a fluorescent product which is monitored in a microflow fluorimeter at 360 nm (excitation) and 470 nm (emission). Nanomolar amounts of the compounds can be detected (Table 5). The effluent ($\sim$1.2 ml/min) from the automated ion exchange system (Technicon amino acid analyzer or a similar apparatus) is mixed (1:1) with the reagent [100 mg of o-phthalaldehyde in 200 ml of 2-methoxyethanol (methyl cellosolve) and 200 ml of distilled water] and (1:2) with concentrated hydrochloric acid. When a run is complete, the hydrochloric acid line is flushed with air in order to preserve the tubing (standard pump tubing). The buffers used to elute the indole derivatives are citrate (pH 4, 1 and 2.0, 0.05 M) containing 0.5 N Li$^+$ and 0.3 N Li$^+$, respectively. The combined effluent and reagents pass through a mixing coil and then a heated bath (65°C) before being determined fluorimetrically.

TABLE 5. Fluorescence of Substituted Indoles after Reaction with o-Phthalaldehyde

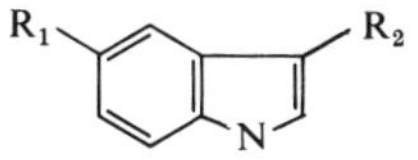

Compound	R$_1$	R$_2$	Fluorescence units per nmol
N-Acetyl-5-methoxytryptamine	CH$_3$O	CH$_2$CH$_2$NHCOCH$_3$	287
5-Methoxytryptamine	CH$_3$O	CH$_2$CH$_2$NH$_2$	275
5-Methoxyindole-3-acetic acid	CH$_3$O	CH$_2$COOH	120
N-Acetyl-5-hydroxytryptamine	HO	CH$_2$CH$_2$NHCOCH$_3$	114
5-Hydroxytryptophan	HO	CH$_2$CH(NH$_2$)COOH	71
5-Hydroxytryptamine	HO	CH$_2$CH$_2$NH$_2$	50
5-Hydroxyindole-3-acetic acid	HO	CH$_2$COOH	47
α-Methyl-5-hydroxytryptophan	HO	CH(CH$_3$)CH(NH$_2$)COOH	36
5-Methoxygramine	CH$_3$O	CH$_2$N(CH$_3$)$_2$	14
5-Methyltryptophan	CH$_3$	CH$_2$CH(NH$_2$)COOH	1.6
N,N-Dimethyltryptamine	H	CH$_2$CH$_2$N(CH$_3$)$_2$	1.5
Tryptophan	H	CH$_2$CH(NH)$_2$COOH	<1
Tryptamine	H	CH$_2$CH$_2$NH$_2$	<1
5-Hydroxyindole	HO	H	<1
5-Methoxyindole	CH$_3$O	H	<1
N-Acetyltryptophan	H	CH$_2$CH(NHCOCH$_3$)COOH	<1
Tryptophol	H	CH$_2$CH$_2$OH	0
Indole-3-acetic acid	H	CH$_2$COOH	0
5-Bromogramine	Br	CH$_2$N(CH$_3$)$_2$	0
Indole	H	H	0

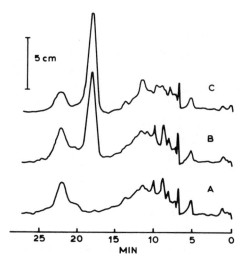

Fig. 27. High-performance liquid chromatography of 5-hydroxytryptophan.[108] Trace *A*: 20 µl of extracted serum from an untreated patient; trace *B*: 20 µl of extracted serum to which 5-hydroxytryptophan was added to a concentration of 0.4 µmol/l; trace *C*: 10 µl of extracted plasma from a patient treated with 5-hydroxytryptophan and carbidopa.

The selectivity of this reaction permits assaying of these compounds in biological samples with a minimum of sample pretreatment.

A modern version of this technique but carried out with second generation autoanalyzer equipment and an up-to-date HPLC unit was reported by Enbaek and Magnussen[108] for 5-hydroxytryptophan in plasma. The combined separation and reaction time was 25 min. With only a simple extraction step of the plasma involved. The sensitivity of the method was indicated as 0.05 µmol/l of sample with 750-µl plasma samples used. Chromatograms for plasma samples of treated and nontreated patients are shown in Figure 27.

The separation of the 5-hydroxytryptophan from metabolites is carried out on cation exchange resin Aminex A-5 (Bio-Rad Labs., Richmond, California) with a solvent system consisting of 100 mM $NaClO_4$ and 100 mM Na-acetate per liter adjusted to pH 5 with acetic acid.

The procedure for plasma analysis is given below:

Mix 750 µl plasma or serum with 4 ml of acidified butanol. Remove the precipitate by centrifugation. Mix the supernate with 6 ml of *n*-hexane and centrifuge. Obtain about 400 µl of the resulting aqueous phase (containing 5-hydroxytryptophan) and keep it for column chromatography. Discard the organic phase. Apply, at most, 25 µl of extracted plasma to the column and develop the chromatogram with the column buffer at 70°C with a flow rate of 0.3 ml/min. React the effluent of the column with fluorogenic reagent at 70°C for 8 min before measuring the fluorescence.

In a recent publication[107] the OPA reaction had been used for the automated analysis of peptides for simultaneous assay and purity test in pharmaceutical dosage forms. Although there is some doubt as to the necessity of using an air-segmented stream in the reaction detector for such a fast reaction (with most peptides, i.e., nonapeptides, it goes to completion in about 1 min), this particular work has to be mentioned since it demonstrates the feasibility of reaction detectors in fully automated and computer-controlled analytical systems.

A schematic of the apparatus used is shown in Figure 28. It consists of a microcomputer-controlled solvent delivery system with gradient elution, a computer-controlled automatic liquid sampler,[111] a reversed phase column, an uv detector, a postcolumn reaction detection system, and a laboratory computer system. The goal of the reaction detection system is to get a selective detection for primary amines. The reactor consists of an air-segmented liquid flow system, and the reaction detector layout is based on conventional Technicon autoanalyzer equipment. The reaction coil provides a 1-min delay. After this time the reaction is almost complete. The band broadening contribution of the reaction detector is about 20%, but this can certainly be optimized.[103] The minimum detectable amount of peptide primarily depends on the number of active primary amino groups and the overall structure of the particular peptide. For Lys⁸-Vasopressin with only one reacting group the detection is about 20 pmol.

A comparative chromatogram and the conditions used to obtain it are shown in Figure 29. With the relatively long time of analysis only a fully automated system such as this can be used adequately for routine monitoring of large sample systems. The repeatability of multiple injections was determined 1.27% relative s.d. ($n = 5$) for the fluorescence signal after gradient elution and derivatization, as determined by the computer system. Schwedt[106] has used a phthalaldehyde reactor for biogenic amines having primary amino groups such as noradrenaline, dopamine, and 3,4-dihydroxyphenylalanine (Dopa) as well as histamine, serotonine, tyramine, and tryptamine. A self-assembled analyzer unit with air segmentation was used. The kinetics of the OPA reaction with these compounds is sufficiently rapid (up to 2-min residence times) that nonsegmented stream technology could be applied without undue band broadening. Schwedt has also demonstrated the feasibility of the phthalate reactor as coupled to reversed phase chromatography with eluents containing small portions of organic polarity modifiers (i.e., acetonitrile, methanol). The disadvantage of this approach for catecholamines is the fact that adrenaline cannot be determined jointly with noradrenaline and hence other more universal fluorigenic reaction detectors have been proposed by the same author.[112–115] One system[112] involved the

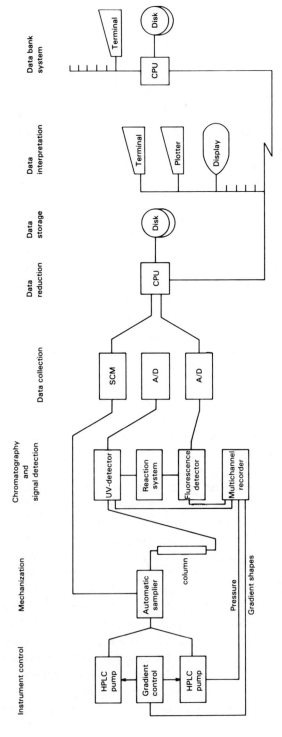

Fig. 28. Scheme of the automated HPLC system.[107]

Fig. 29. Chromatogram of a Lys⁸-Vasopressin solution with the uv signal ("uv"), the reaction detection signal of the OPA derivatization ("fluo"), and the gradient shape.[107] Column: reversed phase RP-18, 7 µm, 250 × 4.6 mm (Knauer, Oberursel, GFR). Mobile phase: step (1) 20% B → 44% B in 15 min; step (2) 44% B → 100% B in 0.5 min; step (3) 100% B for 10 min. A: aqueous KH_2PO_4–solution (1/15 M) pH adjusted to 2.3 with conc. phosphoric acid. B: 1:1 (v/v) mixture of phase A and acetonitrile, pH adjusted to 2.3 with conc. phosphoric acid. Flow 1.0 cm³/min, temperature 22°C; injection volume: 1 cm³; concen-

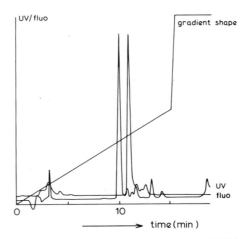

tration: 2 µg/cm³; pump: Altex model 100 with gradient programmer; Model 420 (Altex Scientific Inc., Berkeley, California); uv monitor at 210 nm, 0.2 aufs (chromatogram reconstructed by the Laboratory Data System).

use of redox reactions to produce fluorescent derivatives of compounds such as dopa, dopamine, noradrenaline, adrenaline, methanephrine, and normetanenphrine. A special application of this approach also known as the trihydroxy indole technique[116] has been worked out for adrenaline in the presence of a large excess of noradrenaline[113].

The catecholamines were separated by ion exchanger and reversed phase techniques prior to the reactions. Although these reactions are relatively fast (~2 min) it is the chemical aggressiveness of the reaction system which makes it advisable to use an all-glass–synthetics analyzer system.

This technique has later been applied to the determination of adrenaline and noradrenaline in urine samples.[114] Another approach proposed by Schwedt[115] involved the use of an ethylenediamine reaction detector using condensation with ethylenediamine. Here reaction times are longer (up to 20 min) but the reaction is more universal, hence providing a more widely applicable detection mode. A loss in resolution of up to 30% for such reactor systems was reported, whereby the major contribution to band broadening can again be attributed to the debubbler unit and mixing tees. Considering the selectivity of this detection technique, which permits direct injection of up to 200-µl urine samples, the band dispersion encountered really is not too serious. It is somewhat complex from these publications[106,112–115] to assess the relative merits of the various techniques. Detection limits are all in the low nanogram range and the selectivity make them suitable for complex matrices. The quantitative potential seems to be satisfactory. None of these methods have been reported for application to plasma samples although work in this direction is in progress and should conceivably meet with success.

Amino acids. An interesting approach for the fluorescence detection of amino acids is based on a metal complexation phenomenon.[117] This technique involves separation of the amino acids prior to fluorimetric reaction and determination. As the amino acids are eluted from the column, they are mixed with the pyridoxal–zinc(II) reagent to produce a highly fluorescent zinc chelate. Amounts of as low as 1 nmol of amino acid may be detected. The first reaction involved is the formation of the pyridoxyl–amino acid (Schiff base):

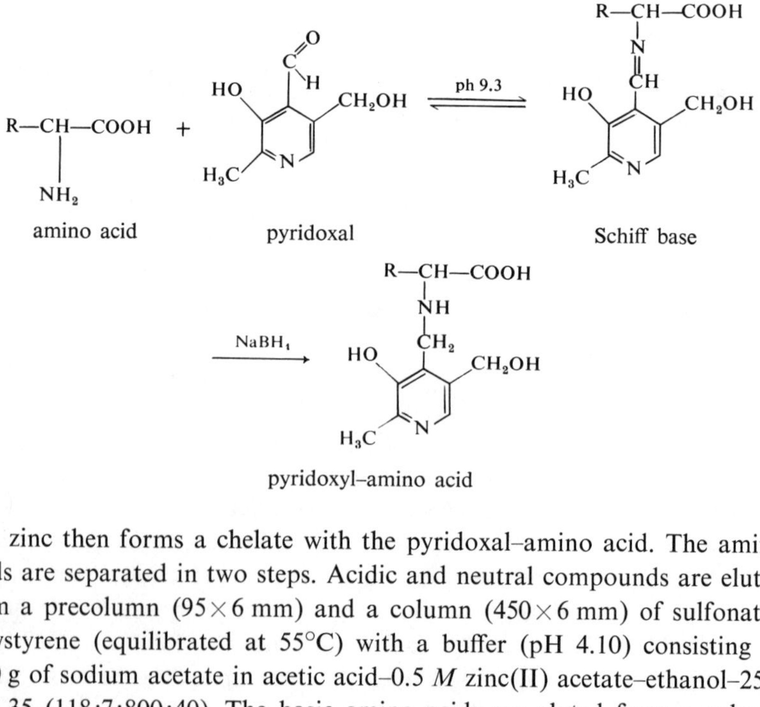

amino acid		pyridoxal	Schiff base

pyridoxyl–amino acid

The zinc then forms a chelate with the pyridoxal–amino acid. The amino acids are separated in two steps. Acidic and neutral compounds are eluted from a precolumn (95×6 mm) and a column (450×6 mm) of sulfonated polystyrene (equilibrated at $55°C$) with a buffer (pH 4.10) consisting of 41.0 g of sodium acetate in acetic acid–0.5 M zinc(II) acetate–ethanol–25% Brij 35 (118:7:800:40). The basic amino acids are eluted from a column (22×6 mm) equilibrated at $55°C$ with a buffer (pH 5.10) consisting of 738 g of sodium acetate in acetic acid–0.5 M zinc(II) acetate–benzyl alcohol–25% Brij (185:20:110:40). The flow rate is 30 ml/hr. The effluent is mixed with the pyridoxal–zinc(II) reagent [1.0 g of zinc(II) acetate and 0.10 g of pyridoxal hydrochloride in 1.0 liter of 2.0% pyridine–methanol] at a flow rate of 120 ml/hr. The mixed effluent is then allowed to react at $65-75°C$ for 10 min in a reaction coil of PTFE capillary tubing (3.0 m × 1.0mm). After reaction, the effluent is examined at 365 nm (excitation) and 485 nm (emission). The limit of detection is ∼1 nmol of each amino acid.

Obviously for true HPLC conditions the systems would have to be miniaturized somewhat and flows reduced at least tenfold. A prime argument for this reaction is again its high selectivity besides being sensitive. Fluram or OPA might well prove to be excellent competitors particularly as regards reaction time and temperature.

Organic acids and sugars. The oxidation reaction of reducible organics with cerium(IV),

$$Ce(IV) + e \longrightarrow Ce(III)$$

nonfluorescent   fluorescent

to yield the fluorescent Ce(III) has been used for several groups of compounds primarily with segmented flow systems. Katz and Pitt[118,119] have been the first group to propose such an approach for HPLC detection and have built a reaction detector. An overview on the effects of reaction temperatures, pH, salt concentration, and possible catalytic effects have been given for this reaction by several authors.[120,121]

The reagent and apparative conditions chosen by Katz and Pitt[118,119] for a group of organic acids are as given in Figure 30, which also indicates the flow rates used. No catalyst was needed for this reaction. A 10^{-4} N cerium(IV) sulfate in 2 N sulfuric acid solution was added.

PTFE tubing (8 m × 0.75 mm i.d.) was coiled and immersed in a bath of boiling water. This provided a residence time of 12 min at a total flow rate of 20 ml/hr in order to ensure completion of the reaction. The limits

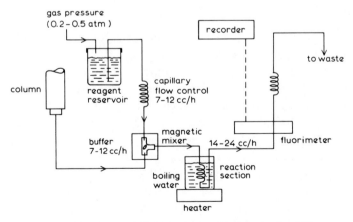

Fig. 30. Diagram of the cerium oxidation monitor.[119]

TABLE 6. Fluorescence and Ultraviolet Sensitivities of Substituted
Aromatic Acids[a]

Acid	Fluorescence sensitivity, units/μg	Fluorescence sensitivity relative to uv response
4-Hydroxymandelic	12	20
4-Hydroxy-3-methoxymandelic	10	>50
4-Hydroxyphenylacetic	19	>50
4-Hydroxy-3-methoxyphenylacetic	25	>50
2-Hydroxybenzoic	8	0.8
4-Hydroxybenzoic	12	4
4-Hydroxy-3-methoxybenzoic	15	2
4-Hydroxy-3,5-dimethoxybenzoic	14	5
4-Hydroxycinnamic	7	4
4-Hydroxy-3-methoxycinnamic	5	4
4-Hydroxyphenyllactic	10	20
4-Hydroxyphenylpyruvic	7	>50

[a] Reference 119.

of detection are of the order of 100–500 ng (fluorescence: excitation, 260 nm; emission, 350 nm). The relative fluorescence responses of a number of aromatic acids in this detection system are listed in Table 6 and compared with their uv responses.

The same approach and a similar type of apparatus and set of conditions were used by the same research group to analyze carbohydrates in blood and urine.[122]

Pollutants (phenolics). Reducing pollutants, for example, phenolics[123] or polythionates[124] were determined via the Ce^{4+}–Ce^{3+} reaction. The principle of this approach is again that oxidizable compounds are detected and then quantitated via the created fluorescent Ce^{3+}. After separation of phenols, for example, by reversed phase chromatography, the compounds were reacted in a Technicon autoanalyzer with a 2.5×10^{-5} N Ce(IV) sulfate solution in 23 N sulfuric acid. The reaction time was 4 min at 25°C. Somewhat longer reaction times were used for the polythionates. Detection conditions were $\lambda_{ex} = 260$ nm and $\lambda_{em} = 350$ nm. Detection limits were in the sub-ppm region. Wolkoff and Larose[123,124] reported negligible band broadening although, according to our own experience, about 10% could be expected from the kind of system used by these workers. Aside from the relatively long reaction times the choice of an all-glass–Teflon autoanalyzer system seems to be indicated for such aggressive chemicals as used for the Ce^{4+}–Ce^{3+} reaction. A filter tube is used on the feed in a practical run in the reaction detector. The column effluent was div ded into small portions by means of a Technicon proportionating pump. To these portions was added a mixture of the cerium(IV) sulfate reagent and the 23 N sulfuric acid. The combined flows were passed through a short coil and then into the microflow cell of the fluorimeter. The relative responses of a number of phenols to this method of detection are listed in Table 7.

Cardiac glycosides. Another example of relatively long reaction times (10 min, 45°C) coupled with an aggressive reaction medium was studied by Gfeller *et al.*[125] for the fluorescence detection of cardiac glycosides. A slightly modified autoanalyzer system was used to react cardiac glycosides with concentrated HCl to yield a fluorescent derivative via a dehydration process. Ascorbic acid and H_2O_2 were added to act as catalysts. The schematic diagram for the postcolumn reaction detector is shown in Figure 31.

An Altex model 100 pump (Altex Scientific, Berkeley, California) was used for chromatography. The reagent pumping and mixing module of this apparatus was a conventional autoanalyzer except for the mixing and reaction spirals, which had to be miniaturized to 1-mm i.d. (Portex Ltd., Hythe, Great Britain). Other parts such as fittings, mixing units, and the peristaltic four-channel pump, were conventional Technicon accessories. The reaction was thermostatted at 45°C and the mixing spirals and the fluorimeter cell at 20°C. The apparatus was laid out for an eluent flow rate of 0.4 ml/min and a total reagent flow rate of 0.58 ml/min (see Figure 31).

The band broadening for this system was around 10% and the corresponding loss in resolution for two critical pairs is shown in Figure 32 for comparative uv and fluorescence scans. The detection limits were <0.1 ng or at least two orders of magnitude better than for uv detection at 220 nm. The selectivity and sensitivity of this reaction should permit adaptation of this method to pharmacokinetics studies of the glycosides in plasma samples.

Pesticides. Moye and Wade[126] determined carbamate pesticides in a fluorimetric enzyme inhibition detector. The principle is as follows: the column effluent is incubated with cholinesterase after which the partially inhibited cholinesterase is reacted with a nonfluorescing substrate to produce a fluorophore. When a cholinesterase inhibition is present in the effluent, the base-line fluorescence drops. Detection takes place at $\lambda_{ex} = 430$ and $\lambda_{em} = 501$ nm. In this paper the influence of the reaction conditions, the substrate concentration, the bubble pattern and the reaction coil length,

TABLE 7. Relative Response of Phenols with Cerium(IV)
Sulfate-Fluorimetric Detection[a,b]

Compound	Weight basis	Mole basis
Phenol	1.000	1.000
o-Chlorophenol	0.50	0.68
p-Chlorophenol	0.50	0.69
2,4-Dichlorophenol	0.24	0.42
2,4,6-Trichlorophenol	0.10	0.22
Pentachlorophenol	0.04	0.12
o-Cresol	0.64	0.74
m-Cresol	0.78	0.90
p-Cresol	0.50	0.57
2,4-Dimethylphenol	0.44	0.57
2,6-Dimethylphenol	0.43	0.55
Catechol	0.47	0.55
Resorcinol	0.88	1.03
Hydroquinone	0.47	0.55
Pyrogallol	0.41	0.55
Orcinol	0.52	0.79
2,4-Dihydroxybenzaldehyde	0.40	0.58
4-Hydroxybenzoic acid	0.42	0.62
2-Methoxyphenol (guaiacol)	0.69	0.89
3-Methoxyphenol	0.59	0.78
4-Methoxyphenol	0.42	0.55
o-Vanillin	0.85	1.37
Eugenol	0.47	0.81
Isoeugenol	0.52	0.92
Syringic acid	0.20	0.41
1-Naphthol	0.56	0.86
2-Naphthol	0.53	0.80
2-Nitrophenol	0.25	0.37
3-Nitrophenol	0.25	0.37
4-Nitrophenol	0.13	0.19
2,4-Dinitrophenol	0.002	0.004
2,4-Dinitro-o-cresol	0.004	0.01
Picric acid	0.01	0.02
4-Aminophenol	0.51	0.59
3-Chloro-7-hydroxy-4-methylcoumarin	0.23	0.51

[a] Reference 123.
[b] Mobile phase, acetonitrile–water (2:3).

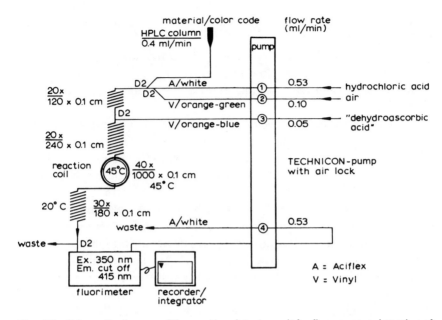

Fig. 31. Schematic diagram of the reaction detector unit for fluorescence detection of cardiac glycosides.[125]

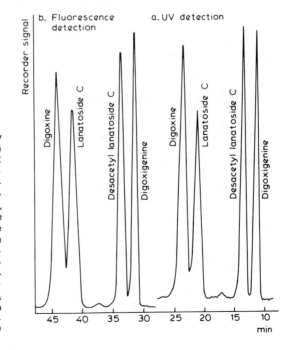

Fig. 32. Comparison of uv and fluorescence detection for the separation of the C group of lanatosides. Column: reversed phase, Nucleosil C_{18}, 10 nm (Macherey-Nagel, Düren, G.F.R.); length, 25 cm; i.d., 0.4 cm; mobile phase, 40% (v/v) acetonitrile —dioxane 1:1) in water at a flow rate of 0.4 ml/min. Detection: uv at 220 nm (Perkin-Elmer LC-55). Fluorescence: λ_{ex} = 350, λ_{em} = 485 nm. Secondary 7-2. A Wratten filter with Aminco Fluoromonitor. Injection volume, 20 μl (Valco loop).[125]

detector flow rate, and detector dead volume on band broadening was studied.

The band broadening contribution by the reaction detector unit was about 35% for the various flow rates tested. An optimal sensitivity was obtained for a 5-min reaction time at 40°C. Detection limits were of the order of 0.2–2 ng.

Compared to the colorimetric cholinesterase technique, discussed earlier,[100] the method by Moye and Wade[126] offers the higher sensitivity inherent in fluorescence detection and a somewhat more sophisticated detector design. Both show elegantly the potential of the highly selective enzymatic techniques for residue analysis in complex matrices.

Another interesting method for carbamates has also recently been published by Moye et al.[127] It involves a two-step derivatization procedure with dynamic hydrolysis of N-methyl carbamates to the phenol and the methylamine following reversed phase HPLC. The methylamine is reacted with o-phthalaldehyde (OPA) to yield a fluorescence derivative. The apparatus used for this work is seen in Figure 33. Band broadening of about 40% was reported due to this OPA detector. A long hydrolysis step of several minuted has to be used, which contributes a major partion to the band broadening reported.

The OPA reaction on the other hand is rapid and goes to completion in less than 1 min. As can be seen in Figure 33, no segmentation was used in the tubular reactor although this is a borderline case with regard to total residence time in the reactor (5 min) and one wonders whether the loss in

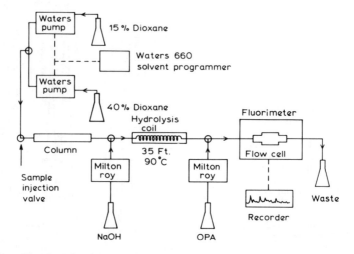

Fig. 33. Modular dynamic fluorogenic labeling liquid chromatograph.[127]

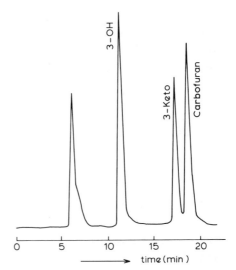

Fig. 34. Chromatogram of carbofuran, 3-hydroxycarbofuran, and 3-ketocarbofuran on *p*-chlorophenylsilane column, modular liquid chromatograph. Program: 15–40% dioxane/water, 20 min, 1 ml/min.[127]

resolution could not be substantially reduced by adapting a bed reactor or a segmented stream device. A typical chromatogram for carbofuran and metabolites is shown in Figure 34. It also shows the feasibility of coupling a gradient with postcolumn reaction techniques[40] which accounts for the relatively small dependence of this two-step derivatization on solvent composition and the low fluorescence background obtained under the experimental conditions used.

Detection limits in the order of 0.1 ng/injection were reported and a 2.7% relative s.d. for 50-ng samples of Lannate ($n = 7$) was obtained. The linearity of calibration plots was excellent over three orders of magnitude.

3.3.2. Solvent Segmentation (Extraction Detectors)

In the previous section it was shown that segmentation principles can be used to advantage for relatively slow reactions. In recent work[128,129] it has been demonstrated that the same principles can also be used favorably for rapid reactions mainly in situations where the excess of reagent interferes with the signal of the product. This comprises the majority of well-known simple and fast reactions such as simple substitution or esterification reactions, well known in precolumn derivatization[1,2] and since the excess of reagent used is usually high (up to several-hundred-fold) this can be quite a problem. Other fast processes which suffer from the same drawback but otherwise would be very useful for postcolumn derivatization are complex formation or ion pair formation.

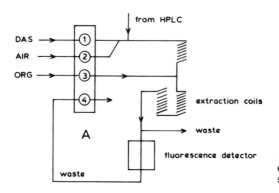

Fig. 35. Postcolumn ion pair extraction system using air segmentation.

Since upon derivatization the polarity of the product is usually significantly lower than that of the reagent, dynamic microextraction procedures well known in classical autoanalyzer knowhow have been adopted to eliminate the excess of reagent from the less polar product.

The feasibility of such an approach has been tested with an ion pair system consisting of tertiary amines of pharmaceutical and agricultural importance, and dimethoxyanthracene sulfonate (DAS) proposed earlier by Westerlund and Borg[130] as a fluorescent counterion for ion pair formation. The schematic of the first detector construction coupled to reversed-phase separation and based on a three-phase segmentation system[127,128] is shown in Figure 35. The reagent is added to the column effluent right after the column. Air segmentation is being used to efficiently reduce band broadening.

The high-density organic extractant, usually chloroform, dichloro, or chloroethane, is then added and after extraction of the ion pair into the organic phase the layers are separated in a conventional Technicon phase separator. Fluorescence detection of the ion pair then takes place in the organic solvent. The model compounds investigated in the course of this study are listed in Table 8. The first five are pharmaceutically active compounds, SAN is a new pesticide, and HA and HT are the major metabolites of the triazine herbicides atrazine and terbutylazine. Other successful tests were carried out with ergot alcaloids[129,131] and with hyoscyamine[128,132] and secoverine.

The detailed outlay of the extraction detector used in these experiments[128,129] is shown in Figure 36.

Standard HPLC pumps (Waters, Perkin-Elmer, Altex, Orlita) were used. The separation was carried out in primarily aqueous media on ion exchangers or polar chemically bonded silica gel material (i.e., diol or CN-bonded phases) or on reversed-phase material RP-8 or RP-18 with organic

TABLE 8. Some Structures of Tertiary Amines Tested

Structure	Name	Abbreviation
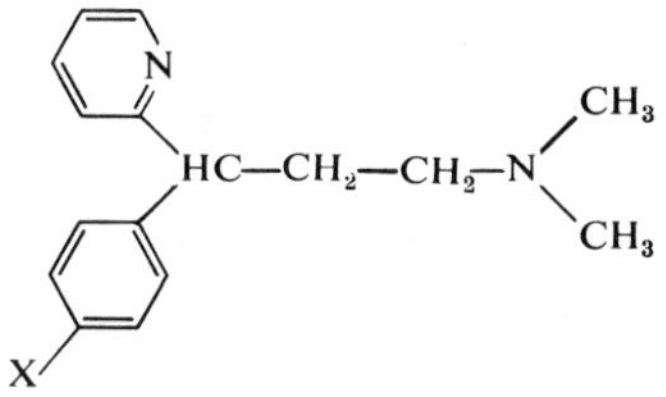	X = Cl Chlorpheniramine	CPA
	X = Br Bromopheniramine	BPA
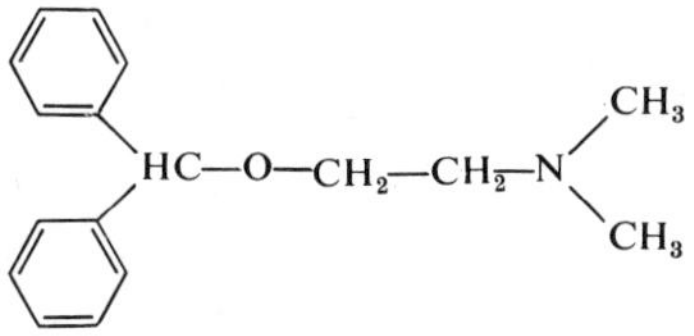	Diphenhydramine	Diph
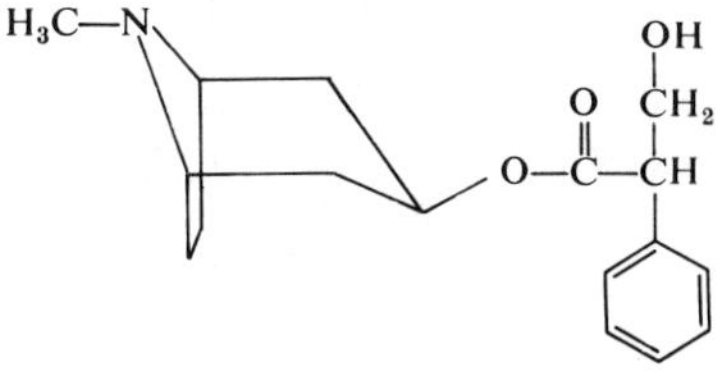	Atropine	Atrop
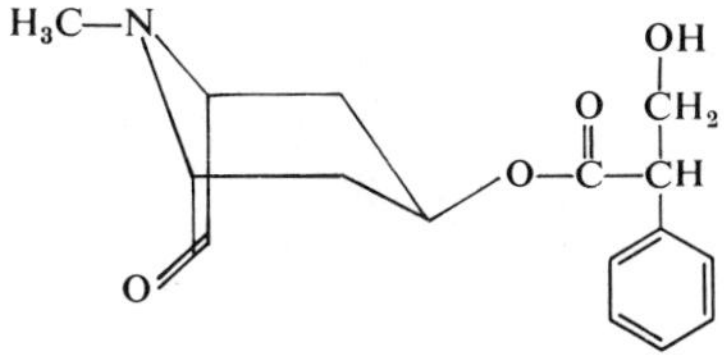	Scopolamine	Scop
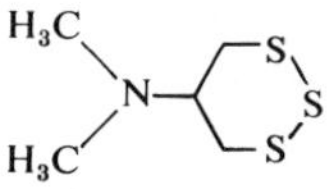	SAN	SAN
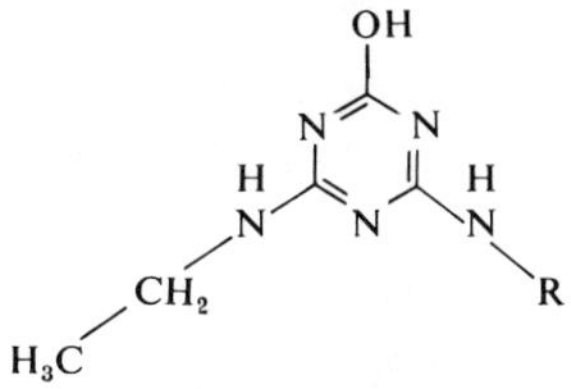	R = CH_3-CH_2-	HA
	R = $CH_3-\overset{\displaystyle CH_3}{\underset{\displaystyle CH_3}{C}}-$	HT

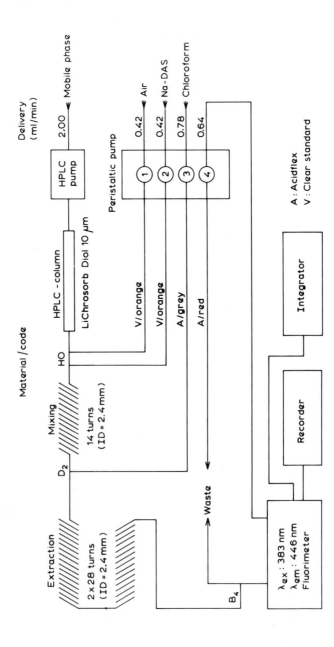

Fig. 36. Flow scheme of the postcolumn reactor. All material from Technicon (Tarrytown, New York) with a Perkin-Elmer fluorimeter model 204.[128]

polarity modifier concentrations (MeOH, acetonitrile) of up to 25% (v/v). The reaction and extraction part of the detector including the phase separator are standard Technicon material. Filter or monochromator fluorescence detectors such as the Aminco fluoromonitor or the Perkin-Elmer model 204 instrument were hooked up.

Besides the one to two orders of magnitude improvement in detection limits it was the selectivity of this fluorescent ion pair formation that was particularly attractive for trace determinations in a complex matrix. This point is amply demonstrated in Figure 37, which shows a comparison of uv and fluorescence detection for 1 ppm of chlorpheniramine in urine samples. 20 μl of the spiked urine were directly injected on the CN column without necessitating a sample pretreatment. In the uv trace (Figure 37) no difference between blank and spiked sample was observed under the test conditions due to an interfering peak (see arrow) and the lower sensitivity of uv detection. The cross-hatched peak in the fluorescence trace corresponds to the 1 ppm of chlorpheniramine which is sufficiently isolated from interferences to assure quantitation.

The results of this work, also with regard to reproducibility and quantitative potential of the extraction detector, were encouraging, and band broadening was kept below 20% for this system relative to the uv signal and using standard Technicon equipment. On the other hand, one is still dealing with a relatively complex three-phase system, which makes the phase separator unit a critical part of the extraction detector; it was, therefore, in a further study decided to drop the air bubble.[133] Continuous extraction without air segments has been used successfully by Karlberg and Thelander[134] for microextraction of caffeine in a continuous flow system. Band broadening was, of course, not as critical in this system as it would be in HPLC detection, but it was reported to be small.

The above authors did not recognize the potential of this segmentation of the aqueous carrier stream with nonmiscible solvent plugs for the reduction of band broadening. It also seems misleading to classify this approach as a flow injection technique such as the above authors[133] have done, since the flow dynamics as described by Ruzicka and Hansen are not applicable for a segmented flow system.

Another recent step in this direction has been made by Tsuji[135] for the fluorimetric determination of erythromycin and its ethylsuccinate in serum[135] and in simulated gastric fluids.[136] He also described an on-stream derivatization dynamic extraction technique using only a two-phase system. The reaction is based on the ion pair formation of the fluorescent naphthatriazole disulfonate with the readily protonated form of erythromycin

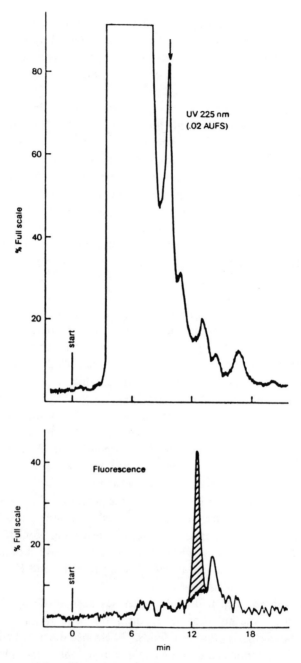

Fig. 37. Comparison of ion pair fluorescence and uv (225-nm) chromatograms of 20 μl urine containing 1.0 ppm chlorpheniramine. Arrow indicates interfering peak in uv results.[129]

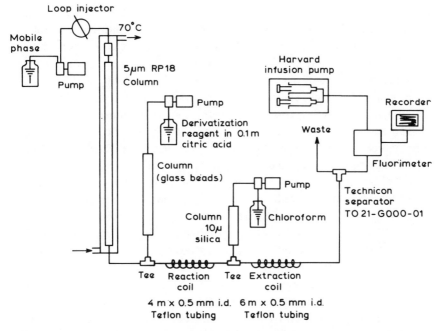

Fig. 38. Schematic diagram of HPLC fluorimetric system for the determination of erythromycin and erythromycin ethylsuccinate.[135]

and its esters.[137] The excitation and emission characteristics for these ion pairs and the free triazole are given with λ_{ex} 360 nm $\lambda_{em} > 440$ nm. Less than 10 ng/ml serum were detected by this approach. The apparatus used by Tsuji is depicted in Figure 38.

The separation was done at a 0.9-ml/min flow rate with a LDC pump. The mobile phase was acetonitrile–0.2 M ammonium acetate–water (65–60:10:25–30), pH 7. The derivatization reagent was added at 0.9 ml/min and the glass bead column (Figure 38) served to give enough backpressure for efficient pulse damping.

The selectivity of the technique permitted analysis with a considerably simplified plasma sample treatment requiring only two extraction steps for all the components. The reproducibility of the technique as applied to serum samples was still in the order of between 4.5% and 6.0% relative s.d. ($n = 7$) for three components, and the linearity of fluorescence signals as a function of concentration was observed over a three order of magnitude concentration range.

Still the apparatus designed for this technique is quite complicated. Tsuji reports about a 40% decrease in number of theoretical plates due to

band broadening in the reaction and extraction detector. To be sure the
phase separator as in all extraction-type detectors contributes a major
amount to band broadening,[133,139] but a major portion is also contributed
by the first (reaction) coil (Figure 38), which contains a nonsegmented
stream for about 1/3 of the total residence time. The author could have used
the nonmiscible solvent chloroform also as a segmentation device by adding
the reagent in the solvent-segmented form, thereby eliminating one mixing
tee and suppressing band broadening right at the outset of the chromatographic column. By proceeding in this manner band broadening could
conceivably be cut in half and the apparatus simplified.

The use of the nonmiscible organic solvents as both a segmentation
device and extraction medium (solvent segmentation principle) for the
construction of HPLC reaction detectors has been proposed by Lawrence
et al.[133] for the same model systems described earlier[129] (see also Table 8).
The resulting simplification in apparatus design is seen in Figure 39 (as
compared to Figure 35). Three channels of the Technicon peristaltic pump
were used for the total operation and only one mixing unit required to mix
the column effluent with the reaction–extraction medium. In this convenient
two-phase system the ion-pairing process occurs instantaneously so that
no reaction coil is needed. The DAS solution is therefore segmented with
the extraction solvent prior to adding it to the column effluent immediately
after the column. The extraction takes place in a ten-turn glass coil (2-mm
i.d.) and is of about the same efficiency as the previously used 20-turn coil
in the three-phase system.[129] This corresponds to about a 1.4-min extraction
or residence time. Doubling this extraction time by using a 20-turn coil
resulted in a less than 5% increase in the signal. A reduction of the extraction
unit to a five-turn coil under otherwise identical flow conditions resulted
in a 25% drop in the fluorescence signal. The phase separation was carried
out with a conventional Technicon phase separator with a PTFE insert.

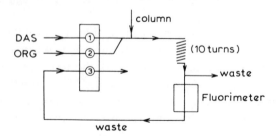

Fig. 39. Postcolumn ion pair extraction system using solvent segmentation. Flow
rates (ml/min): DAS, 0.32; organic solvent, 0.92 through the flow cell, 0.34; HPLC,
1.0.[133]

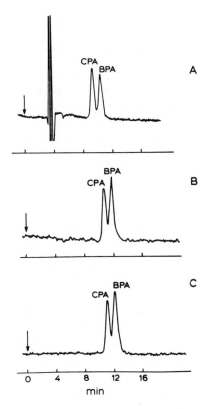

Fig. 40. Comparative chromatograms. (A) uv dete tion of chloro- and bromopheniramine (CPA and BPA, respectively) at 225 nm. (B) Fluorescence, air segmentation. (C) Fluorescence, solvent segmentation. Organic solvent, chloroform. HPLC conditions: CN column, 25% methanol–0.1 M NaH$_2$PO$_4$ as mobile phase; flow rate 1.0 ml/min.[133]

In the present setup, only about one-third of the organic phase is drawn into the detector cell, the reason for this being that the capillary inlet to the cell causes too much backpressure at higher flowrates. This can be changed with a redesign of the cell compartment.

The influence of the extraction detector units on band broadening was also investigated. In Figure 40, the use of uv and fluorescence (for ion pairs) detection in the separation of chloro- and bromopheniramine is compared.

The band broadening observed for the extraction detectors in comparison with direct uv detection was of the order of 12 sec, or less than 20%, for the two pheniramine peaks. As band broadening was shown to be less than 2 sec per ten-turn coil, it can be assumed that the phase separator is still the critical unit in the extraction detector. In addition, the band broadening caused by the ion pair extraction technique was found to be identical for the present solvent-segmented two-phase and the previously used air-segmented three-phase system, the former moreover seemingly giving a more consistent phase separation. As a consequence, the warmup time in the two-phase system, which is about 10 min, was considerably shorter than in the three-phase system, where about 1 hr was needed.

Regarding the dependence of band broadening and extraction efficiency on the ratio of aqueous to organic phase, a ratio close to unity was found to be optimal. Hence, most of the work was carried out under such conditions, which corresponds to a rate of about one extractant segment per second. However, we have observed that the ratio of aqueous to organic phase can be varied from about 0.5 to 1.5 without seriously increasing band broadening (12 ± 2 sec) or reducing the extraction efficiency. The quantitative potential of this technique was again satisfactory with 3.5% relative s.d. for multiple assay ($n = 4$).

A further simplification was possible by adding the ion-pairing reagent (DAS) to the mobile phase prior to the column and hence possibly separating the amines (Table 8) as ion pairs. This dual function of the ion-pairing process both for chromatographic selectivity and improvement of detection properties has been investigated.[138] The resulting reaction detector design is seen in Figure 41. A two-channel peristaltic pump can be used for this modification, or if work is carried out under high-pressure conditions a single conventional HPLC pump can be adapted to deliver the organic solvent since a solvent-segmented system is essentially not compressible.

In the same study the influence of the mobile phase composition (conc. of DAS, buffer, pH, MeOH) on separation and detection of the model compounds was studied and the possibility of using different chemically bonded stationary phases such as diol, CN, RP-2, RP-8, and RP-18 was investigated. Of particular interest was the influence of the polarity modifier

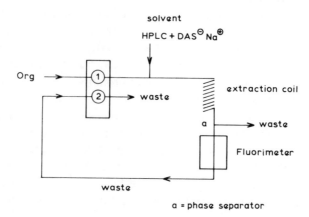

Fig. 41. Schematic apparatus for extraction detector with reagent added to the mobile phase.[138]

(MeOH) since high concentrations of MeOH (or acetonitrile) tend to increase the fluorescence background because of increased solubility of the DAS in the organic extractant.

A drastic influence on retention is exerted by the percentage of methanol in the mobile phase on all supports tested except on the diol column. Here, essentially nonretention conditions ($k' < 0.5$) prevail for all model compounds with and without DAS added.

Results for the nitrile and RP-2 supports are shown in Table 9. The addition of 10^{-4} M DAS to the mobile phase does not produce a significant change in k' values under any condition (with the single exception of SAN or RP-2 with pure water). It is interesting to note that the k'_{DAS}/k' values for RP-2 are all between 1.2 and 1.6, while significantly lower values of 0.9–1.2 are found for the nitrile support.

The essence is that with the flexibility in choice of mobile phase and of polar and nonpolar chemically bonded phases, the principle of the extraction detector should be applicable to a wide range of compounds. The flexibility of this technique is further demonstrated by an adaption of the extraction detector principle to normal phase chromatography.[139] The apparatus used under these circumstances looks just like the one in Figure 41 except that the eluent is now the organic phase (i.e., chloroform with a low percentage of MeOH as polarity modifier) and is segmented with an aqueous stream containing the ion-pairing reagent (DAS). The ion pair formation takes place at the organic–aqueous interface and the ion pair is transferred into the organic segment. Since the extraction equilibrium is established within a few seconds an extraction spiral is not needed and an efficient mixing unit is sufficient.

Since the two phases are immiscible, a regular pattern of organic and aqueous segments flows from the mixing tee, through the extraction coils. The volume of each segment was found to be about 1–2 μl. The two phases are then separated by means of a standard Technicon phase separator modified as shown in Figure 42 to minimize dead volume and assure a smooth separation. All of the aqueous phase plus a portion (about 50%) of the organic phase go to waste; the remaining organic phase is directed through the fluorometer for measurement. The considerable loss of organic extract is necessary for optimum detector performance (low noise levels).

The distribution coefficient of HA—as its DAS–HA ion pair—experimentally determined in the detector system is $D = 0.44$. Because of this relatively low value, the ratio of organic to aqueous phase could be expected to have a significant influence on the extraction of the ion pairs. When varying the flow rate of the DAS (aqueous phase) at a constant HPLC

TABLE 9. Effect of the Per Cent Methanol on Capacity Factors k' and on k'_{DAS}/k' [a] on RP-2 and CN-Stationary Phases

Compound tested	RP-2 k'			RP-2 k'_{DAS}			RP-2 k'_{DAS}/k'			CN k'		CN k'_{DAS}		CN k'_{DAS}/k'	
% MeOH	25	10	0	25	10	0	25	10	0	25	0	25	0	25	0
CPA	4.71	16.61	—	6.86	—	—	1.46	—	—	1.67	6.67	1.67	7.17	1	1.07
BPA	5.71	22.9	—	8.29	—	—	1.45	—	—	1.83	9.33	1.83	10.00	1	1.07
Diph	9.00	37.71	—	14.00	—	—	1.56	—	—	2.67	12.5	2.83	11.67	1.06	0.93
HT	2.14	7.29	—	3.14	11.3	—	1.47	1.55	—	0.67	1.67	0.67	2.00	1	1.20
HA	1.00	2.86	17.4	1.29	3.71	—	1.29	1.30	—	0.67	1.00	0.67	1.00	1	1
Scop	0.57	1.57	4.5	0.77	2.00	5.5	1.35	1.27	1.22	0.67	1.17	0.67	1.33	1	1.14
Atrop	1.14	4.57	20.0	1.71	8.00	—	1.50	1.75	—	0.67	2.00	0.83	1.83	1.24	0.92
SAN	0.57	0.71	0.86	0.71	1.00	7.14	1.25	1.41	8.3	0.67	0.83	0.67	1.00	1	1.20

[a] k'_{DAS} capacity factor for 16^{-4} M DAS in mobile phase.

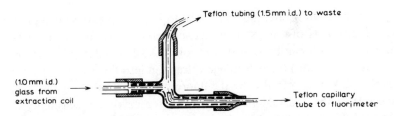

Fig. 42. Modified phase separator. 2.0-mm-o.d. Teflon tubing (dashed lines) was inserted through the lower portion of the glass phase separator. A hole was cut into it to permit the aqueous phase to escape. A small length of 1-mm-i.d. Teflon tubing (o.d. 1.5 mm) was inserted into the first tubing on the entrance side of the phase separator. Teflon capillary tubing to the fluorometer was inserted into the other end of the 2.0-mm-o.d. Teflon tubing as shown in the diagram.[139]

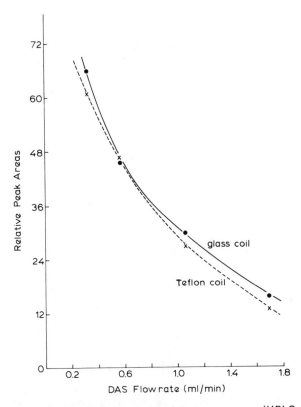

Fig. 43. Influence of DAS flow rate on detector response (HPLC flow rate 1.0 ml/min).[139]

(organic phase) flow rate, this was found to be true. Figure 43 compares the relative sensitivity of the system under such conditions, for 1-m-long Teflon and glass coils. It can be seen that at low DAS flow rates (high organic to aqueous phase ratio) the percentage extraction into the organic phase—and thus the peak area recorded by the fluorometer—increased significantly, irrespective of the coil material used. Thus, for maximum sensitivity a low flow rate of DAS should preferably be used (0.32 ml min^{-1} in the present study)—a requirement which was compatible with that of our band broadening studies (see also Figure 5).

Just as observed in the reversed phase mode the percentage of methanol in the mobile phase had a profound effect on the background signal. Figure 44 shows the increase in base line and noise when the methanol content varied from 0 to 30%. The increase in noise was not accompanied by a corresponding increase in extractability of the ion pairs of any of the com-

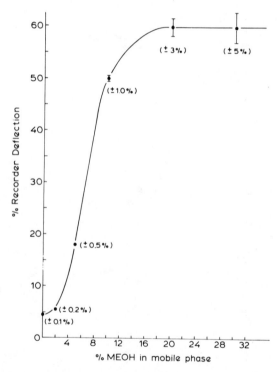

Fig. 44. Influence of methanol content of the mobile phase on noise and background fluorescence. DAS flow 0.32 ml/min. Mobile phase flow 1.0 ml/min. Numbers in parentheses indicate peak-to-peak noise. Measurements made at (or converted to) 10X attenuation.[139]

pounds studied. Thus, the signal-to-noise ratio went down considerably at higher methanol concentrations.

The most useful range was found to be 0–15% of methanol. Hence the conclusion must be that the alcohol content of the mobile phase should be kept as low as possible for best detector sensitivity. Evidently, this will somewhat limit one's choice of mobile phase for HPLC systems and, more importantly, make even step gradient elution in a relatively polar (greater than 15% alcohol) solvent–mixture range problematic. Increasing the concentration of DAS was also found to increase base-line noise. For routine operation, $10^{-4}\,M$ DAS solutions were used. For the rest, background noise was shown not to be affected by varying the flow rate at any of the DAS concentrations used, while varying the pH of the aqueous phase from 2.5 to 6.5 had no effect on either the signal of the compounds studied or noise. For routine use, pH = 3.5 was used.

The detection limits for HA, atropine, and ergotamine (as the DAS ion pairs) are 30, 40, and 100 ng, respectively. They are hence of the same order of magnitude as the values previously reported[137] for some related compounds in a reversed phase system when using a filter fluorimeter.

An improvement of at least one order of magnitude can be expected by using a monochromator instrument of the type PE-204.

To show the utility of the postcolumn ion pair extraction principle with normal phase chromatography a urine sample spiked with 2.0 ppm of HA was analyzed. Figure 45 compares uv detection and postcolumn ion pair extraction with fluorescence detection. As can be seen, the fluorescence result is much superior to that of uv analysis in selectivity, and in ultimate detection limit calculated to be less than 0.1 ppm by fluorescence and greater than 1.0 ppm by uv, at a 3:1 signal:peak-to-peak noise ratio. Recovery was found to be about 85% with the described extraction technique.

The large peak following the DAS–HA peak in both chromatograms is due to free DAS extracted form the sample, which partitions to a certain extent into the organic phase.

From the foregoing it could be concluded that the extraction–detector principle is just as feasible and can be as useful for normal phase chromatography as it is for the reversed phase mode.[133,138] Although in the present study, only one type of mobile phase (methanol in chloroform) and of reagent (DAS) have been used, recent experiments have shown that it can also be used for other ion-pairing reagents.

It should be realized that the prime value of the extraction detector coupled to normal phase chromatography is its high potential for the

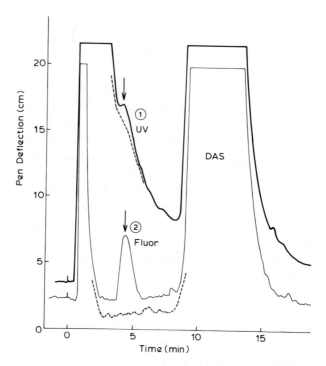

Fig. 45. Comparison of uv and fluorescence–ion-pair extraction for the chromatographic detection of hydroxyatrazine spiked in urine at 2.0 ppm. (1) uv results. 0.01
absorbance units full scale. Dotted line indicates blank. (2) Fluorescence–ion-pair
results. 10X attenuation. Dashed line indicates blank. DAS = response due to counter
ion. Mobile phase was 8% methanol in chloroform, 0.1 M in butyric acid. Arrow shows
hydroxyatrazine peak.[139]

handling of organic extracts such as obtained in residue and trace analysis
of complex matrices.

As to the solvent–segmentation approach one can conclude that it
is just as applicable to postcolumn reaction systems as an air-segmented
system.

The advantages of the present approach over the three-phase system
used for HPLC detection earlier[128,129] are obvious. The design is simpler
and the phase separation can be effected more conveniently. The residence
time per unit spiral volume is longer, as no volume is occupied by air bubbles.
The system is less temperature sensitive and more flexible with regard to
flow variations owing to the absence of compressible gas bubbles. For the
same reason, it should be possible to work at higher pumping pressures, so
that standard HPLC pumps and lines could conceivably be used.

Variation of the segmentation rate and the ratio of organic to aqueous phase can be flexible. The possibility for choosing an optimal organic solvent with regard to, e.g., viscosity, solubility, density, vapor pressure, and spectral characteristics seems to introduce another interesting aspect. As a consequence of all of these advantages it should be possible to minimize band broadening and to reduce background noise and accordingly, also to improve the detection limits and the reproducibility of this detection mode.

Finally, it should be realized that the solvent–segmentation principle offers many interesting possibilities in the development of reaction detectors for HPLC including, in addition to ion pairing and complex formation, the whole range of relatively simple and fast chemical reactions where the excess of reagent would interfere in classical reactor approaches.

In regard to actual chemical reactions with reaction times of several minutes the proof for the efficiency of a solvent-segmented system to supress band broadening was recently given for reactions (residence times) of up to 23 min.[140]

Dansylation (see Table 10) of secondary and primary amines has been taken as a fluorigenic model system. The plugs of organic solvent introduced in order to minimize band broadening also acted as reagent carrier, and as product carrier after the reaction had taken place at the interface.

Dansylations reactions have been well known for many years for prechromatographic fluorescence labeling of compounds with primary and secondary amino groups and phenolic OH groups.[1,141] For postcolumn reactions they were never considered since for complete reactions even at elevated temperatures reaction times on the order of 10 min were required for the type of compounds studied,[142] and the excess reagent in its hydrolyzed form gives strong fluorescence interference. Nevertheless, it would be desirable to adapt this reaction for a reaction detector since artifact formation can also be very troublesome with dansylation procedures[143] and since a good reaction detector for secondary amines and phenolics would be helpful.

The possibility of carrying out dansylations at an aqueous–organic interface was proposed earlier in connection with batch reactions[144] and was therefore adopted in this study to reduce the problem of fluorescence background. Unfortunately these interface reactions do not proceed as rapidly as reactions for similar compounds in a homogeneous medium,[145] as can be seen for example in Figure 46; except for β-histidine no plateau has been reached even after over 20-min reaction times at elevated temperatures.

In order to reach a reasonable detection limit, one therefore had to use as long a residence time as possible in the reactor. Still reasonable

TABLE 10. Structures of Model Compounds for Postcolumn Dansylation

β-Histine

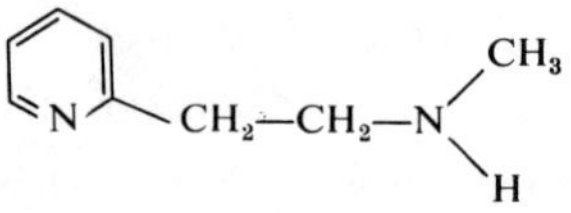

Clovoxamine

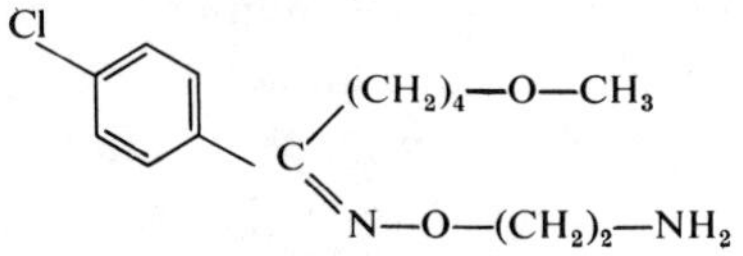

Fluvoxamine

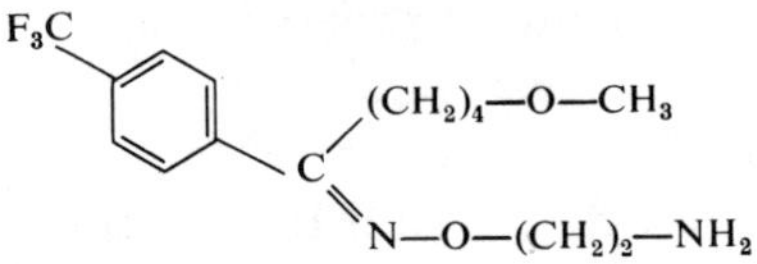

Dansylation of amines

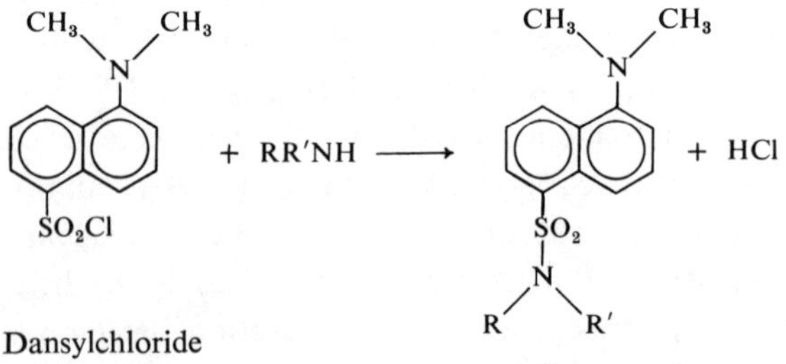

$+ RR'NH \longrightarrow$

$+ HCl$

Dansylchloride
nonfluorescent

fluorescent

Hydrolysis:

$$\phi - SO_2Cl + H_2O \longrightarrow \phi - SO_3H + HCl$$

(DNS-Cl)   Fluorescent

detection limits were obtained with practicable residence times for most compounds tested (Table 11) with the experimental setup described below. The apparatus used consisted again of a Technicon autoanalyzer system as set up for solvent segmentation (Figure 39). Dichloroethane was invariably used as the organic solvent which contained 350 μg/ml Dans-Cl. An Aminco

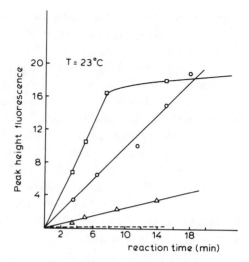

Fig. 46. Dependence of peak height of fluorescence signal on reaction time. Equimolar amounts of the amines were used. The peak heights are in arbitrary units.[140] □, β-Histine; ○, clovoxamine; △, fluvoxamine; - - -, emetine.

fluoromonitor was employed for the detector of the dansylated amines (excitation filter, 365 nm; emission cutoff filter, >450 nm). The T-piece used was a Technicon part (No. 1168034-01A10); the all-glass phase separator was homemade and had a PTFE insert similar to that described in

TABLE 11. Detection Limit of Amines (as Their Dansylamides) in a Two-Phase Flow System[a,b]

Compound	Reaction time, min	Reaction temperature, °C	Detection limit, ng
β-Histine	5	Ambient	1
Fluvoxamine	16	56	3
Clovoxamine	16	55	0.5
Ergotamine	6.5	54	8
Cephaeline	16	56	30
Emetine	16	56	30
Ephedrine	16	56	60
Histamine	6.5	54	600
Di-n-butylamine	10.5	54	18

[a] Reference 142.
[b] For structures see Table 10 and reference 145.

Figure 42. A Perkin-Elmer liquid chromatograph was employed to pump the carrier stream, which consisted of an aqueous 0.05–0.1 M sodium bicarbonate solution, containing 0–30% (v/v) of acetonitrile; the flow rate was 0.8–1.0 ml/min. A Rheodyne injection valve with 175-μl loop was used for plug injections.

In band broadening studies, a variable-wavelength uv detector (LC 55, Perkin-Elmer) was inserted between the injection valve and the reactor.

It is again evident that postcolumn band broadening is chiefly due to the phase separator and, possibly, the mixing tees. Even with residence times of below 1 min, band broadening, calculated as the standard deviation, σ_t, was in the order of 4–7 sec. Besides, it was also observed that relatively minor changes in T-piece and phase separator design or in the connecting parts effected substantial differences in dead volume and, consequently, band broadening.

For an example, on decreasing the inner volume of a glass phase separator by inserting PTFE tubing (Figure 42), thereby simultaneously effecting a smoother phase separation, the total band broadening of the system decreased from 11.1 to 9.4 sec; that is, the improvement due to the use of the modified glass phase separator was 6 sec. In another experiment, substituting a glass T-piece for a PTFE version having a reagent inlet with a somewhat wider internal diameter, caused a reduction of σ_t from 11.6 to 9.3 sec; this corresponds to an *additional* contribution of the PTFE T-piece of about 7 sec. The dependence of band broadening on reaction coil design was studied for different reaction times, such as are required for the dansylation reaction (see Table 12). The main conclusion to be drawn from these data is that varying the residence time from 2 to some 20 min does not result in a noticeable increase in band broadening. Neither does increasing the internal diameter of the capillary tubing from about 1 to 2 mm. Further, similar results were observed for glass, stainless steel, and PTFE coils. However, the use of PTFE as construction material is not recommended because of unfavorable wetting conditions.[139] In such cases phase separation was rather poor, with part of the water being drawn through the flow cell of the fluorescence spectrophotometer.

Another important observation was that working with tightly wound PTFE coils having a small internal diameter did lead to an increase in band broadening (cf. Table 10). This is due to secondary mixing effects which cause successive organic segments partly to merge, as could be observed visually. Calibration curves were linear over a three order of magnitude concentration range. Regression coefficients for clovoxamine and β-histine, for example, were 0.9957 and 0.9997, respectively. The repeatability was

TABLE 12. Dependence of Total Band Broadening on Reaction Coil Design. Model System: Dansylation of Fluvoxamine[140]

Coil material	Internal diameter, mm	Coil diameter, cm	Residence time, min	$\sigma_{t,\text{total}}$,[a,b] sec
Glass	2.0	2	2.5	9
			10.5	9
			12	9
			15	9
Stainless steel	1.1	25	2	9
			7	9
			16	9
			23	9
	2.0	11	5	10
PTFE	0.9	0.7	4	13
		12.5	4	9
		12.5	9	9
	1.1	6	4	9
	2.0	6	5	9

[a] Relative s.d., $\pm 5\%$.
[b] Measurements carried out at flow rate of 0.8 ml/min through detector flow cell.

better than 2% (relative s.d.) ($n = 6$), which again renders this type of detector a truly quantitative tool also for this system.

In conclusion it can be said that the present study demonstrates the feasibility of the solvent-segmentation approach for actual chemical reactions with relatively slow kinetics.

The solvent segments, besides acting as reagent carrier and extraction medium for the reaction product, also effectively suppress band broadening, hence rendering the approach suitable to detection in HPLC. The principle should also be of interest in autoanalyzer work where relatively long reactions are involved, since dilution is suppressed and hence sensitivity will be high; besides, sample throughput can be increased when no separation is necessary. The large contribution of mixing tees and phase separators to band broadening will necessitate further improvement and possible miniaturization of their design.

Electronic desegmentation principles could also be a feasible approach.

4. SPECIAL APPLICATIONS AND TRENDS

4.1. Coupling with Other Detection Modes

4.1.1. Electrochemical Detectors

Next to uv and fluorescence detection electrochemical detection techniques are gradually becoming more widely accepted as valuable alternatives for detection in HPLC. A recent review has given the current state of the art.[146] The same author has also presented an overview on chemical derivatization techniques coupled to electrochemical detection[147] in which the possibility of postcolumn reactions was discussed. The philosophy is again similar to the one applicable for other detection modes mainly to improve eluted species in their electrochemical detection properties. Another approach to overcome some of the limitations of coulometric and amperometric detection is to measure a change in the reagent which has been added. This aspect has been explored for the first time by Takata and Muto[148] by determing reducing sugars. The underlying reaction is given below:

$$\text{red. sugars} + [Fe(CN)_6]^{3-} \longrightarrow \text{ox. sugars} + [Fe(CN)_6]^{4-}$$

$$[Fe(CN)_6]^{4-} \longrightarrow [Fe(CN)_6]^{3-} + e^-$$
$$\longrightarrow \text{electrochemical detection step}$$

The same principle has also been reported by Kissinger et al.[149] for sulfhydryl compounds which are known to be difficult to oxidize directly on a graphite electrode but are easily oxidized by the ferricyanide complex in a homogeneous medium.

A recent account of this technique has been given by Deelder et al.[18] with the consequences for reactor design being discussed. Since the reaction occurred in alkaline solution at high temperature (reaction time = 2 min at 90°C), glass bead columns could not be used for this reaction because of the alkaline solutions attack on the packing material.

For this reaction a tubular reactor was used. The scheme of the reaction system is given in Figure 47. The sugars were separated on a strongly acidic cation exchange resin (Aminex A-9, Biorad, Richmond, Virginia) in the K^+ form. Deionized water was used as the eluent at a flow rate of 0.35 cm^3 min^{-1}. The column effluent was mixed with the reagent consisting of a solution of 0.01 M $K_3Fe(CN)_6$ in 0.3 M NaOH, flow rate 0.3 cm^3 min^{-1}. A coiled glass capillary tube 25 m × 0.254 mm i.d. was used as a reactor; the coil diameter was 2.5 cm. A coulometric detector as described by Lankelma

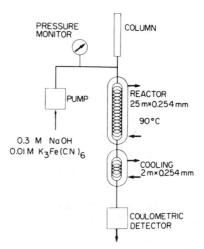

Fig. 47. Schematic diagram of the system for separation and detection of reducing sugars.[18]

and Poppe[150] was used for measuring the concentration of the reduced ion $Fe(CN)_6^{4-}$. The particular reaction mixture made it necessary for the detector to be slightly modified. The glassy carbon electrodes were replaced by platinum sheets. Moreover, an external calomel electrode was used instead of the internal reference electrode of the original design. The working electrode had a constant potential of $+0.5$ V. Figure 48 shows a chromatogram for a mixture of sugars. The additional band broadening in the reaction system was 5 sec and the pressure drop 18 atm. The band broadening was measured by injecting fructose directly into the reactor.

Such homogeneous redox reactions by the general equation of

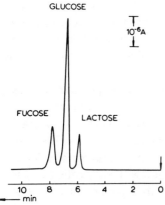

Fig. 48. Chromatogram for a mixture of sugars. Column 25 cm × 0.46 mm packed with Aminex A-9 cation exchanger (K^+ form). Eluent: water. Eluent flow rate, 0.35 cm³ min⁻¹; reagent flow rate, 0.3 cm³ min⁻¹. The fucose peak corresponds to 1.6 μg.[18]

followed by either

(a) reagent$_{red}$ $\longrightarrow$ reagent$_{ox}$ $+$ ne^- (solid state detector)

or

(b) analyte$_{ox}$ $\longrightarrow$ analyte$_{red}$ $-$ ne^- (Hg—drop detector)

should be generally applicable with a large number of compounds and would help to overcome the difficulty with many organic compounds reacting slowly at the electrode surface and exhibiting irreversible voltammetric waves with potentials clearly in excess of the conventional potentials. Since usually a large excess of reagent is used, for equation (b) to be applicable the excess reagent should not interfere in the detection. It is clear that redox reactions in which the reagent is the reducing agent should also be feasible.

Work along these lines are in progress in the author's laboratory. Another feasible procedure of postcolumn reaction was also mentioned by Kissinger[146] regarding the bromination of phenolics or unsaturated compounds such as pheromones, prostaglandins, etc. via bromination in a two-stage electrochemical reaction detector. Coulometrically generated bromine was used followed by an amperometric detection of the bromine consumed.

The coulometric generation is in general an interesting approach to reagent production in reaction detectors which for its advantages (microtechnique, good control, clean, use of unstable reagents, on-line working mode) would be well worthwhile in pursuing further.

On-line nitration or nitrosation is another potential approach.[151] These reactions carried out on aromatic nuclei give rise to large and well-defined reduction curves well suited for quantitative analysis. A large number of compounds have been reported for easy conversion to electroactive groups such as nitrosation of penicillins, nitration of drugs, i.e., glutethimide, phenobarbital, and diphenylhydantoin.[152] However, artifact formation poses particularly serious problems in nitration procedures carried out in a precolumn mode, hence postcolumn derivatization would be a feasible approach since the excess reagent does not interfere as others have shown.[151-153]

Electrochemical detection has also become quite popular recently in conjunction with enzymatic reactions and reactors.[147] For example, the product produced in the enzymatic assays for isoenzymes as discussed by Regnier et al.[84-86] is the fluorescent NADH, which is also well suited for electrochemical detection. The important place that amperometric detection takes in continuous-flow analysis in conjunction with enzymatic assay techniques is also clearly seen in the state-of-the-art report by Snyder et al.[91]

TABLE 13. Amperometric Methods Adaptable to Continuous Flow Analysis[a]

Analyte	Enzyme or substrate	Species monitored
Glucose	Glucose oxidase	O_2, H_2O_2
Urea	Urease	NH_4^+, HCO_3^-, $NH_3(g)$, $CO_2(g)$
Creatinine	Creatininase	NH_4^+, $NH_3(g)$
Uric acid	Uricase	O_2, H_2O_2
Cholesterol	(Cholesterol oxidase) (cholesterol hydrolase)	O_2, H_2O_2
Lactatic acid	Lactic dehydrogenase	NAD^+/NADH
Inorganic phosphorus	Alkaline phosphatase	O_2, H_2O_2
Lactate dehydrogenase	Lactic acid	NAD^+/NADH
Glutamic–pyruvic transaminase	Lactic dehydrogenase	NAD^+/NADH
Creatine phosphokinase	Hexakinase, glucose-6-phosphate dehydrogenase	O_2, $H_2O_2(I^-)NADP^+$/NADPH
Alkaline phosphatase	Phenyl phosphate polyphenoloxidase	O_2

[a] Reference 91.

in which the authors have given a tabulation of systems that fall into this category (see Table 13).

The enzyme reactor principle can be turned around and the substrate taken as the compound to be analyzed (analyte). Since enzymes are expensive and enzyme solutions rather unstable, the use of immobilized enzymes either in a fixed bed version on glass beads or in an open tubular version have been proposed as continuous-flow enzyme reactors. The potential of such an approach for HPLC detection has been studied and discussed recently by Schifreen et al.[154] and also by Adams and Carr.[155] They used an immobilized urease bed reactor as model system. The same publications[154,155] give a comprehensive overview of work done and demonstrate the potential of coulometric and other electrochemical detection devices in this area.

Although this work has not been done in conjunction with LC its potential for reaction detector design is undisputed. Enzymatic reactors would obviously represent a most sophisticated line of reaction detection in LC.

The simplest versions of postcolumn reagent–reactant interactions for electrochemical detection could be a modification of the mobile phase after separation has taken place. As mentioned earlier the most serious drawback of reaction detectors in HPLC is the limiting influence they exert on the choice of an optimal mobile phase; this influence is often even more pronounced in electrochemical systems where complex surface reactions are involved. Nevertheless when it is difficult to make a good compromise between separation and detection requirements postcolumn addition of suitable electrolytes, buffers, and solvents can drastically improve the situation, as has been demonstrated by several authors,[147–149] and may go to the extent of even permitting normal phase HPLC to be coupled to EC detection.[156] In a strict sense one could define essentially all coulometric and amperometric detection processes as chemical reactions and such a detector as a reaction detector with the reaction (oxidation, reduction, complexation, etc.) taking place at the electrode surface. It would certainly go beyond the scope of this chapter to discuss the field in this sense.

On the other hand, some very selective reactions which take place at electrode surfaces could well be described in this context. One example is the complexation reaction of sulfur compounds (i.e., thiourea compounds) on the surface of a mercury drop, to name just one. Such a method has recently been proposed by Hanekamp et al.[157]

A mercury drop detector described in detail in another publication[158] and depicted in Figure 49 has been used for this purpose. The most important criteria for this detector are summarized below:

The detector has a favorable time constant for all flow rates higher than 0.5 ml/min; at these flow rates the cell seems to behave as a mixing chamber. For nitrobenzene as a test compound the calibration curve computed via linear regression revealed an excellent linear dynamic range of three to four orders of magnitude. The detection limit was computed for a

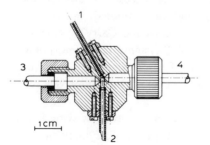

Fig. 49. DME detector.[158] 1, Inlet; 2, outlet, counter electrode; 3, reference electrode; 4, DME.

TABLE 14. Structures of the Thiourea Compounds Studied

Compound	Structure	Molecular weight
Allylthiourea	NH_2—$\overset{\overset{S}{\|\|}}{C}$—NH—$CH_2$—CH=$CH_2$	116.20
N,N'-Diphenylthiourea	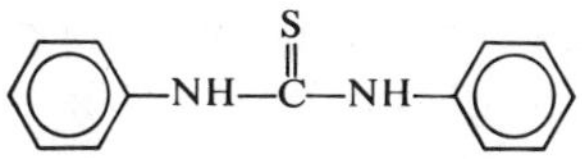 ⬡—NH—$\overset{\overset{S}{\|\|}}{C}$—NH—⬡	228.32
Ethylene thiourea	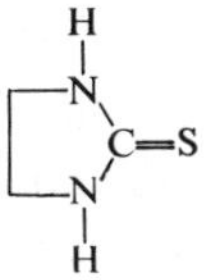 (cyclic) $\overset{H}{N}$—C=S, $\overset{H}{N}$	102.16
N-Methylthiosemicarbazide	CH_3—NH—$\overset{\overset{S}{\|\|}}{C}$—NH—$NH_2$	105.17
N-Phenylthiourea	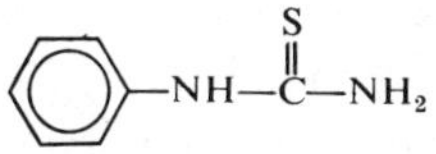 ⬡—NH—$\overset{\overset{S}{\|\|}}{C}$—$NH_2$	152.23
Thiourea	NH_2—$\overset{\overset{S}{\|\|}}{C}$—$NH_2$	76.12

signal-to-noise ratio of 2 and was 3 ng with no damping applied and 1 ng with a time constant of 0.3 sec.

The compounds studied are shown in Table 14. Ethylene thiourea (ETU), a suspected carcinogen, was of particular interest.

For this special application of the detector to thiourea (TU) compounds advantage was taken of the specific complexation of the thiocarbonyl functional group with the mercury drop surface. The overall reaction is as follows:

$$\text{anode:} \quad Hg^0 + 2TU \longrightarrow Hg(TU)_2^{2+} + 2e$$

As a result of this reaction a polarographic wave is obtained of which the limiting current is proportional to the concentration of the complexing species. A relatively low pH is required. The half-wave potential of the resulting polarographic wave is shifted towards more negative potentials compared to the free formation of mercuric ions.

The extent of this shift is dependent on the stability of the complex formed. All the compounds of the thiourea group which were tested gave a good wave and hence lend themselves nicely to this mode of detection.

As a favorable working potential $+190$ mV was chosen. This low applied positive potential has the advantages of not being critical with regard to oxygen, of giving a low noise, and of providing good selectivity.

A comparison of this detector (Figure 49) with other detectors for this group of compounds showed the mercury drop detector to be about equal or slightly superior to the uv detector but at least one order of magnitude less sensitive than a glassy carbon detector used in the oxidation mode. For ETU, for example, it was ~ 10 ng with the DME, 20 ng with the uv detector, and 1.3 ng with a glassy carbon detector.

Although the sensitivity of the DME detector is not as attractive as in the solid state detector, it is the selectivity of this detection mode which seemed particularly worth pursuing. Monitoring of ETU in urine samples, which is an important problem area and which so far has not been solved satisfactorily, was hence chosen as a practical example.

ETU was directly determined in urine under the chromatographic conditions shown in the caption to (Figure 50). Thanks to its high selectivity, the DME detector permits the measurement of ETU in urine without sample preparation, while with the wall jet electrode detector under otherwise the same conditions it is hardly possible to determine ETU in spite of its high sensitivity. Another problem with the solid state detector is caused by the decrease in its sensitivity that results from the poisoning of the electrode surface after multiple injections of urine. Already after one injection the response was observed to drop significantly. The linearity, sensitivity, and detection limit of the DME detector for a calibration curve for urine samples spiked with ETU are comparable to those in artificial solutions.

The detection limit was 10 ng for a S/N ratio of 3 with a noise of 0.1 nA.

A regression coefficient of 0.999 for a linear range from 0.092 to 1.108 μg and with a sensitivity of 0.032 (nA/ng) was observed. The reproducibility for this detection mode was 2.1% relative s.d. ($n = 10$) and the linear range was usually observed over two or three orders.

The selectivity of the complexation reaction of the thioureas with mercury can hence make it an interesting detection principle for many other organo sulfur compounds where the sulfur exhibits good complexation properties, i.e., proper oxidation state and steric configuration. A further advantage is the lack of interference of oxygen since we are working in the positive voltage range.

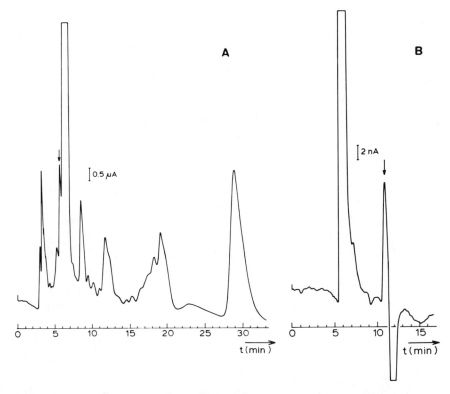

Fig. 50. (A) Wall jet electrode detector: chromatogram of urine (20 µl injected) spiked with 1.0 µg ETU (indicated peak). E_{Ind} = +1050 mV; chromatographic conditions: 1.2 ml/min; 1% (v/v) methanol in water, 10^{-2} M KNO_3, 5 × 10^{-3} M HNO_3. Column: 28-cm length, 4.6-mm i.d. RP-8, 10 µm.[157] (B) DME detector: chromatogram of urine (20 µl injected) spiked with 1.1 µg ETU (indicated peak). E_{Ind}: + 190 mV; chromatographic conditions: 0.9 ml/min; 1% (v/v) methanol in water, 0.1 M KNO_3, 0.02 M HNO_3.[157]

The principle for selective electrode surface reactions as demonstrated above can be extended in many directions as stated also by Heineman and Kissinger[147,159]:

There is a considerable effort at present to prepare chemically modified electrodes (CMEs) in which various "ligands" or potential electron transfer catalysts are chemically bound to a solid electrode surface (often graphite, or metal oxide semiconductors). The technology used is in most cases very similar to that employed for preparation of chromatographic stationary phases or chemically bound enzyme reactors. It would appear likely that this very active area of electrochemical research may provide significant improvements in the selective electrochemical detection of solutes which ordinarily would require a large overvoltage.

4.1.2. Other Miscellaneous Detectors

A number of other techniques have been developed for detection in HPLC[5] which might also lend themselves to postcolumn reaction technology but which are at the moment less easily available either for cost reason or lack of commercialization.

Mass spectroscopy (mass detector principle) is probably the highest up on the line and closest to a wider commercialization. A state-of-the-art report has recently been given by Arpino and Guiochon.[160]

One of the disadvantages so far of this detection mode has been the difficulty of introducing normal flow rates of aqueous streams into the MS source without splitting and the difficulty in having high buffer concentrations such as frequently encountered in ion exchange or reversed phase chromatography. Karger *et al.*[161] have adopted the extraction detector principle described earlier[129] to overcome some of these difficulties by coupling a Technicon microextraction unit between the column and the MS detector. It is easy to see the extension possibility of this principle to actual derivatization reactions suitable for mass detection by adopting, for example, the techniques described in references 137–140 to this end.

2,4-Dinitrophenyl (DNP) derivatives are useful for the separation and the subsequent MS identification of amines.[162] Most of the derivatives give reasonably intense molecular ions and an ion at m/e 196, which is usually the base peak formed by α-alkyl fission. A similar TLC separation and MS identification is also possible by converting aliphatic amines into 4-nitroazobenzene-4-carboxylic acid amines.[163] The molecular ions are relatively intense (the second most abundant ion) and the base peak corresponds in all cases to a fragment at m/e 254.

Dansyl (DNS) derivatives can also be used for MS identification and have been employed for the confirmation of a large number of biogenic amines.[164] The molecular ions of these derivatives are usually intense and are often accompanied by an ion at m/e 170 or 171 as the key fragment.

The mass spectra of DNS derivatives of chlorophenols[165] and hydroxybiphenyls[167] have been recently reported, and intense ions similar to those from the amine derivatives were observed at m/e 170.

The use of electron donor–acceptor complexes has been reported for the visualization and MS identification of compounds separated by TLC.[166,167] The mass spectra of the donor and of the acceptor can be separated by using different probe temperatures.

Radiochemical detection could be another mode for convenient adoption in reaction detectors. Very little work has been carried out on radiochemical derivatization for analysis of trace amounts of materials in combination with LC. The technique has the advantage of being both selective and sensitive. The main advantage is that the sample background does not cause interference in the detection as it does in most other methods and which necessitates some degree of cleanup. Also, the reactions used are those for normal derivatization procedures, the only difference being that the reagent is radiolabeled and that appropriate precautions are required for radioactive substances. Consequently extraction detector principles[129,137] could also here be adapted to detect the radiolabeled compounds following a post-column labeling process. A continuous-flow scintillation counter would be best suited for this purpose.[168] Several known reactions would be suitable for this approach, i.e., the iodination of unsaturated compounds with radioactive iodine (^{125}I and ^{131}I).[169] Fatty acids and barbiturates with unsaturated side chains were used as model systems. The detection limits obtained with this procedure at a signal-to-noise ratio of 3:1 are 20 ng of butalbital per spot using ^{125}I and 5 ng per spot with ^{131}I. The reproducibility of these techniques is $\sim 2\%$ (relative standard deviation) for concentrations greater than 0.5 µg per spot and $\sim 4\%$ or higher for amounts less than 0.1 µg per spot.

The method is generally applicable to a number of unsaturated compounds; others, such as ergotamine and codeine, exhibit some uncontrollable secondary reactions and therefore cannot be measured. In general, the reaction is additive, i.e., the number of double bonds determines the sensitivity of the method. Derivatization of amino acids with (^{14}C)DNS-Cl is very useful for protein sequencing.[170,171] The reactions involved are the same as described for the fluorimetric derivatization techniques.[2]

Low concentrations of thiol groups can be detected by reaction with $CH_3\ ^{203}HgNO_3$.

The removal of excess of reagent from the labeled thiol can be achieved again via dynamic microextraction. This technique has been applied to the analysis of proteins which contain thiol groups.[172]

The quantitation of compounds by acetylation with (^{14}C)acetic anhydride may be another useful approach, particularly where sample matrices cause problems in other common methods. Acetylation reactions with ^{14}C reagents are well known and have been applied to many types of compounds.[173] There could still be a long list of other potential applications let alone the possibility of using fast complexation or ion pair formation reactions by simply adding radioactive metals or radiolabeled ion-pairing reagents, i.e., ^{14}C–picric acid, to suitable groups of compounds. The benefits of such techniques to the bioanalytical field cannot be doubted.

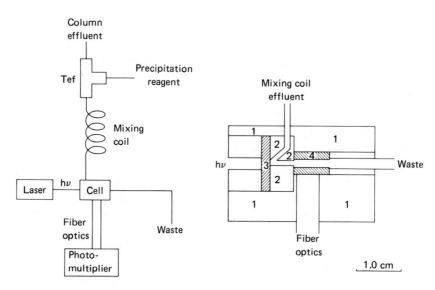

Fig. 51. Schematic diagram of detection system and simplified schematic of light-scattering detection cell. (1) Stainless steel detector block; (2) stainless steel entrance aperture; (3) Pyrex window; (4) Pyrex scattering cell.[174]

Nephelometry has been another principle suggested for a reaction detector by Jorgensen *et al.*[174] This novel detection system for liquid chromatography works on the basis of a postcolumn solute precipitation and light-scattering measurement of the formed precipitate. Its efficiency has been demonstrated by the sensitive detection of nonpolar lipids of biological origin separated by reversed phase chromatography. A linear relationship between the square root of the detector response and sample concentration was found for several lipid solutes. Detection limits in the submicrogram range were reported.

The schematic of the apparatus used and the detector design is shown in Figure 51, which also shows a more detailed schematic diagram of the detection cell assembly.

Solutes enter a glass light-scattering cell (approximately 17-μl volume) through a stainless steel aperture. A 0.5-mW helium–neon laser (model 155, Spectra Physics, Mountain View, California) operating at 633 nm provides the incident light beam. A laser was chosen as the light source to provide easy alignment with the cell and thus minimize background scatter from the inner walls of the cell. Scattered light is transmitted from the cell to the photomultiplier tube (type R 372, Hamamatsu, Japan) via fiber optics (Edmund Scientific, Barrington, New Jersey). Although the fiber optics are arranged at 90° to the incident laser beam, the cell geometry and fiber optics allow collection of scattered light from a larger solid angle. Associated electronic equipment is a high-voltage power supply (model 244, Keithley, Cleveland, Ohio) for the photomultiplier tube, and an electrometer (model 616, Keithley) for signal amplification.

Detection parameters were investigated with nonpolar lipids using a 30-cm × 3.9-mm-i.d., μBondapak C_{18} reversed phase column (Waters Assoc., Milford, Massachusetts). The mobile phase consisted of a 2:1 (v/v) mixture of acetone and methanol. Postcolumn precipitation of solutes was accomplished by adding an ammonium sulfate solution to the column effluent. The resulting change in solvent composition was sufficient to precipitate nonpolar lipids. Cholesterol, cholesterol esters, and triglycerides were used as standards.

High-pressure reciprocating piston pumps (model M 6000, Waters Assoc.) were employed to deliver both the mobile phase and the precipitation agent. Complete degassing of both solvents was found necessary to prevent the formation of bubbles upon their mixing. All investigations were carried out using a mobile phase flow rate of 1.0 ml/min.

It can be expected that the detector response is affected by both solvent composition and dynamic conditions of detection. Hence a compromise must be sought between separation and detection conditions. Optimalized conditions are shown in Figure 52, which gives a chromatogram for an extract of human serum according to Folch et al.[175]

While in this work response factors are about the same as for refractive index detectors, one can definitely assign advantages to this approach. The instrument design can likely be improved considerably, for example, by using pulseless pumping and better cell geometry, to yield at least a one order of magnitude better detection limit. The selectivity of reaction detectors of this type will permit a minimum of sample handling. As to the general potential of this detection mode Novotny[174] gives the following quotation, which the author, based on his own experience in the field, can fully support:

The present study gives an example of precipitation based on a decreased solute solubility due to an altered solvent composition. However, this represents just one mode of forming precipitates, and further applications of the light-scattering detection principle in conjunction with solute chemical alterations should be feasible. For example, many classical analytical precipitations could be used to detect various ions. Another method of precipitation may involve the reaction of solute leading to an insoluble product or by-product. An example is the reaction of permanganate with an oxidizable solute to produce insoluble manganese dioxide.

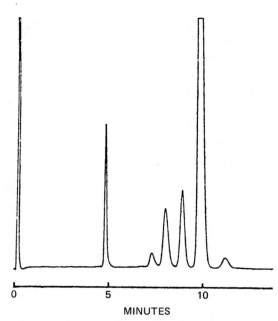

MINUTES

Fig. 52. Chromatogram of nonpolar lipids of human serum (sample size equivalent to 50 µl serum). Conditions: column flow rate, 1.0 ml/min; precipitation agent concentration, 0.025 M ammonium sulfate; precipitation agent flow rate, 0.5 ml/min. Detected peaks are various triglycerides. The first peak (retention time, 5 min) is cholesterol, the other peaks are triglycerides and cholesterol esters.[174]

In conclusion, the light-scattering detector appears to be a valuable addition to the family of LC detectors, and it may serve well in cases where other existing detection techniques fail. Through the adaption of many of the qualitative precipitation reactions generally known from the literature, various levels of selectivity can be attained. These may range from very specific and functional group detections to "selectivity" in a broader sense. It should be emphasized that a maximum utilization of the light-scattering detector will always be strongly dependent on proper design of postcolumn chemistry.

Flame, oven, and plasma techniques form another group which the author considers of some future interest in conjunction with derivatization techniques. Several groups have pioneered the use of atomic absorption spectroscopy (AAS) for detector purpose in HPLC. A short overview of the status has been given by Fernandez.[176] Applications, for example, to chelating agents[177] as chromium organometallics and lead alkyls[178] or copper complexes of amino acids[179] have been discussed. For reaction detector technology there remains of course the handicap that when, for example, adding metal ions for complexation and detection of chelating species the excess of reagent (i.e., metal ions) will interfere. Application of the extraction detector principle[137–139] will again provide a solution. Complexing species can then be separated either by reversed phase or normal phase chromatography followed by addition of an excess of suitable metal ions prior or after the column and the chelates selectively extracted into a suitable organic extractant. Chelating agents, amino acids, peptides, complexing N-, O-, or S- containing drugs or pesticides would be just some of the systems that could be adapted. Commercial AAS burners will have to be adapted somewhat in order to handle organic phases as has been suggested by Jones et al.,[180] but if this is done aspirator efficiencies and consequently sensitivity can be better than with aqueous media. The use of gradient elution techniques should pose no problems with this extraction detector combination.

Flameless AAS was also recently propagated for HPLC detection.[181] The use of this commercial flameless (graphite furnace) atomic absorption detector (GFAA) automatically coupled to a high-pressure liquid chromatograph (HPLC) was demonstrated to provide element-specific separation and detection of organometallic compounds at nanogram concentrations in both protic and nonpolar solvents using conventional columns. Relative sensitivities of the HPLC–GFAA system for compounds of arsenic, lead, mercury, and tin were shown to be mainly functions of LC flow rate and relative AA sensitivity for each element. Separation of mixtures of organometal ions with both isocratic and gradient elution on reverse phase columns was possible for achieving complete resolution of different organometal species. The GFAA detector was operated in either a rapid sampling mode, providing higher resolution of the HPLC effluent, or in a batch survey mode. Brinkman et al.[181] demonstrated the possibility of full automation of such a system. The sensitivities being one to two orders of magnitude better than for a corresponding flame detector would further enhance the usefulness and feasibility of an AAS–extraction detector combination. Typical chromatograms obtained with this type of a detector as compared to uv signals can be seen in Figure 53 for organomercurials.

Fortuitously, for mercurial analytes, concurrent operation of the uv detector at 254 nm provides a selective, moderately sensitive indication of mercury-containing species, probably by a photodecomposition reaction occurring in the uv cell, which preceded the GFAA unit.

Flame emission offers a further alternative in the same vein and complementary to AAS. The emission technique is particularly attractive for the detection of phosphorus- or sulfur-containing compounds.[182] The

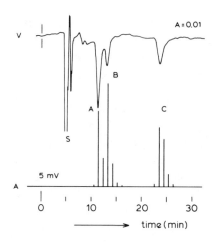

Fig. 53. HPL chromatogram run under isocratic conditions shows speciation of a 20-µl three-component solution (in water–methanol, 1:1) of trace mercurial ions: peak A, Hg^{2+} (235 ng as element); B, CH_3Hg^+ (235 ng); C, $C_2H_5Hg^+$ (260 ng); S solvent. Column: "Lichrosorb" C_8; mobile phase: 0.01 M ammonium acetate in water (94%) plus methanol (4%) containing 25 ppm 2-mercaptoethanol; flow rate, 0.30 ml min^{-1}; column pressure, 470 p.s.i.[181]

principle is based on flame excitation in a hydrogen-rich flame and has been widely used for detection in gas chromatography.[183]

In the design discussed by Julin *et al.*[182] up to 5 ml/min of the total column effluent is nebulized and directed into a cool hydrogen–nitrogen flame. Phosphorus- and sulfur-containing compounds emit characteristic emissions at 526 and 383 nm, respectively, which are measured by a simple bandpass filter–photomultiplier system. In favorable situations, the apparatus permits the detection of about 2×10^{-8} and 2×10^{-7} g/ml of phosphorus and sulfur, respectively, in the column effluent. However, with optimization of the system, sensitivity may be increased by an order of magnitude.

The flame emission detector can be used for compounds containing phosphorus and sulfur in different forms. Here the detector was utilized for the assay of 5'-monophosphate nucleotides by reversed phase chromatography. Excellent short- and long-term quantitation has been experienced in trace analysis. In this work, application of the detector has been limited to systems with aqueous mobile phases since significant concentrations of organic materials affect detector sensitivity; but again this is a technical problem which can conceivably be solved by redesigning the burners.

The introduction of suitable groups responding well with such a detector would again be feasible in a postcolumn reaction mode coupled to the microextraction principle. The problem with nonmiscible organic solvents to be introduced into a flame emission detector can also be solved by adapting a transport-type detector such as marketed by Pye Unicam (Cambridge, Great Britain) to this end. This possibility has been described by Compton and Purdy,[184] who described the modification of a Pye Unicam flame ionization detector via conversion of the FID to a thermal ionic detector (TID) by adding a rubidium silicate glass bead above the FID unit.

The transport detector principle is in general an interesting aspect in conjunction with the microextraction technique since a much larger portion of the effluent or carrier stream consisting of volatile organics can be taken up by the moving wire (band) (see also reference 159).

Plasma emission technology, for an extension of what has been said here, can also prove interesting in this regard. The principle of inductively coupled plasma in optical emission spectroscopy has been reviewed by Fassel and Kniseley.[185] The possibility of using this principle for HPLC detectors has recently been suggested by Kraak et al.[186] with a number of organometallics. The advantage of a plasma over a flame would be the enhanced detection limits and the broader spectrum of elements that can still be handled. Other than that all arguments mentioned above are still applicable.

Electron capture detection (ECD) widely used in GC has recently also been proposed for HPLC work.[187,188] Very little use has yet been made of this detection principle in LC for which a commercial instrument (Pye Unicam) is now available. An appraisal of its usefulness has been given by Brinkman et al.[188] One of the major drawbacks is obviously its limitation to highly pure, nonpolar organic solvents such as octane or heptane with only small quantities of polarity modifier. Adaption of this principle to reversed phase or ion exchange LC would mean an enormous widening of its potential, and simple postcolumn reactions to introduce electron capture active groups such as halogens or nitro groups could then be effected easily to improve detection properties. In analogy to previous work[161] the extraction detector principle[137] can again provide interesting solutions. Work in this direction is in progress.

4.2. Coupling with Other Techniques

4.2.1. Preconcentration and Precolumn Technology

In many trace analytical procedures particularly in chromatography a true preconcentration step is unavoidable prior to quantitative analysis to attain the necessary concentration of compounds for the detection step. The classical techniques such as freeze drying, extraction and/or evaporation, steam distillation, etc. have their serious limitations in terms of recovery, capacity, and possible loss of sample during processing, particularly for temperature-sensitive compounds. Preconcentration techniques via adsorption on solid surfaces have therefore been used for some time in conjunction with chromatographic techniques, particularly GC. The best known is probably the carbon adsorption method (CAM), reviewed recently by Suffet and Sowinski.[190] The use of chromatographic-grade untreated silica gel or possibly also alumina for trace enrichment purposes has been known for some time already in conjunction with classical column chromatography and cleanup procedures as well as with thin-layer chromatography. In the latter the process can be further enhanced by using intermediate drying steps. Huber and Becker[191] have recently treated this aspect specifically for modern column liquid chromatography. They named their approach enrichment by displacement LC, but then one has to transfer the enriched sample to another column for actual separation. Nevertheless some of the considerations in their paper are of general validity for the preconcentration process on solid substrates. The major conditions are that an adequate adsorbent be used and that the sample be applied from a suitable solvent assuming a high initial adsorption.

A high enrichment factor will hence be obtained if the chemical difference between solvent and solute is quite large. With proper conditions chosen, i.e., p-cresol in cyclohexane, up to a 10,000-fold concentration factor could be obtained.[192] The potential of such preconcentration or trace-enrichment techniques has been considerably expanded with the emergence of chemically bonded silica gels. Particularly hydrophobic surfaces such as obtained on the commercially available reversed phase materials (C_8, C_{18}) lend themselves nicely to the collection of a wide range of organics. In 1974 Kirkland[193] discussed the possibility of using preconcentration phenomena on the top of analytical HPLC columns for the purpose of trace enrichment. Little and Fallick[194] reported on the use of very large injection volumes of up to 200 ml for relatively nonpolar compounds applied from an aqueous solution to a C_{18} reversed phase column. Similar results were reported for peptides and alkaloids injected from aqueous solutions.[195-197] The following can serve as an explanation.

When relatively nonpolar organic species are injected from an aqueous solution onto a hydrophobic surface they will become immobilized until the elution strength of the solvent mixture is increased. This means that the compounds are concentrated into a very small zone on top of a reversed phase chromatographic column (trace-enrichment effect). The components can then be eluted with a suitable eluent with very little band broadening. This phenomenon can be observed nicely in Figure 54. Different nonapeptides ranging in retention from $k' = 1.5$ to $k' = 11$ were injected in 340-μl volumes, and the influence of this large injection volume on peak

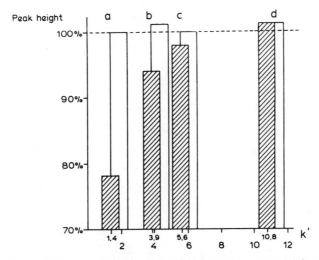

Fig. 54. Effect of injection volume on peak height (the amount injected is kept constant). Peak height: - - - -, resulting from a 34-μl injection (= 100%); ▨, resulting from a 340-μl injection (solution tenfold diluted with mobile phase); ☐, resulting from a 340-μl injection (solution tenfold diluted with water). *a*, Lypressin + ornipressin; *b*, oxytocin; *c*, felypressin; *d*, demoxytocin.

height was investigated. No reduction in peak height was observed when aqueous samples were injected, but when the peptide is dissolved in the mobile phase prior to injection the ones with low retention start moving during the injection step and a drastic reduction in peak height (large band broadening) is observed.

The trace-enrichment phenomenon was studied in more detail[198] with larger injection volumes. Figure 55 shows the injection of 150 ml of "pure" distilled water (*A*) and again a 150 ml water sample containing 6 ppb of dihydroergotoxine (*B*). Volumes of this size had to be injected with a pump. Noticeable is the large shift in base line and the concentrating effect for impurities contained in the distilled water. The disadvantage of such an extreme preconcentration technique is also obvious from Figure 55 in that not only the compound of interest but also many interferences are preconcentrated; hence, for example, the dihydroergocryptine peak contains also the impurity peak from water and could not be evaluated. Besides, a drastic displacement of the base line will often occur.

In such situations one can adopt the selectivity of a reaction in a postcolumn reaction detector to eliminate many interferences. This possibility of combining the preconcentration approach with postcolumn derivatization has been mentioned.[40] A typical example is shown in Figure 56 for the injection of a nonapeptide. With 20 μl of a oxytocine solution injected the

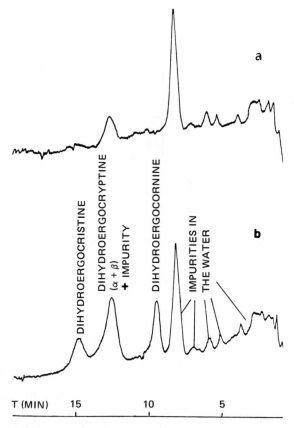

Fig. 55. Chromatogram for dihydroergotoxine mesilate. (*a*) Injection of 165 ml of distilled water; (*b*) injection of 165 ml of distilled water containing 6.7 ppb of dihydro-ergotoxine mesilate. Chromatographic conditions: Nucleosil C_{18}, 5 µm; column, 10-cm × 3-mm i.d.; acetonitrile–0.1 M ammonium carbonate solution (40:60); flow rate, 0.97 ml/min; detection, 280 nm.[197]

advantage of the Fluram derivatization is already quite obvious since upon sufficient amplification even by-products can be quantitated. With an injection volume of 1.77 ml (∼90-fold concentrating factor) the selectivity aspect with the use of the postcolumn reactor is clearly observable. While in the uv trace a quantitation of oxytocine becomes difficult owing to uv active interferences, the fluorescence trace still looks identical to the one obtained for a 20-µl injection.

Precolumn technology. The direct injection of large amounts of sample and (or) very dirty samples onto an analytical column can exert much strain on this column. Frequently it is observed that after a few injections of, for example, biological fluids or extracts, the separation performance of the

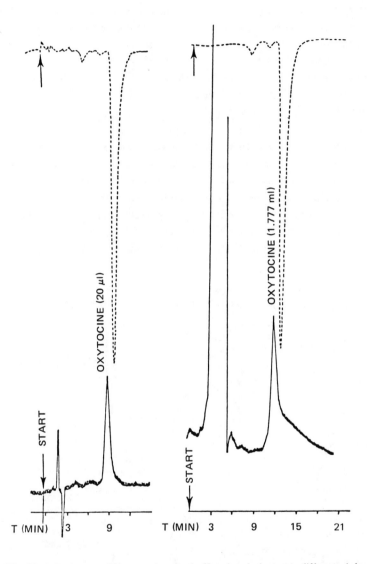

Fig. 56. Chromatograms of oxytocine in buffered solutions at different injection volumes (loop 20 μl and 1.777 ml) on a column of 5-cm length. ———, uv detection at 210 nm prior to derivatization; - - -, fluorescence detection after derivatization with Fluram. Conditions: solution of 30 mg of Fluram in 100 ml of acetonitrile; flow rate, 0.02 ml/min; detection conditions $\lambda_{ex} \simeq 360$, $\lambda_{em} \simeq 470$ nm. Chromatographic conditions: Merck RP-8, 5 μm; column, 4-mm i.d.; mobile phase, acetonitrile–water 20:80 (pH 7); flow rate, 0.2 ml/min; column thermostated at 24°C. The absolute amounts of oxytocine injected are the same for both injection volumes (0.5. I.U.).[40]

column drops drastically. HPLC columns are expensive and high-efficiency columns are valuable. The use of short precolumns has therefore been recommended for protection of the analytical column.[199] These arguments are even more valid when trace enrichment is the ultimate goal. The precolumn can in addition serve in a cleanup function, for preconcentration and as a sampling and storage device. Such an approach has recently been discussed in conjunction with pollution methodology.[199] The instrumental setup is described in Figure 57. Another application of this precolumn concept has been proposed by Lankelma and Poppe[200] for the analysis of the cytostatic drug methotrexate in plasma. Preconcentration was done on C_8 reversed phase material and the separation was carried out on a silica-gel-based chemically bonded anion exchanger. Contrary to the approach discussed in Figure 57 backflushing was used in the latter technique. A somewhat different approach was chosen by Ishii et al.[201–203] for the pre-concentration of phthalates from aqueous samples. Miniaturized HPLC

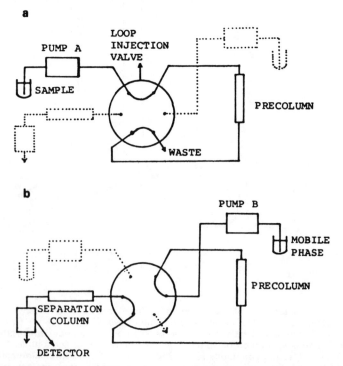

Fig. 57. Apparatus scheme for precolumn trace enrichment technique: (a) Precon-centration step; (b) separation step. Pumps A and B Altex model 100 piston pump (Altex, Berkely, California); loop injection valve Altex; precolumn: 3-cm length, 4-mm i.d., slurry packed with reversed phase RP-18 5 μm material (Merck, Darmstadt, West Germany).[198]

equipment was used[202] and correspondingly the column dimensions were 0.5-mm i.d. and 18-mm length. This limited the sampling rate to <100 µl/min. Sampling occurred off-line and the precolumn was dried with air prior to connection to the separation column. The long sampling times required with such a device for the detection of actual concentrations of phthalates in polluted river water would make this miniaturized version rather unsuitable for such problems but might render it valuable for bioanalytical applications where smaller sample volumes are used. This has been demonstrated by Ishii *et al.*[202] with an example of corticosteroid analysis in serum samples in the 20–130 ppb concentration range. 200 µl of serum containing the corticosteroids were preconcentrated on Hitachi gel No. 3010 20 µm and after air drying separated on a silica gel column with dichloromethane-methanol (97:3 v/v). The air-drying step has distinct advantages when using such widely different phases for the preconcentration and the separation step. In the situations described before[199,200] a drying procedure is not needed. It should be emphasized that both techniques can lend themselves to automation and routine analysis.

In all these cases relatively long precolumns packed with small particles have been used. This is in many instances a waste since with very high k' only the upper few millimeters of the sorbents are used for the loading process. Long columns have also the disadvantage of a slow sample throughput due to high backpressure. In a recent study therefore an attempt was made to work with very short precolumns having sorbent layers of only 2 mm. The idea was to actually go toward a filter concept useful for field sampling or, for example, pollutants. The work was again carried out with phthalates as a model system and samples of up to 1 liter of such water samples containing between 0.1 and 2 ppb of the phthalates were preconcentrated and analyzed with excellent recoveries and reproducibility.[204]

A new precolumn design has been described. The same type has also been used later for trace analysis in biological samples, i.e., chlorpheniramine in urine. Chlorpheniramine had been preconcentrated on a 2-mm layer C_{18} 10 µm material from an untreated urine sample of 200 µl. The precolumn acted at the same time as a cleanup device and the drug was transferred on-line to the analysis column (see Figure 58).

Although the precolumn has an appreciable cleanup effect which permits the detection of the drug with uv detection (compare Figures 37 and 58), it is again the combination of these described technologies with reaction and extraction detectors which mutually will amplify the advantages of both techniques. Details on the detection procedure and a schematic of the apparatus (Figure 35) have been given earlier (see also reference 129). In

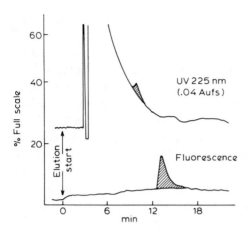

Fig. 58. Comparison of ion pair fluorescence and uv (225-nm) chromatograms of a 100-µl urine sample containing 1.0 ppm chloropheniramine, after preconcentration on a LiChrosorb RP-18 precolumn. The cross-hatched peaks illustrate the response due to the presence of chloropheniramine.[129]

some cases up to 10 ml undiluted and untreated urine samples have been analyzed with this procedure[138] for traces of a triazine metabolite (see Table 8).

Preconcentration of 10-ml samples of relatively polar compounds such as HA on 2-mm LiChrosorb RP-18 layers can already be quite critical with regard to recoveries due to breakthrough of the HA.

The reason for this probably is not an overloading phenomenon. HA and many other metabolites are often significantly more polar than their parent compound and would have a higher mobility (lower k'). With large sample volumes this mobility will permit them to move through the entire precolumn during the injection step, and eventually this results in a loss of the component. The polarity can be reduced by forming an ion pair and preconcentrating the compound as an ion pair. This was done in the case of HA by adding DAS to give a 10^{-3} M solution. The effect of this phenomenon on an actual chromatogram is shown in Figure 59 with a more than 50% gain in recovery when injecting the HA in the DAS ion pair form.

On-line coupling of a preconcentration on an ion exchange bed and a reaction detector has been studied by Adler et al.[205] for polyamines.

Putrescine, spermine, and spermidine were analyzed in urine and whole blood. The samples were hydrolyzed with barium hydroxide and neutralized with sulfuric acid. The polyamines were concentrated and separated from amino acids on a small bed of ion exchange resin that then served to load the samples on a two-channel, automated ion exchange chromatography apparatus with detection based on the ninhydrin reaction. As many as 110 samples can be analyzed in a 24-hr period.

The particle size of the acid-type ion exchanger was >12 µm but band broadening on this medium pressure analyzer was not critical. The selectivity of this combination permitted also a relatively simple sample treatment.

A similar technique was discussed for amino acids[206] in up to 1-liter water samples with preconcentration of the acids on a 30–60-mesh particle ion exchanger bed (Lewarit S-100 cation exchanger). Transfer of samples was off-line by necessity and the acids were analyzed on a Biotronic amino acid analyzer with orthophthalaldehyde labeling and fluorescence detection. Needless to say the on-line operation could easily be done with the technique described earlier.[204]

The selectivity of such a preconcentration effect can also be further enhanced by using on-column derivatization techniques, i.e., preconcentration of the compounds of interest on a precolumn and selective derivatization and elution of the derivatives on the same column. Such an approach has been proposed, for example, by Maitra et al.[207] for the development of a gentamicin assay in serum. O-phthalaldehyde was used as a selective fluorigenic reagent.

The principle of on-column trace enrichment and chemical derivatization followed by a postcolumn reaction offers again many possibilities. Chao et al.[208] have described such a technique for the trace determination of nitrite in water. The nitrite was converted to an alkyl nitrite by passing an acidified solution through a bed of packed beads (XAD-2) coated with 1-decanol. The latter acted both as a reactant and as an extractant. The

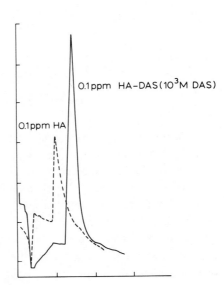

Fig. 59. Resulting chromatograms for the preconcentration and analysis of 25 ml of urine. - - -, Direct preconcentration; ——, preconcentration after the addition of 10^{-3} M DAS. HA spiked at 0.1 ppm. Chromatography conditions; column RP-2,10 μm, length 10 cm, i.d. 4.6 mm. Mobile phase consisting of 25% MeOH–0.1 N NaH$_2$PO$_4$, pH 3.5. Flow rate, 1.0 ml/min.[138]

0.1ppm HA–DAS(10^3M DAS)

0.1ppm HA

reaction proceeds by rate-limited diffusion of nitrous acid into the alkanol phase where it is rapidly esterified. Esterification by this pathway was quite rapid and efficient. The product, decyl nitrite, was retained in the alkanol phase where it is resistant to hydrolysis because of the limited solubility of water in the alkanol phase. Under optimized conditions, 68% of the nitrite present in the original water solution was retained on the column as decyl nitrite. Decyl nitrite was eluted from the column with acetone and then converted to highly colored azo dye by sequential reaction with sulfanilamide and N-(1-naphthyl)ethylenediamine. The chromophore was monitored spectrophotometrically. Acetone limits for nitrite quantitation by this method were on the order of <200 pg/ml.

Another example where the precolumn not only serves as an enrichment tool but also can act directly as a reaction medium has been suggested by our own research group. The reactions involved concerned urea herbicides which can be preconcentrated from an organic extract on a precolumn filled with silica gel (SI-100 10 μm or TLC-grade material). The precolumn is heated to $\sim160°C$ and catalytic hydrolysis of the urea on the silanol groups of the silica gel surface occurs within 10–15 minutes without reagents added. The proposed reaction is shown below:

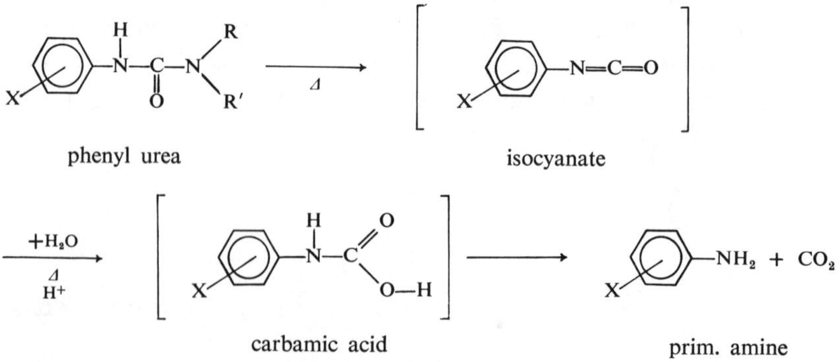

phenyl urea   isocyanate

carbamic acid   prim. amine

This technique was first worked out for thin-layer chromatography[209] whereby the aniline was *in situ* derivatized with Dans-Cl followed by separation and fluorescence detection. The same approach was later developed for column technology[210] with on-column dansylation similar to Maitra et al.[207] and separation of the Dans derivatives by reversed phase HPLC and fluorescence detection.

In an other technique[210] the substituted anilines were transferred to a reversed phase column for separation and reacted in a Fluram reaction detector similar to one described earlier.[39,40] The selectivity of this method permits residue analysis of, for example, linuron, diuron, and metoxuron in soil to be carried out without a Florisil cleanup.

On-line chemical derivatization can also occur between two separation steps, so-called two- or multidimensional technology, such as proposed by Erni and Frei[211] and by Johnson et al.[212] with an on-line coupling of one column system with another. In order for this transfer of zones to occur with a minimum of band broadening it is necessary to have a preconcentration on the second or subsequent column. In the above situations gel permeation chromatography as a first separation step was combined with a reversed phase system. The schematic of such an apparatus is shown in Figure 60.

Other combination possibilities such as GPC with organic solvents followed by adsorption chromatography or ion exchange separations cou-

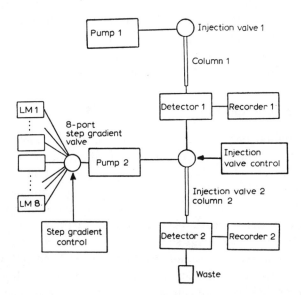

Fig. 60. Schematic diagram of on-line two-dimensional HPLC with step gradient elution and automatic injection control of loop 2. Pump 1: Lewa model FL1 (Lewa, Herbert Ott G, Leonberg, Stuttgart, West Germany). Injection valve 1: Valco loop, 7000 p.s.i., 50 μl (Valco Instruments, Houston, Texas). Column 1: stainless steel, 200 × 0.4 cm, filled with CPG (controlled pore glass), 200–400 mesh, 88A/113/A 170 A'240A (Electro-Nucleonics, Fairfield, New Jersey). Recorder 1: W + W model 600 recorder (Kontron, Zurich, Switzerland). Injection valve 2: Valco loop, 7000 p.s.i., 1.777 ml. Control unit for the injection valve: home-made, two type RDF time relays (Summerer, Zürich, Switzerland). Pump 2: Altex model 100 (Altex, Berkeley, California). Step gradient valve: Labotron eight-port valve, No. 2581 (Kontron). Step gradient control unit: home-made, RS 21 × PG time relays (Comatelectric, Worb, Switzerland). Column 2: stainless steel, 25 × 0.4 cm, filled with Nucleosil C_{18} reversed phase material, 5 μm (Machery, Nagel & Co., Düren, West Germany). Detector 2: L 55 uv detector (Perkin-Elmer). Recorder 2: W + W model 600 recorder (Kontron).[24]

pled with reversed phase chromatography could also be feasible and the introduction of an on-line reaction step between the two chromatographic systems to change the detection and chromatographic properties of the compound would be technically feasible and enhance the information content.

An example for the feasibility of such a two-column approach with a reaction step between the two columns was proposed via an enzymatic technique by Usher and Rosen[213] for the analysis of oligoadenylates differing at the 3'-terminus. The HPLC support was RPC-5 reversed phase material prepared according to an earlier publication.[214] The first column separated $A_{\parallel}Ap$ plus $A_{\parallel}A > p$ from $A_{\parallel}A$ and at the start of the second column a layer of bacterial alkaline phosphatase enzyme converted the $A_{\parallel}Ap$ into $A_{\parallel}A$. Hence this $A_{\parallel}A$ emerged separately from the original $A_{\parallel}A$ and from the $A_{\parallel}A > p$. The technique can be used to analyze a three-component mixture for a single chain length or a mixture of $A_{\parallel}Ap$ and $A_{\parallel}A > p$ of mixed chain lengths ($n = 3$–7).

Although the technology used was not of the newest origin, this example supports the feasibility of such an idea.

4.2.2. Gradient Elution

By definition one would consider it difficult to couple reaction detectors with gradient elution. The reaction medium would be continuously varied and it would be difficult to imagine a good reproducibility for the detection step. This is certainly true for continuous gradients and with reactions which are sensitive to small changes in eluent composition, and a large portion are. With relatively insensitive reactions and (or) reactions with low reagent background signal even a continuous gradient may be applicable, as shown for example by Moye et al.[127] for OPA reactions. If problems indeed exist then an alternative is given by the adoption of step gradient technology where it can be assumed that a reaction equilibrium is established relatively quickly for each step and that the signals are at least reproducible for each individual step.

This possibility has been demonstrated for the Fluram reaction,[39,40] which has quite a high fluorescence background, increasing rapidly with increasing acetonitrile concentrations. The result is shown in Figure 61 for the reversed phase separation of a complex mixture of peptides, using a step gradient, described by Erni et al.[215] and depicted in Figure 60. The reproducibility of the chromatographic pattern is excellent and this combination gives truly a quantitative technique for the system described. It

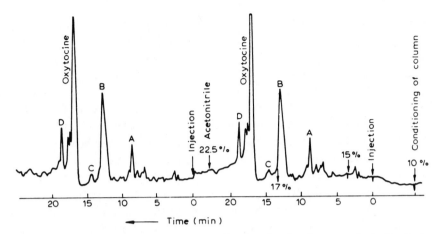

Fig. 61. HPLC of a peptide mixture (purification fraction for oxytocine). Duplicate chromatogram. Step gradient: acetonitrile, 10–22.5% (pH 7). Other conditions as in Figure 8.[56]

can be certainly assumed that many of the other reactions described would be just as amenable to coupling with step gradients. In certain cases, i.e., routine analysis of large series of samples one can even see the possibility of automating a system such that reagent is only added during the gradient step(s) in which the compound(s) of interest appear. This can result in significant time and reagent saving.

4.2.3. Catalysis

The use of elevated temperatures to enhance the kinetics of a reaction is quite common in reaction detector know-how. Some authors have even proposed the use of a pressure cooker effect for significant acceleration of reactions.[25,26] Heterogeneous and homogeneous catalysis can also provide elegant solutions to the improvement of the reaction kinetics. For heterogeneous catalysis an example has been cited with the catalytic hydrolysis of compounds such as ureas on silica gel surfaces.[209] Besides cutting down the time requirement from a few hours to a few minutes the absence of agressive reagents was another advantage. This procedure is, however, hardly feasible for a postcolumn reaction since the solvent medium would likely prevent the catalytic action of the silanol group.

Vratny et al.[216] have published an alternative approach by using a reaction postcolumn packed with a small-diameter strongly acidic cation exchanger in the H+ form (14.6 ± 2-μm i.d.) for the hydrolysis of sugars on a solid surface without addition of reagents. The separation of the non-reducing oligosaccharides was done on an ion exchange column containing the same material in the Ca^{2+} form with deionized water as a mobile phase. Both separation and reaction column were operated at 80°C. Hydrolysis of these sugars occurred at different rates, but a residence time of 48 sec in the reactor column resulted in, for example, a 97% conversion for sucrose, 67% for raffinose, and 40% for stachyose, while essentially no hydrolysis occurred with trehalose. Longer residence times can be achieved by lowering the flow or by using longer reaction columns which if well packed contribute relatively little to band broadening (see Section 2 on Theoretical Aspects). The hydrolyzed sugars were then reacted with p-hydroxybenzoic acid hydrazide according to Lever[218] and detected colorimetrically.

It should be mentioned that this is one of the first studies which demonstrates that a bed reactor can also be used as a reaction column and directly participate in the reaction either in a catalytic fashion or as a reagent donor (H^+ donor) or both, such as may well be the case with this hydrolysis column. The efficiency of this column remained unaltered even after 50 hr of operation.

One can see many more such applications, i.e., with metal-loaded supports to catalyze redox reactions.

An example of the use of homogeneous catalysis in postcolumn derivatization where the analyte acts as the catalyst has been shown by Nachtmann et al.[217] Zero-order kinetics conditions prevailed during a fixed detection time. The analytes in this case were the hormones tetra- and triiodothyronine, which catalyze the redox reaction between cerium(IV) and arsenic(III) to yield a decoloration due to disappearance of Ce(IV) at 365 nm. This detection principle could be of general use for organoiodine compounds and is selective and sensitive to the subnanogram level:

$$2Ce^{4+} + As^{3+} \xrightarrow{\quad I \quad} 2Ce^{3+} + As^{5+}$$

The reaction is pseudomonomolecular.[217] The kinetics are described by the equation

$$-\frac{d[Ce^{4+}]_t}{dt} = k[Ce^{4-}]_t \cdot [I] \tag{20}$$

Integration gives

$$\ln[Ce^{4+}]_t = \ln[Ce^{4+}]_{t=0} - k[I]t \tag{21}$$

The concentration of Ce^{4-} can be monitored by a spectrophotometer. One can then write

$$\ln A_t = \ln A_0 - k_A[I]t \tag{22}$$

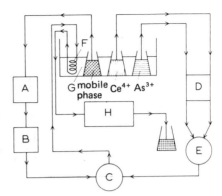

Fig. 62. System for optimization of the catalytic reaction. (A) Pump for mobile phase; (B) injection system; (C) mixing device; (D) pump for reagent solutions; (E) mixing device; (F) reaction capillary; (G) thermostat; (H) spectrophotometer.[217]

where A is absorbance. If all of the parameters in a continuous system are kept constant, the concentration of iodine is proportional to the change in absorbance:

$$\ln A = b\,\mathrm{I}$$
$$\ln A = \ln A_0 - \ln A_t \qquad (23)$$
$$k_A t = b$$

Equation (23) is the basis of this investigation. The term b represents the sensitivity of the reaction, if it is used for the determination of iodine.

Conditions that are important in this method are the use of constant flow pumps (no pulsation), a thermostatted system, and working with good reproducible timing.

A schematic of the apparatus is shown in Figure 62. For the addition of the reagent a pulseless infusion pump was recommended and a mixing unit made of PTFE with a geometry as recommended earlier[39,40] was used to keep band broadening low. The limiting aspect for the method was the band broadening which had to be accepted in the reaction vessel (a capillary of 15-m length made of PTFE with 0.5-mm i.d.). Much could be gained by using a segmented stream approach to reach still higher sensitivities at lower zone dispersion. The technique is limited to aqueous mobile phases and reversed phase systems containing not more than 50% organic phase. Addition of acetonitrile caused a reduction in sensitivity but gave a longer linear dynamic range. In spite of all these limitations the technique is by far the most sensitive yet obtained for these thyroid hormones (subnanogram region) and can still further be optimized. A chromatogram of three of the hormones is shown in Figure 63 at 3-ng concentrations; the limiting factor here is the noise from the reagent delivery and HPLC pumps. The

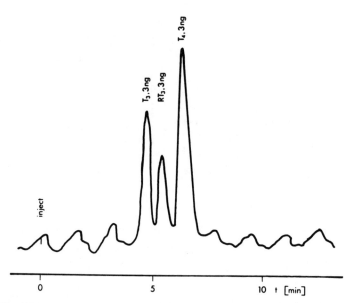

Fig. 63. HPLC separation and catalytic detection of three iodinated thyronines. Reagent pump: Ismatec PMP 10 Duo. Solvent system: water–acetonitrile (5:2) + 1% of acetic acid. Column: Nucleosil C_{18}, 5 μm, 15-cm × 3.2-mm i.d. Flow rates: mobile phase, 1 ml/min; reagents 1 ml/min. Injection volume: 100 μl. Temperature: 30°C. Length of reaction capillary: 15 m. Detector: Schoeffel SF 770 at 365 nm, range 2.[217]

selectivity of this catalytic function should render the method suitable for the monitoring of plasma samples with only a simple handling step.

This study by Nachtman *et al.*[217] shows the feasibility of this approach. Many of the known catalyzed reactions[219] in metal trace analysis could probably be adopted for detection purposes in dynamic systems. Since zero-order reaction conditions prevail and the systems are quite sensitive to flow and temperature fluctuations, much of the know-how gained from enzyme reaction systems can be adopted for such catalytic detection techniques.

Obviously the compound to be detected does not always have to be the catalyst. The use of a catalyst added, for example, with the reagent to speed up a conventional reaction can also help much to facilitate a reaction detector design and to improve a detection method.

4.2.4. Photochemical Reactions

Another attractive approach in postcolumn derivatization is the use of photochemical reactions. This can be done with different goals in mind, one of which is the improvement of the reaction time of conventional reactions. Iwaoka and Tannenbaum[220] have used a photochemical reaction for the detection of *N*-nitroso compounds. In their work the nitrosamines were hydrolyzed photohydrolytically by irradiation with a long-wave mercury lamp. The liquid stream was contained in Pyrex glass capillary tubing placed in the irradiator. The hydrolysis product was nitrite which was determined on line by the Griess reaction (see also reference 221). The residence times here were of the order of several minutes, hence considerable band broadening was observed. The use of segmentation principles would thus also be interesting for relatively slow photochemical reactions. The same principle of a photochemically induced hydrolysis was again discussed recently by Snyder and Johnson[222] with a review of work done in this line. A xenon arc lamp was used.

The nitrite produced was concentrated by adsorption on an anion-exchange column, stripped with 0.01 M HClO$_4$ and detected on a flow-through platinum electrode after mixing with a stream of 9 M HCl.

Nitrite produced by photolysis of nitrosamines is reduced at a platinum electrode in 4.5 M HCl at an electrode potential in the range 0.7–0.4 V by a cathodic process limited by convective-diffusional mass transport. The reaction involves one electron per molecule of HNO$_2$ and NO is produced:

$$R_2N - NO + OH^- \xrightarrow[\text{5 mM NaOH}]{h\nu} NO_2^- + R_2NH$$

$$H_2O + NO^+ (NOCl) \xleftarrow{\text{4.5 } M \text{ HCl}}$$

$$NO^+ + e \longrightarrow NO$$

In order to prevent poisoning of the Pt surface by coadsorption of organic by-products the electrodes were treated with iodine ions, which adsorb preferentially.

For the photolytic conversion to nitrite within a 30-sec residence time about 70–90% of *N*-nitrosodimethylamine and *N*-nitrosodiphenylamine were produced, respectively, in 5 mM NaOH. Calibration curves and detection limits were investigated for this detection mode and were typically in the 5-ng to 1-µg linear range for *N*-nitrosodipropylamine with a detection limit of 1 ng. The technique was also applied to food samples but without the use of a chromatographic separation system.

The above study shows the potential of coupling photochemical reactions with electrochemical detection as a reaction detector principle in dynamic systems.

Although it has not been used here in conjunction with HPLC, there is little doubt that this approach can have an excellent potential for such an extension.

Another goal for the use of photochemistry for detection is the production of derivatives with favourable detection properties which cannot be produced easily by other means. This can be done with or without reagent addition to the eluent stream, and if reagent has to be added it can often be added prior to the column directly to the eluent, hence avoiding band broadening from the reagent mixing step.

Twitchett et al.[223] were the first group to use this principle without reagent addition to develop a sensitive and specific technique for the detection of cannabinoids in body fluids. Cannabinol, for example, converts within a few seconds irradiation time to a highly fluorescent photoproduct. The medium is isooctane–dioxane (3:2). No reagent is added. Detection limits in the subnanogram range can be obtained but one of the major features is the high selectivity which permits quantitation of such compounds in body fluids without cleanup.

In this work a 100-W medium-pressure mercury arc was used for an intensive irradiation of the HPLC effluent stream. The cooling medium was water, and a quartz capillary conducted the carrier with the compounds of interest around the lamp. The conversion of cannabinol (I) according to the following equation to the fluorescent product (II) occurs within 1–5 sec and goes to completion:

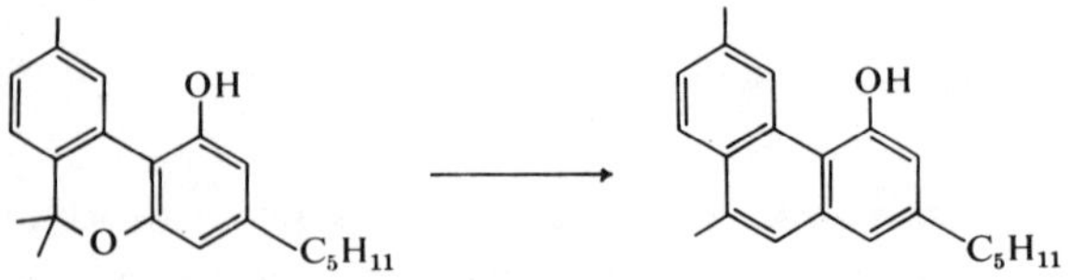

The technique was used with normal phase separation on Partisil PAC with isooctane–dioxane (3:2) as mobile phase. A urine hexane extract was injected and the photoproduct measured at $\lambda_{ex} = 258$ nm and $\lambda_{em} = 362$ nm.

The same authors[223] have mentioned briefly the possibility of using this technique for selectively destroying the fluorescence of LSD and the use of this method for positive identification of this drug in plasma samples. Although the selective loss of fluorescence as a function of a photochemical process is obviously less interesting from an analytical point of view, it can be a powerful and fast qualitative tool.

This aspect has recently been explored further for the group of ergot-alkaloids.[224]

The fluorescence signal of alkaloids decreased within about 20 sec of irradiation and disappeared selectively from complex chromatograms. Stoll and Schlientz[225] have investigated the photochemical reaction of ergot-alkaloids and proposed the reaction as seen below:

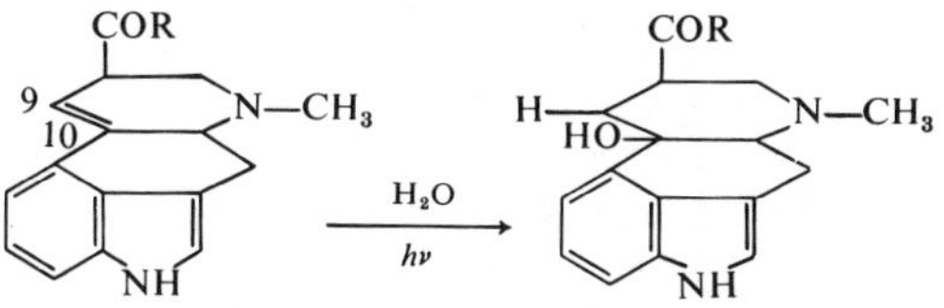

ergotamine   lumiergotamine

whereby no fluorescence is observed for the lumi derivatives of ergotalka-loids. This loss of the fluorescence upon irradiation with uv light has been investigated for ergot and 9,10-dihydroergotalkaloids amd seems to be a specific reaction for this group of compounds. It can therefore be used for identification.

The photochemical reactor was somewhat altered from previous designs.[223] An air-cooling principle was adapted for greater ease of exchanging different reaction spirals.

The reactor consisted of a lamphouse and power supply (Siemens, 5 NS 1102) with a XBO 150 W/1 xenon high-pressure lamp. Around this lamp a coiled quartz capillary of 1-m length, 0.5-mm inner diameter, and coil diameter of 6 cm was placed. Under the lamphouse an aluminium foil enhanced the reflection of the emitted light.

Cooling was done with pressurized air, which was previously cooled by solid CO_2 in a Dewar vessel.

For selective identification of the ergotalkaloids a decrease of at least 90% in fluorescence signal was necessary. This was best obtained with a low flow rate (e.g., 0.5 ml/min), a low percentage of acetonitrile, and a high pH. However, the amount of acetonitrile and the pH strongly influenced the separation; therefore the variation possibilities were limited. A good compromise was made using the following conditions: 58:42 (v/v) 0.01 M $NaHCO_3$/acetonitrile at pH 2.2 at a flow rate of 0.5 ml/min. At these conditions the fluorescence decrease was 90–99% for 17 ergotalkaloids and dihydroergotalkaloids.

The technique was then applied to a spiked urine sample for dihydro-ergotoxine (see Figure 64).

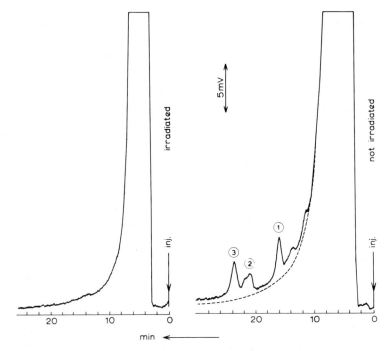

Fig. 64. Urine sample, spiked with dihydroergotoxine λ_{em} = 280 nm, λ_{ex} = 340 nm; flow rate, 0.5 ml/min; recorder 5 mV; inj. 100 μl. (1) Dihydroergocornine (46 ng/ml); (2) (α + β)-dihydroergocryptine (46 ng/ml); (3) dihydroergocristine (46 ng/ml); - - -, blank urine.[224]

A similar selective identification was possible for all the ergotalkaloids tested. Although it is suspected that the mechanism proposed earlier is not necessarily followed under the high-density irradiation conditions used here, one can conclude that the described technique is of general applicability to ergotalkaloids.

Preliminary results suggest that the drastic fluorescence decrease may be at least partly due to a change on the indole ring, since indole exhibits the same behavior upon irradiation. The technique is suitable for qualitative work in complex samples, as can be seen in Figure 64. It may also be interesting for cross-contamination checks in the pharmaceutical production processes. Nevertheless what one is really interested in is the production of signals rather than their suppression.

In an attempt to further explore the potential of photochemical reaction detectors several model compounds were investigated by our group (see Table 15).

Clobazam, a 1,5-benzodiazepine, is a new antianxiety agent[227,228];

TABLE 15. Structures of the Investigated Mode Compounds for Photochemical Reaction Detection

		Before irradiation, nm		After irradiation, nm	
		λ_{ex}	λ_{em}	λ_{ex}	λ_{em}
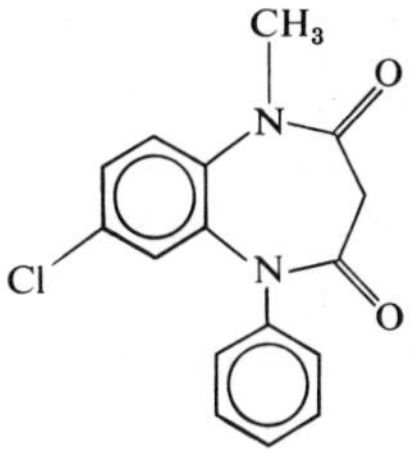	Clobazam	—	—	356	395
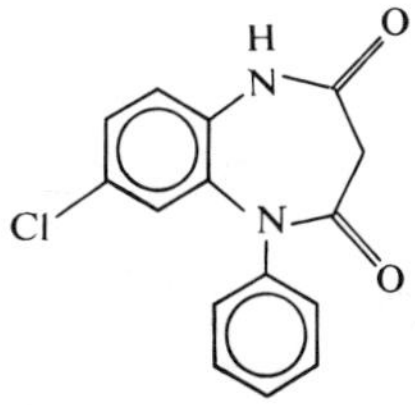	Desmethylclobazam	—	—	356	395
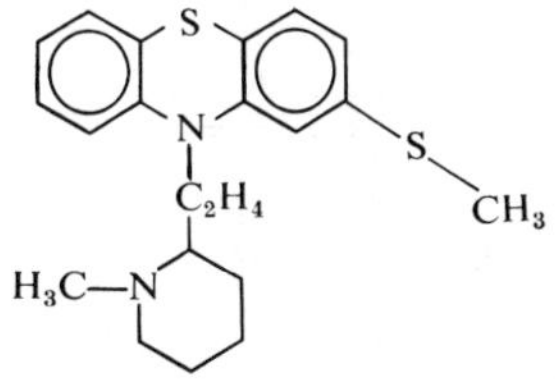	Thioridazine	325	455	340	378
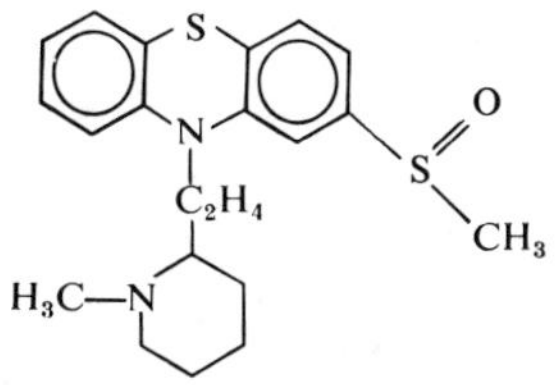	Mesoridazine	325	495	345	385
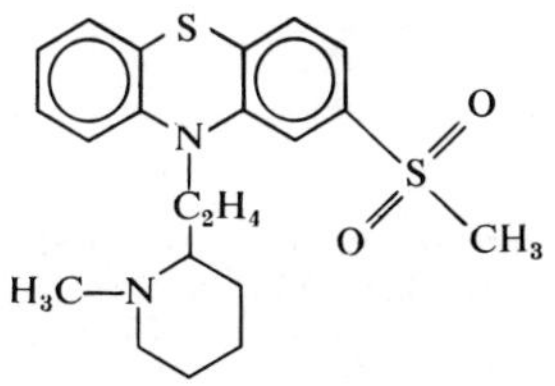	Sulforidazine	340	505	355	405

desmethylclobazam is its major plasma metabolite, which possesses psychosedative and anticonvulsant activity itself. The phenothiazines used were thioridazine, mesoridazine, and sulforidazine, a group of well-known psychopharmaca.[229] It was the aim of this investigation (i) to study some basic parameters such as band broadening of photochemical reaction detectors coupled on-line to a high-performance liquid-chromatographic (HPLC) system, and (ii) to demonstrate the potential of photochemical detection by developing sensitive and selective analytical methods for the determination of some of the model compounds in urine and plasma. The first step consisted of a design improvement of the photochemical reactor in comparison to previously used models.[223,224] The design of the reactor (Figure 65) was slightly different from that used in an earlier paper.[224] The main improvements were the use of (i) a 200-W Xe–Hg lamp as the light source, (ii) a vacuum cleaner, which sucks air through the reactor and thus effects better and more reproducible cooling, and (iii) an aluminium

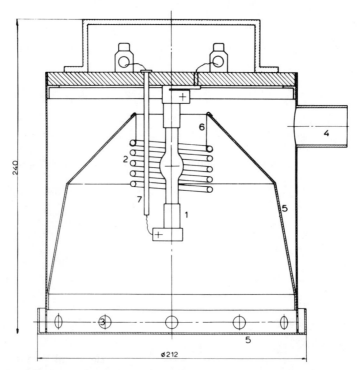

Fig. 65. The photochemical reactor. (1) 200-W Xe–Hg high-pressure lamp (Hanovia, Nema Electonics, Amsterdam, the Netherlands); (2) quartz capillary 0.5-mm i.d.; (3) inlet for air cooling; (4) outlet for air cooling; (5) aluminum reflection shield; (6) clip; (7) glass isolation.[226]

TABLE 16. Comparison of Fluorescence Intensity of Phenothiazines Before and After UV Irradiation[a] [(226)]

Compounds	Fluorescence intensity		Gain factor
	Nonirradiated	Irradiated	
Thioridazine	17.5	204	11.5
Mesoridazine	70	160	2.3
Sulphoridazine	102	327	3.2

[a] Conditions: mobile phase, methanol–0.01 M pH 5 acetate buffer (1:1); ϕ, 0.5 ml/min; reaction time, 50 sec; solute conc., 10 μg/ml; plug-injection volume, 20 μl; detection at optimal wavelengths.

reflection shield which encloses the lamp and capillary. Uv irradiation of clobazam and desmethylclobazam, which themselves are nonfluorescent, resulted in the formation of highly fluorescent compounds within minutes (see Table 15).

The phenothiazines studied exhibit native fluorescence. Upon irradiation, a photochemical oxidation process ensued, as reported earlier,[(230,231)] which resulted in a shift of the emission maxima to shorter wavelengths and an increase of the fluorescence yield (Table 16). From Table 16 one also reads that the native fluorescence increases from thioridazine to sulforidazine, i.e., with increasing oxidation state. As is to be expected, irradiation produced the largest gain in fluorescence in the case of thioridazine, which is in the lowest oxidation state; however, even with sulforidazine a distinctly higher signal was obtained, which suggests that the sulfur atom in the ring system can be oxidized relatively easily.

The influence of the carrier stream on the fluorescence signal was also studied with a view toward preferable mobile phase systems for HPLC; three organic solvents, viz. methanol and acetonitrile, and their mixtures with water, were used. The general trend observed was an increase in fluorescence yield from acetonitrile to methanol, that is, with increasing polarity of the solvent; as is to be expected, the effect is less pronounced at lower concentrations due to the lower optical density of the irradiated solution. Although organic nonpolar phases, as has been shown earlier,[(223)] can also be used in photochemical reaction detectors, reversed phase systems with polar solvents seem more ideal for this purpose. These results had to be verified in an actual chromatographic system. Since the carrier stream now

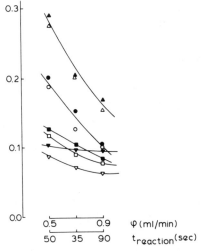

Fig. 66. Influence of the water content in the mobile phase. For clobazam: △, (1:1, v/v) methanol–0.01 M sodium acetate (NaAc); ▲, (6:4, v/v) methanol–0.01 M NaAc; ○, (4:6, v/v) acetonitrile–0.01 M NaAc; ●, (1:1, v/v) acetonitrile–0.01 M NaAc; for desmethylclobazam: ▽, (1:1, v/v) methanol–0.01 M NaAc; ▼, (6:4, v/v) methanol–0.01 M NaAc, □, (4:6, v/v) acetonitrile–0.01 M NaAc; ■, (1:1, v/v) acetonitrile–0.01 M NaAc. Conditions: gain 3, sens. range 10, recorder 2 mV, concentration 70 ng/ml, 20-μl injection.[226]

also acts as mobile phase, suitable mixtures of water and an organic solvent had to be selected in order to obtain appropriate retention behavior of clobazam and desmethylclobazam. The results are shown in Figure 66. Obviously, over the range of mobile phase compositions studied, varying the water content has a relatively minor effect. Low water contents are preferable on account of the slightly higher fluorescence signals. For the rest, the trend observed with regard to changes in organic solvent followed the earlier observations in batch.

Figure 66 also gives an indication of the influence of reaction time (irradiation time) on the fluorescence signal in the mobile phases used. For clobazam, increasing the residence time from 28 to 86 sec resulted in a two- to threefold increase in peak height. For desmethylclobazam, the effect was much smaller; this suggests distinctly slower kinetics for the fluorophore formation with the metabolite. It should be added that with both clobazam and desmethylclobazam an irradiation time of 86 sec was not nearly sufficient to reach plateau conditions. That is, the use of longer residence times could well be beneficial and this system could serve as a model for a study of other photochemical reactor designs which permit longer irradiation times with a minimum of band broadening. Work along this line is in progress. For the phenothiazines, the general trend was similar to that observed for clobazam and its metabolite. Results of batch experiments suggest that a more rapid oxidation can be obtained by adding an oxidizing agent such as ammonium persulfate to the carrier stream. The uv radiation would then catalyze the oxidation process in an example of using photochemical re-

TABLE 17. Dependence of Band Broadening on Reactor Conditions[a (226)]

Conditions	Band broadening, sec		Asymmetry factor
	$\sigma_t{}^b$	$w_{t,10\%}{}^c$	
Without reactor	6	29	2.2
With reactor: nonirradiated	12	68	2.5
nonirradiated, 60°C	11.5	55	1.5
irradiated	12.5	57	1.5

[a] HPLC conditions: mobile phase, methanol–0.01 M pH 5 acetate buffer (1:1); flow rate, 0.5 ml/min; reaction time, 50 sec; model compounds, sulforidazine and mesoridazine.
[b] Measured at front of peak.
[c] Width measured at 10% peak height.

actions to accelerate already known chemical processes (e.g., see references 68 and 229). The dependence of band broadening in the photochemical reactor on residence time was studied with the phenothiazines as model compounds, since these exhibit native fluorescence and therefore can easily be detected without prior irradiation. Experiments were carried out with and without the photochemical reactor inserted in the line. In the former case, one series of data was collected under normal operating conditions, i.e., with irradiation, while a second series was obtained with the reactor shut off. The results are summarized in Table 17. From these results the band broadening due to the reactor, σ_{tr}, was calculated to be 10.4 sec (at $\phi = 0.5$ ml/min). The data indicate that the peak shape of the irradiated solutes is more symmetrical and narrower than the shape recorded with the reactor shut off. This is believed to be largely due to the higher diffusion rates prevailing under irradiation (high-temperature) conditions, as was demonstrated in a final series of experiments carried out at high temperature, but with the reactor shut off; these brought about a similar decrease in peak width and asymmetry. As an application of this detection principle, 1 ml of human serum was spiked with 50 µl of methanol containing 0.7 µg of each clobazam and desmethylclobazam. After the addition of 1 ml of methanol and shaking, the sample was centrifuged for 3 min at 7 g; 20-µl aliquots of this serum sample were injected on the top of the analytical column; the HPLC chromatogram is shown in Figure 67. The recovery for clobazam was 95% and for desmethylclobazam 70%. Analysis of spiked samples of human urine also was successful. Here, it is interesting to note that Sticht

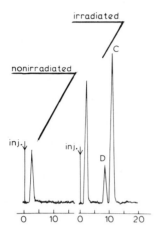

Fig. 67. Chromatogram of human serum sample spiked with 0.7 ppm of clobazam and desmethyl-clobazam, before and after irradiation. HPLC conditions: mobile phase, methanol–0.01 M sodium acetate solution (1:1, v/v); flow rate, 0.9 ml/min. Detector conditions: gain, 3; sensitivity range, 10; slit width, 10 nm; response, slow; recorder at 5 mV. C, clobazam; D, desmethylclobazam.[226]

and Kaeferstein[232] have reported that clobazam is poorly excreted in urine. These authors, however, calculated a detection limit for (hydrolyzed) clobazam of 50 ng, i.e., some three orders of magnitude higher than that found in the present study (30–50 pg/injection). The repeatability of the dynamic photochemical process was in all cases better than 1.0% relative s.d. ($n = 8$). The linearities of calibration curves extended over two to three orders of magnitude concentration ranges.

What made this study quite useful is the fact that many of the insights gained here can be of general utility in photochemical reactor design and choice of reaction conditions. A photochemical reactor can be constructed at moderate cost and be coupled to different detection modes, as has been shown in this brief overview of the field.

In contrast with many other detection principles, with photochemical detection postcolumn addition of reagent generally is not necessary. This implies that the design of the reaction detector can be rather simple and no additional band broadening due to reagent addition is introduced. The relatively fast kinetics of reactions such as those studied so far have permitted the use of reasonably short irradiation times. This in turn has enabled the study groups to work with nonsegmented flow without undue band broadening. However, there seems to be a need for reaction detectors which permit longer irradiation times on the order of several minutes. Segmentation techniques could well offer solutions in this direction. In the future, work will also have to be carried out to study the reaction mechanisms of the model systems. Hopefully, this will permit optimization of the design and use of the photochemical reaction detectors and drawing general conclusions as to structure requirements for useful photochemical reactions to occur.

Many potentially interesting reactions could certainly be elucidated from the photochemical literature and from batch and precolumn techniques already in use for analytical purposes. To name just one example: the analysis of tamoxifen, which is a phenylstilbene derivative and is currently being used as a metastatic breast cancer agent. Tamoxifen and its 4-hydroxylated derivatives can be converted to a highly fluorescent product via uv irradiation, and this fact has been used for clinical analysis in urine and plasma samples after separation of the fluorescent photoproducts by HPLC.[233] The photochemical conversion seems to proceed from the original non-fluorescent phenylstilbene structure to highly fluorescent phenanthrenes with $\lambda_{ex} = 256$ nm, $\lambda_{em} = 320$ nm. The reaction goes to completion after about 30 min irradiation with relatively low radiation, and it seems highly probable that the same reaction can be adapted with advantage to a postcolumn-mode photochemical reaction detector.

4.2.5. Chemiluminescence

Chemiluminescence and also likely bioluminescence are areas which could become of interest in the design of new reaction detectors.[234] Only very few studies hint at such a possibility and most of the work that has been done in coupling of HPLC with a chemiluminescence detector is in the field of metallics. Chemiluminescence reactions offer excellent selectivity for postchromatographic detection of many metal ions. Luminol (3-amino-phthalyhydrazide) reacts with hydrogen peroxide when catalyzed by certain metal ions to produce luminescence. For application to HPLC, the column effluent is mixed with a luminol–hydrogen peroxide mixture at a cell where light is produced. The advantage of this system is that no light source is required, only a photomultiplier to monitor luminescence. This approach has been evaluated for the detection of metal ions at parts per 10^{12} levels.[235,236] The experimental approach such as used by Neary et al.[234] is described below and gives an idea of a typical chemiluminescence reactor.

The effluent from an ion exchange column is mixed in a 100-µl chamber such as shown in Figure 68. The components are mixed by means of a gas turbine mixer which is driven by compressed air. The reactants are loaded into 50-ml plastic syringes and driven by Harvard model 975 infusion pumps. All connective tubing is made of PTFE. Luminescence is detected with a photomultiplier. The reagent solutions consist of 1.4×10^{-4} M luminol in 0.1 M lithium hydroxide and 0.1 M boric acid buffer (pH 10.5). Hydrogen peroxide is diluted with deionized water to 2.5×10^{-4} M. The flow rates on the column range from 0.1 to 2 ml/min depending on the desired separation.

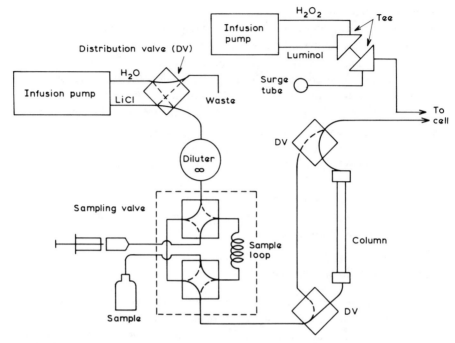

Fig. 68. Diagram of the luminol reaction detector and chromatograph for analysis of trace amounts of metals.[234]

The eluting solvents are distilled deionized water or 2.4 M lithium chloride (pH 2.8).

An organic analytical application of the chemiluminescence principle in conjunction with HPLC has recently been described by Fine *et al.*[238-240] for the detection of nitrosamines with the so-called TEA (thermal energy analyzer) detector. The basic principle is the chemiluminescence detection of nitrogen oxides by addition of ozone. The interaction of nitric oxide and ozone results in a chemiluminescent emission which is detected with a photomultiplier tube. More recently this principle has been applied to the detection of *N*-nitrosamines after catalytic cleavage of the N–NO bond to produce nitric oxide.[238] These detectors incorporate an optical filter to eliminate emissions occurring below 600 nm, such as those between ethylenic compounds and ozone.

Combination of this detector principle with GC[239] has been the first chromatographic application and has considerably enhanced its utility.

Chemiluminescent detectors thus provide substantially stronger evidence for the presence of an *N*-nitrosamine than detectors which are only nitrogen- rather than NO selective. For the analysis of foodstuffs and other commodities in which nitrosamines may occur it is therefore a valuable technique.

The extension of this principle to HPLC[237,240] is desirable for handling the many aromatic nonvolatile or poorly volatile nitrosamines. A typical design is shown in Figure 69. In this version, the HPLC solvent stream was delivered directly into the furnace using 1/16-in.-o.d., 0.010-in.-i.d. stainless steel tubing.

The vaporized solvent was carried through the system using nitrogen at 50 ml/min (standard temperature and pressure). The furnace consists of a nonglazed ceramic tube (1/8-in.-i.d. × 22 in.), maintained at 350°C by a concentric heating element. The vaporized solvent and nitrogen gas carrier were carried into a trap system maintained at a temperature slightly above the freezing point of the HPLC solvent. The nitrogen gas carrier was mixed with ozone in a chamber directly in front of the phototube assembly. The output of the ozone generator was adjusted so that the pressure in the chemiluminescence chamber was 1.0 Torr above the background pressure (0.2 Torr, typically). In actual operation, with the HPLC solvent flowing, the nitrogen passing at 50 ml(STP)/min and the ozone properly adjusted, the pressure in the chemiluminescence chamber was between 2.0 and 4.0

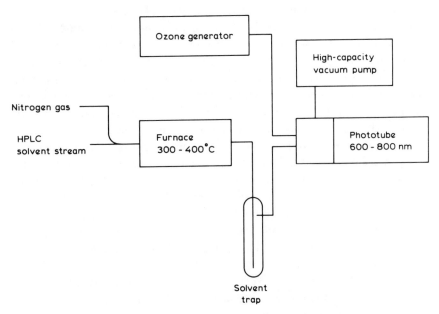

Fig. 69. Block diagram of the TEA–HPLC system.[240]

Torr. A vacuum pump with a free air pumping capacity of 150 l/min was used to evacuate the chemiluminescence chamber.

The characteristics of this detector such as time variation, nitrogen carrier flow rate, and furnace temperature were studied.[240] The loss of resolution that has to be accepted due to the TEA reaction tract is on the same order as for previously discussed reactors ($\sigma \simeq$ <10 sec).

Considering the specificity and the sensitivity ($\sim$20-µg detection limit) of the TEA technique, which permits trace determination of these nitrosamines in foods with a minimum of cleanup, then the increased zone spreading is a fair trade-off.

5. CONCLUSIONS

In spite of the relatively recent nature of this field a respectable list of achievements has accumulated in the past few years demonstrating that postcolumn reaction detectors can help in solving many demanding analytical tasks.

The table of applications (Table 18) will serve to demonstrate this point and may aid the reader in finding a solution to his or her own problems.

One of the more remarkable features of reaction detectors is their ability to fit into automated analytical systems. Their selectivity and sensitivity routed in the reaction itself will often permit one to drastically cut down the amount of sample preparation or the requirements for a powerful separation. They are, in other words, an important link in a totally integrated system and a necessary step in the direction of tailor-made systems. With the further development of reactor technology and the better understanding of their operation parameters many more reactions and other detection modes will be made accessible to this field of HPLC detectors.

To conclude, a word should again be said on the aspect of band broadening, since one would like to maintain the resolution obtained in the chromatographic system as much as possible. Proper design of the reaction detector is the major criterion to assume minimal band broadening. The choice of the optimal reactor mode is another important aspect. As a rule of thumb we can say that for very short reaction times, i.e., up to 30 sec, a tubular reactor system may be used to advantage. For longer reaction times of up to 5 min one might preferably adopt a bed reactor, and for reaction periods above this the segmentation principle can be recommended. Besides reaction kinetics other considerations have to be taken into account, such as aggressiveness of chemicals, high-temperature or high-pressure reaction conditions,

TABLE 18. List of Applications for Postcolumn Reaction Detectors in HPLC

Compounds	Reagent(s)	Detection[a]	Reference(s)
Amines	Iodine	uv	77
	Ninhydrin	vis	95
Amines (tertiary)	Dimethoxyanthracene sulfate (DAS)	F	127–133, 138, 139
Amino acids	Fluram	F	38, 52–54
	NBD–Cl	F	58, 62
	Orthophthalaldehyde (OPA)	F	18, 19, 57–61, 64, 204
	Ninhydrin	vis	25, 26, 60, 92, 206
	Pyridoxal–Zn	F	117
Amino acids (secondary)	N-chloro-succinimide–Fluram	F	52
Acids (organic)	Ce^{4+}, H_2SO_4	F	118, 119
	O-nitrophenol–Na	vis	78, 79
Barbiturates	Borate buffer (pH 10)	uv	80
Biogenic amines	Ethylenediamine	F	115
	OPA	F	106
	Miscellaneous	F, M	112–114, 164
Catecholamines	$[Fe(CN)_6]^{3-}$ and others	F	112–115
Cannabinoids	Photochemical	F	223
Carbamates	NaOH, OPA	F	127
	Choline esterase	vis, F	100, 126
Carbohydrates (sugars)	Heat, H_2SO_4	uv	73, 74
	Phenols, H_2SO_4	vis	75, 76
	Cu^{2+} + Neocuproin	vis	97

continued overleaf

TABLE 18. (*continued*)

Compounds	Reagent(s)	Detection[a]	Reference(s)
Carbohydrates (sugars)	Bicinchonate	vis	98
	Tetrazolium salt	vis	71–76
	$Ce^{4+} + H_2SO_4$	F	118, 119, 122
	$[Fe(CN)_6]^{3-}$	E	18, 148
Clobazam + metabolites	Photochemical	F	226
Creatinine	Picric acid	vis	70
Cyclohexanone	Dinitrophenylhydrazine	vis	102
Enzymes	NAD and other substrates	uv, F (E)	69, 84–86
Ergotalkaloids	Photochemical	F	224
Erythromycins	Naphthatriazole disulfide	F	135
Gentamycins	OPA	F	41, 55
Hormones	Fluram	F	38, 54
	Ce^{4+}, As^{3+}	vis	217
Hydroxyperoxides	NaI	vis	23
Indoles	OPA	F	108–110
Lanthanides	Complexing reagent	uv	83
Lipids	Ammonium sulfate	M	175
Metals	Complexing reagent	uv, F	81–83
	Luminol, H_2O_2	F, M	234–236
Nitrite	Sulfanil amide	vis	208
Nitrosamines	Photochemical, Griess reaction	vis, E	220, 222
	Ozone	M	238–240
Oligoadenylates	Phosphatase	uv	213

Oligosaccharides	Hydroxybenzoic acid	vis	216
Penicillins	Imidazole, $HgCl_2$	uv-vis	103
Pesticides	Cholinesterase and diverse	vis, F	100, 101, 126
	DAS	F	133, 138, 139
Peptides	Dans-Cl	F	140
	Fluram	F	38–40
	OPA	F	64, 107
Pharmaceuticals	DAS	F	127–133, 138, 139
	Dans-Cl	F	140
	Photochemical	F	226, 224
	Miscellaneous	F, uv	135, 80, 68, 103
	Fluram	F	38–40, 54
	OPA	F	41, 55, 64, 107
Phenols	Ce^{4+}, H_2SO_4	F	123
Phenothiazines	Photochemical	F	226
	$KMnO_4$	F	68
Pheromones	Br_2	E	146
Phosphates	H_2SO_4, Va–molybdate	vis	18
Polyamines	OPA	F	65
Polythionates	Ce^{4+}, H_2SO_4	F	124
Prostaglandins	Br_2	E	146
Sugars (see Carbohydrates)			
Sulfhydryl compounds	$[Fe(CN)_6]^{3-}$	E	149
Thioureas	Hg complexation	E	157
Thyroid hormones	Ce^{4+}, As^{3+}	vis	217

[a] Abbreviations:: OPA, orthophthalaldehyde; DAS, dimethoxyanthracene sulfonate; Dans-Cl, dansylchloride; uv, ultraviolet; vis, visible; F, fluorescence; E, electrochemical; M, miscellaneous.

photochemical aspects, etc. In any case it would be valuable if investigators would report precise band broadening data on their system(s) in order to permit an objective comparison of new designs and to eventually come to conclusions and rules of general validity regarding the use of reaction detectors.

However, one should not be a purist in this regard since the gain in selectivity which one might obtain by using selective chemical derivatization techniques could well compensate for a couple per cent of band broadening.

As a final comment we can certainly state that this line of novel detection devices will make a distinct impact on future HPLC work and will be of considerable usefulness in solving the complex analysis problems that we are confronted with in many areas.

REFERENCES

1. J. F. Lawrence and R. W. Frei, *Chemical Derivatization in Liquid Chromatography*, Elsevier, Amsterdam (1976).
2. K. Blau and G. S. King, *Handbook of Derivatives for Chromatography*, Heyden, London (1977).
3. R. W. Frei and W. Santi, *Z. Anal. Chem.* **277**, 303 (1975).
4. R. W. Frei, *Res. Dev.*, 42 (February, 1977).
5. S. A. Wise and W. E. May, *Res. Dev.*, 45 (October, 1977).
6. R. W. Frei, in *Methodological Surveys in Biochemistry*, Vol. 7, E. Reid, ed., Ellis Horwood Publishers, Chichester, England (1978), p. 243.
7. R. W. Frei, *J. Chromatogr. Rev.* **165**, 75 (1979).
8. J. Drozd, *J. Chromatogr.* **113**, 303 (1975).
9. R. W. Frei and A. H. M. T. Scholten, *J. Chromatogr.* **17**, 152 (1979).
10. D. H. Spackman, W. H. Stein, and S. Moore, *Anal. Chem.* **30**, 1190 (1958).
11. J. Hrdina, in Proceedings of the 6th Technicon Amino Acid Colloquium, London, September 1968, p. 355.
12. G. Ertinghausen, H. J. Adler, and A. S. Reichler, *J. Chromatogr.* **42**, 355 (1969).
13. G. Taylor, *Proc. R. Soc. London Ser. A* **219**, 186 (1953).
14. R. A. Aris, *Proc. R. Soc. London Ser. A* **235**, 67 (1956).
15. L. R. Austin and J. P. Seader, *AIChE J.* **19**, 85 (1973).
16. W. R. Dean, *Philos. Mag.* **4**, 208 (1927); **5**, 673 (1928).
17. R. N. Trivedi and K. Vasuveda, *Chem. Eng. Sci.* **29**, 2291 (1974); **30**, 1317 (1975).
18. R. S. Deelder, M. G. F. Kroll, A. J. B. Beeren, and J. H. M. van den Berg, *J. Chromatogr.* **149**, 669 (1978).
19. J. H. M. van den Berg, dissertation, Technische Hogeschool, Eindhoven, 1978.
20. R. Tijssen, dissertation, Technische Hogeschool, Delft, 1979.
21. R. Tijssen, *Sep. Sci. Technol.* **13**, 681 (1978).
22. J. Ruzicka and E. H. Hansen, *Anal. Chim. Acta* **99**, 37 (1979).
23. R. S. Deelder, M. G. F. Kroll, and J. H. M. van den Berg, *J. Chromatogr.* **125**, 307 (1976).

24. T. D. Schlabach, S. H. Chang, K. M. Goodring, and T. E. Regnier, *J. Chromatogr.* **134**, 91 (1977).
25. K. M. Jonker, H. Poppe, and J. F. K. Huber, *Z. Anal. Chem.* **279**, 154 (1976).
26. K. M. Jonker, H. Poppe, and J. F. K. Huber, *Chromatographia* **11**, 123 (1978).
27. J. W. Hiby, in *Proceedings of Symposium on Interaction Between Fluids and Particles*, P. A. Rottenburg, ed., Institution of Chemical Engineers, London (1962), p. 312.
28. L. J. Skeggs, *Am. J. Clin. Pathol.* **28**, 311 (1957).
29. R. W. Deelder and P. J. H. Hendricks, *J. Chromatogr.* **83**, 343 (1973).
30. G. Ertinghausen, H. J. Adler, and A. S. Reichler, *J. Chromatogr.* **42**, 355 (1969).
31. L. R. Snyder and H. J. Adler, *Anal. Chem.* **48**, 1017, 1022 (1976).
32. L. R. Snyder, *J. Chromatogr.* **125**, 287 (1976).
33. G. M. Singer, S. S. Singer, and D. G. Schmidt, *J. Chromatogr.* **133**, 59 (1977).
34. R. L. Habig, B. W. Schlein, L. Walters, and R. E. Theis, *Clin. Chem.* **15**, 1045 (1969).
35. J. F. Lawrence, U. A. Th. Brinkman, and R. W. Frei, *J. Chromatogr.* **171**, 73 (1979).
36. J. F. Lawrence, U. A. Th. Brinkman, and R. W. Frei, *J. Chromatogr.* **185**, 473–481 (1979).
37. C. E. Werkhoven, U. A. Th. Brinkman, and R. W. Frei, *Anal. Chim. Acta* **114**, 147 (1980).
38. K. Zech and W. Voelter, *Chromatographia* **8**, 350 (1979).
39. R. W. Frei, L. Michel, and W. Santi, *J. Chromatogr.* **126**, 665 (1976).
40. R. W. Frei, L. Michel, and W. Santi, *J. Chromatogr.* **142**, 261 (1977).
41. E. Oelrich and D. Theuerkauf, *J. High Resolution Chromatogr. & Chromatogr. Commun.* **2**, 256 (1979).
42. K. M. Jonker, thesis, University of Amsterdam, 1977.
43. J. H. Knox and J. F. Parcher, *Anal. Chem.* **41**, 1599 (1969).
44. J. H. Knox, G. R. Laird, and P. A. Raven, *J. Chromatogr.* **122**, 129 (1976).
45. S. Udenfriend, S. Stein, S. Böhlen, P. Dairman, W. Leimbruger, and M. Weigele, *Science* **178**, 871 (1972).
46. M. Roth, *Anal. Chem.* **43**, 880 (1971).
47. A. H. M. T. Scholten, U. A. Th. Brinkman, and R. W. Frei, *J. Chromatogr.* (in press).
48. S. Stein, P. Böhlen, and S. Udenfriend, *Arch. Biochem. Biophys.* **163**, 400 (1974).
49. K. Samejima, H. Kawase, S. Sakamoto, M. Okada, and Y. Endo, *Anal. Biochem.* **76**, 392 (1976).
50. P. M. Fröhlich and T. D. Cunningham, *Anal. Chim. Acta* **84**, 427 (1976).
51. P. M. Fröhlich and L. D. Murphy, *Anal. Chem.* **49**, 1606 (1977).
52. A. H. Felix and G. Terkelsen, *Anal. Biochem.* **60**, 28 (1974).
53. J. R. Benson, and P. E. Hare, *Proc. Natl. Acad. Sci USA* **72**, 619 (1975).
54. W. Voelter and K. Zech, *J. Chromatogr.* **112**, 743 (1975).
55. E. Oelrich, H. Preusch, E. Wilhelm, and D. Theuerkauf, *J. Chromatogr. Sci.* **17**, 243 (1979).
56. R. E. Kaiser and E. Oelrich, *Optimierung in der HPLC*, Dr. Alfred Hüthig Verlag, GMBH, Heildelberg (1979).
57. R. Hakanson and A. Rönnberg, *Anal. Biochem.* **54**, 353 (1973).
58. M. Roth and A. Hampai, *J. Chromatogr.* **83**, 353 (1973).
59. J. R. Cronin and P. E. Hare, *Anal. Biochem.* **81**, 151 (1977).
60. E. Lund, J. Thomsen, and K. Brunfeld, *J. Chromatogr.* **130**, 51 (1977).
61. K. Aoki and Y. Kuroiwa, *Chem. Pharm. Bull.* **26**, 2684 (1978).
62. M. Roth, *Clin. Chim. Acta* **83**, 273 (1978).

63. P. B. Ghosh and M. W. Whitehouse, *Biochem. J.* **108**, 155 (1968).
64. E. H. Creaser and G. H. Hughes, *J. Chromatogr.* **114**, 69 (1977).
65. L. J. Morton and P. C. Y. Lee, *Clin. Chem.* **21**, 1721 (1975).
66. (a) E. Ueda, N. Yoshida, K. Nishimura, T. Joh, S. Antoku, K. Tsukada, S. Ganno and T. Kobuku, *Clin. Chim. Acta* **80**, 447 (1977).
    (b) R. J. Merrils, *Anal. Biochem.* **6**, 272 (1963).
67. K. Mori, *Ind. Health* **12**, 171 (1974).
68. R. G. Muusze and J. F. K. Huber, *J. Chromatogr. Sci.* **12**, 779 (1974).
69. R. P. Schroeder, P. J. Kudirka, and E. C. Toren Jr., *J. Chromatogr.* **134**, 83 (1977).
70. N. D. Brown, H. C. Singer, W. E. Neeley, and S. E. Koeritz, *Clin. Chem.* **23**, 1281 (1977).
71. K. Mopper and E. T. Degens, *Anal. Biochem.* **45**, 147 (1972).
72. D. Noel, T. Hanai, and M. D'Amboise, *J. Liq. Chromatogr.* **2**, 1325 (1979).
73. Y. Tijima, *Dempun Kagaku (Starch Sci.)* **21**, 124 (1974).
74. S. Katz and L. H. Thacker, *J. Chromatogr.* **64**, 247 (1972).
75. S. Katz, S. R. Dinsmore, and W. W. Pitt Jr., *Clin. Chem.* **17**, 731 (1971).
76. W. W. Pitt Jr., C. D. Scott, W. F. Johnson, and G. Jones Jr., *Clin. Chem.* **16**, 657 (1970).
77. C. R. Clark, C. H. Darling, J. Chan, and A. C. Nichols, *Anal. Chem.* **49**, 2080 (1977).
78. L. Kessner and E. Muntwyler, *Anal. Chem.* **38**, 1164 (1966).
79. K. W. Stahl, G. Schafer, and W. Lamprecht, *J. Chromatogr. Sci.* **10**, 95 (1972).
80. C. R. Clark and J. Chan, *Anal. Chem.* **50**, 635 (1978).
81. J. S. Fritz and J. N. Story, *Anal. Chem.* **46**, 825 (1974).
82. J. P. Sickafoose, PhD. thesis, Iowa State University, 1971.
83. S. Elchuck and R. M. Cassidy, paper presented at the 176th National American Chemical Society Meeting, Miami Beach, Florida, 1978.
84. S. H. Chang, K. M. Goodring, and F. E. Regnier, *J. Chromatogr.* **125**, 103 (1976).
85. T. D. Schlabach, S. H. Chang, K. M. Goodring, and F. E. Regnier, *J. Chromatogr.* **134**, 91 (1977).
86. T. D. Schlabach, A. J. Alpert, and F. E. Regnier, *Clin. Chem.* **24**, 1351 (1978).
87. R. W. Frei, Proceedings of the 8th Technicon International Congress, London, December 1978, IX-FR-1-16, 1979.
88. D. A. Burns, Proceedings of the 7th Technicon International Congress, New York City, December 1976.
89. G. Schwedt, *Angew. Chem.* **91**, 192 (1979).
90. Bibliography Service, Technicon Inc., Tarrytown, New York.
91. L. Snyder, J. Levine, R. Stoy, and A. Conetta, *Anal. Chem.* **48**, 942A (1976).
92. G. Ertinghausen, H. J. Adler, and A. S. Reichler, *J. Chromatogr.* **42**, 355 (1969).
93. H. Tschesche, C. Frank, and H. Ebert, *J. Chromatogr.* **85**, 35 (1973).
94. C. E. Vander Cook and R. L. Prince, *J. Assoc. Offic. Anal. Chem.* **57**, 124 (1974).
95. H. Hatano, K. Sumizu, R. Robushika, and F. Murakami, *Anal. Biochem.* **35**, 337 (1970).
96. M. Sinner and J. Puls, *J. Chromatogr.* **156**, 197 (1978).
97. M. H. Simatupang and H. H. Dietrichs, *Chromatographia* **11**, 89 (1978).
98. K. Mopper and E. M. Gindler, *Anal. Biochem.* **56**, 440 (1973).
99. G. F. Smith and W. H. McCurdy, Jr., *Anal. Chem.* **24**, 371 (1952).
100. K. A. Ramsteiner and W. D. Hörman, *J. Chromatogr.* **104**, 438 (1975).
101. D. E. Ott, *Bull. Environ. Contam. Tox.* **17**, 261 (1977).

102. R. S. Deelder and P. J. H. Hendricks, *J. Chromatogr.* **83**, 343 (1973).
103. D. Westerlund, J. Carlqvist, and A. Theodorsen, *Acta Pharm. Suecica* **16**, 187 (1979).
104. H. Bundgaard and K. Ilver, *J. Pharm. Pharmac.* **24**, 790 (1972).
105. P. Schauwecker, F. Erni, and R. W. Frei, *J. Chromatogr.* **136**, 63 (1977).
106. G. Schwedt, *Anal. Chim. Acta* **92**, 337 (1977).
107. F. Erni, K. Krummen, and A. Pellet, *Chromatographia* **12**, 399 (1979).
108. F. Engbaek and I. Magnussen, *Clin. Chem.* **24**, 376 (1978).
109. R. P. Maikel and F. P. Miller, *Anal. Chem.* **38**, 1937 (1966).
110. H. H. Brown, M. C. Rhindress, and R. Griswold, *Clin. Chem.* **17**, 92 (1971).
111. F. Erni and H. Bosshard, *Chromatographia* **12**, 412 (1979).
112. G. Schwedt, *Fresenius Z. Anal. Chem.* **287**, 152 (1977).
113. G. Schwedt, *J. Chromatogr.* **143**, 463 (1977).
114. G. Schwedt, *Fresenius Z. Anal. Chem.* **293**, 40 (1978).
115. G. Schwedt, *Chromatographia* **10**, 92 (1977).
116. W. H. Weil-Malherbe, *Methods Biochem. Anal.* **16**, 294 (1968).
117. M. Maeda, A. Tsuji, S. Ganno, and Y. Onishi, *J. Chromatogr.* **77**, 434 (1973).
118. S. Katz, W. W. Pitt Jr., and G. Jones, *Clin. Chem.* **19**, 817 (1973).
119. S. Katz and W. W. Pitt Jr., *Anal. Lett.* **5**, 177 (1972).
120. G. den Boef and H. L. Polak, *Talanta* **9**, 271 (1962).
121. W. H. Richardson, in *Oxidation in Organic Chemistry*, Part A, K. B. Weberg, ed., Academic Press, New York (1965), p. 243.
122. S. Katz, W. W. Pitt Jr., J. E. Mrochek, and S. Dinsmore, *J. Chromatogr.* **101**, 193 (1974).
123. A. W. Wolkoff and R. H. Larose, *J. Chromatogr.* **99**, 731 (1977).
124. A. W. Wolkoff and R. H. Larose, *Anal. Chem.* **47**, 1003 (1975).
125. J. C. Gfeller, G. Frey, and R. W. Frei, *J. Chromatogr.* **142**, 271 (1977).
126. H. A. Moye and T. W. Wade, *Anal. Lett.* **9**, 801 (1976).
127. M. A. Moye, S. J. Scherer, and P. A. St. John, *Anal. Lett.* **10**, 1049 (1977).
128. J. C. Gfeller, G. Frey, J. M. Huen, and J. P. Thevenin, *J. High Resolution Chromatogr. & Chromatogr. Commun.* **1**, 213 (1978).
129. R. W. Frei, J. F. Lawrence, U. A. Th. Brinkman, and I. Honigberg, *J. HRC & CC* **2**, 11 (1979).
130. D. Westerlund and K. O. Borg, *Anal. Chim. Acta* **67**, 89 (1973).
131. J. C. Gfeller and G. Frey, *Z. Anal. Chem.* **291**, 332 (1978).
132. J. M. Huen and J. P. Thevenin, *J. HRC & CC* **2**, 154 (1979).
133. J. F. Lawrence, U. A. Th. Brinkman, and R. W. Frei, *J. Chromatogr.* **171**, 73 (1979).
134. B. Karlberg and S. Thelander, *Anal. Chim. Acta* **98**, 1 (1978).
135. K. Tsuji, *J. Chromatogr.* **158**, 337 (1978).
136. K. Tsuji and J. F. Goetz, *J. Chromatogr.* **147**, 359 (1978).
137. K. Y. Tserng and J. G. Wagner, *Anal. Chem.* **48**, 348 (1976).
138. C. van Buuren, J. F. Lawrence, U. A. Th. Brinkman, I. L. Honigberg, and R. W. Frei, *Anal. Chem.* **52**, 700 (1980).
139. J. F. Lawrence, U. A. Th. Brinkman, and R. W. Frei, *J. Chromatogr.* **185**, 483 (1979).
140. C. E. Werkhoven-Goewie, U. A. Th. Brinkman, and R. W. Frei, *Anal. Chim. Acta* **114**, 147 (1980).
141. N. Seiler and L. Demisch, in *Handbook of Derivatives in Chromatography*, K. Blau and G. Kind, eds., Heyden Publishers, London (1978), p. 349.
142. R. W. Frei, W. Santi, and M. Thomas, *J. Chromatogr.* **116**, 365 (1976).

143. R. W. Frei, M. Thomas, and I. Frei, *J. Liq. Chromatogr.* **1**, 443 (1978).

144. J. F. Lawrence, C. Renault, and R. W. Frei, *J. Chromatogr.* **121**, 343 (1976).

145. F. Nachtmann, H. Spitzi, and R. W. Frei, *Anal. Chim. Acta* **76**, 57 (1975).

146. P. T. Kissinger, Amperometric and coulometric detectors for HPLC, *Anal. Chem.* **49**, 447A–456A (1977).

147. P. T. Kissinger, K. Bratin, G. C. Davis, and L. A. Pachla, *J. Chromatogr. Sci.* **17**, 137 (1979).

148. Y. Takata and G. Muto, *Anal. Chem.* **45**, 1864 (1973).

149. P. T. Kissinger, L. J. Felice, D. J. Miner, C. R. Preddy, and R. E. Shoup, Detectors for trace organic analysis by liquid chromatography; Principles and applications, in *Advances in Analytical and Clinical Chemistry*, Plenum Press, New York (1978), pp. 55–175.

150. J. Lankelma and H. Poppe, *J. Chromatogr.* **125**, 375 (1976).

151. P. Zuman, *Organic Polarographic Analysis*, Pergamon Press, Oxford (1965).

152. W. F. Smyth, *Proc. Analyt. Div. Chem. Soc.*, 187 (1975).

153. Y. Audonard, A. Suzanne, O. Vittori, and M. Porthault, *Bull. Soc. Chim.* **85**, 130 (1975).

154. R. S. Schifreen, D. A. Hanua, L. D. Bowers, and I. W. Carr, *Anal. Chem.* **49**, 1929 (1977).

155. R. E. Adams and P. W. Carr, *Anal. Chem.* **50**, 944 (1978).

156. M. Lemar and M. Porthault, *J. Chromatogr.* **99**, 372 (1977).

157. H. B. Hanekamp, P. Bos, and R. W. Frei, *J. Chromatogr.* **186**, 489 (1979).

158. H. B. Hanekamp, P. Bos, U. A. Th. Brinkman, and R. W. Frei, *Z. Anal. Chem.* **297**, 404 (1979).

159. W. R. Heineman and P. T. Kissinger, *Anal. Chem.* **50**, 166–175R (1978).

160. P. J. Arpino and G. Guiochon, *Anal. Chem.* **51**, 682A (1979).

161. B. L. Karger and P. Vouros, paper presented at the IVth International Symposium on Liquid Chromatography, May 1979, Boston, Massachusetts; *Anal. Chem.* **51**, 2324 (1979).

162. A. Zeman and I. P. G. Wirotama, *Z. Anal. Chem.* **247**, 155 (1969).

163. K. Heyne, H. P. Harke, H. Scharmann, and H. F. Grutzmacher, *Z. Anal. Chem.* **230**, 118 (1967).

164. N. Seiler, H. Schneider, and K. D. Sonnenberg, *Z. Anal. Chem.* **252**, 127 (1970).

165. M. Frei-Häusler, R. W. Frei, and O. Hutzinger, *J. Chromatogr.* **79**, 209 (1973).

166. R. M. Cassidy, D. S. LeGay, and R. W. Frei, *J. Chromatogr. Sci.* **12**, 85 (1974).

167. M. Frei-Häusler, R. W. Frei, and O. Hutzinger, *J. Chromatogr.* **84**, 214 (1973).

168. Brochure on HPLC Radioactivity Monitor LB 503, Berthold Ltd., D-7547 Wildbad 1, Federal Republic of Germany.

169. G. Gübitz, R. W. Frei, W. Santi, and B. Schreiber, *J. Chromatogr.* **169**, 213 (1979).

170. L. Casola, G. DiMatteo, G. Di Prisco, and F. Cervone, *Anal. Biochem.* **57**, 38 (1974).

171. L. Casola and G. DiMatteo, *Anal. Biochem.* **49**, 416 (1972).

172. M. C. Cullen and E. T. McGuinness, *Anal. Biochem.* **42**, 455 (1971).

173. R. H. Benson and R. B. Turner, *Anal. Chem.* **32**, 1464 (1960).

174. J. W. Jorgenson, S. L. Smith, and M. Novotny, *J. Chromatogr.* **142**, 233 (1977).

175. J. Folch, M. Lees, and G. H. Sloane-Stanley, *J. Biol. Chem.* **226**, 497 (1957).

176. F. J. Fernandez, *Chromatogr. Newsl.* **5**, 17 (1977).

177. D. R. Jones and S. T. Manahan, *Anal. Lett.* **8**, 421 (1975).

178. C. Bove, F. Cacace, and R. Cozzani, *Anal. Lett.* **9**, 825 (1976).

179. D. R. Jones and S. E. Manahan, *Anal. Chem.* **48**, 502 (1976).
180. D. R. Jones, H. C. Tung, and S. E. Manahan, *Anal. Chem.* **48**, 7 (1976).
181. F. E. Brinkman, W. R. Blair, K. L. Jewett, and W. P. Iverson, *J. Chromatogr. Sci.* **15**, 493 (1977).
182. B. G. Julin, H. W. Vandenborgh, and J. J. Kirkland, *J. Chromatogr.* **112**, 443 (1975).
183. S. S. Brody and J. E. Chaney, *J. Gas Chromatogr.* **4**, 42 (1966).
184. B. J. Compton and W. C. Purdy, *J. Chromatogr.* **169**, 39 (1979).
185. V. A. Fassel and R. N. Kniseley, *Anal. Chem.* **46**, 1110A (1974).
186. C. H. Gast, J. C. Kraak, H. Poppe, and F. J. M. J. Maessen, paper presented at the Fourth International Symposium on Liquid Chromatography, May 1979, Boston, and *J. Chromatogr.* **185**, 549 (1979).
187. G. Nota and R. Palombari, *J. Chromatogr.* **62**, 153 (1952).
188. F. W. Willmott and R. J. Dolphin, *J. Chromatogr. Sci.* **12**, 695 (1974).
189. U. A. Th. Brinkman, P. M. Onel, and G. de Vries, *J. Chromatogr.* **171**, 424 (1979).
190. I. Suffet and E. J. Sowinski Jr., in *Chromatographic Analysis of the Environment*, R. L. Grob, ed., Marcel Dekker, New York (1975), pp. 435–483.
191. J. F. K. Huber and R. R. Becker, *J. Chromatogr.* **142**, 765 (1977).
192. F. Eisenbeiss, H. Hein, R. Joester, and G. Naundorf, *Chrom. Newsl.* **6**, 8 (1978).
193. J. Kirkland, *Analyst* **99**, 859 (1974).
194. J. N. Little and G. J. Fallick, *J. Chromatogr.* **112**, 389 (1975).
195. K. Krummen and R. W. Frei, *J. Chromatogr.* **132**, 27 (1977).
196. K. Krummen and R. W. Frei, *J. Chromatogr.* **132**, 429 (1977).
197. P. Schwauwecker, R.W. Frei, and F. Erni, *J. Chromatogr.* **136**, 63 (1977).
198. R. W. Frei, *Int. J. Environ. Anal. Chem.* **5**, 143 (1978).
199. W. E. May, S. N. Chester, S. P. Cram, B. H. Gump, H. S. Hertz, D. P. Enagonio, and S. M. Dyszel, *J. Chromatogr. Sci.* **13**, 535 (1975).
200. J. Lankelma and H. Poppe, *J. Chromatogr.* **149**, 587 (1978).
201. D. Ishii, K. Hibi, K. Asai, and M. Nagaya, *J. Chromatogr.* **152**, 341 (1978).
202. D. Ishii, K. Hibi, K. Asai, M. Nagaya, K. Mochizuki, and Y. Mochida, *J. Chromatogr.* **156**, 173 (1978).
203. D. Ishii, K. Asai, K. Hibi, T. Jonokuchi, and M. Nagaya, *J. Chromatogr.* **144**, 157 (1977).
204. H. P. M. van Vliet, Th. C. Bootsman, R. W. Frei, and U. A. Th. Brinkman, *J. Chromatogr.* **185**, 483 (1979).
205. H. Adler, M. Margoshes, L. R. Snyder, and C. Spitzer, *J. Chromatogr.* **143**, 125 (1977).
206. D. Kasiske, K. D. Klinkmüller, and M. Sonneborn, *J. Chromatogr.* **149**, 703 (1978).
207. S. K. Maitra, T. T. Yoshikawa, J. L. Hansen, I. Nilsson-Ehle, W. J. Palin, M. C. Schotz, and L. B. Gruze, *Clin. Chem.* **23**, 2275 (1977).
208. M. K. Chao, T. Higuchi, and L. A. Sternson, *Anal. Chem.* **50**, 1670 (1978).
209. A. H. M. T. Scholten, C. van Buuren, J. F. Lawrence, U. A. Th. Brinkman, and R. W. Frei, *J. Liq. Chromatogr.* **2**, 607 (1979).
210. A. H. M. T. Scholten, B. J. de Vos, J. F. Lawrence, U. A. Th. Brinkman, and R. W. Frei, *Anal. Lett.* **13**, 1235 (1980).
211. F. Erni and R. W. Frei, *J. Chromatogr.* **149**, 561 (1978).
212. E. L. Johnson, R. Gloor, and R. E. Majors, *J. Chromatogr.* **149**, 571 (1978).
213. D. A. Usher and J. A. Rosen, *Anal. Biochem.* **92**, 276 (1979).

214. R. L. Pearson, J. F. Weiss, and A. D. Kelmers, *Biochim. Biophys. Acta* **228**, 770 (1971).
215. F. Erni, R. W. Frei, and W. Lindner, *J. Chromatogr.* **125**, 265 (1976).
216. P. Vratny, J. Cuhrabkova, and J. Copikova, Proceedings of the Second Danube Symposium on Progress in Chromatography, Carlsbad, Czechoslovakia, 1979, Elsevier (1979).
217. F. Nachtmann, G. Knapp, and H. Spitzy, *J. Chromatogr.* **149**, 693 (1978).
218. M. Lever, *Anal. Biochem.* **47**, 273 (1972).
219. K. B. Yatsimirskii, *Kinetic Methods of Analysis*, Pergamon Press, Oxford (1966).
220. W. Iwaoka and S. R. Tannenbaum, *IARC Sci. Publ.* **14**, 51 (1976).
221. G. M. Singer, *J. Chromatogr.* **133**, 59 (1977).
222. B. G. Snyder and D. C. Johnson, *Anal. Chim. Acta* **106**, 1 (1979).
223. P. J. Twitchett, P. L. Williams, and A. C. Moffat, *J. Chromatogr.* **149**, 683 (1978).
224. A. H. M. T. Scholten and R. W. Frei, *J. Chromatogr.* **176**, 349 (1979).
225. A. Stoll and W. Schlientz, *Helv. Chim. Acta* **38**, 585 (1955).
226. A. H. M. T. Scholten, U. A. Th. Brinkman, and R. W. Frei, *Anal. Chim. Acta* **114**, 137 (1980).
227. Clinical Investigator Literature, Hoechst-Roussel Pharmaceuticals Inc., Sommerville, New Jersey, May 1973.
228. R. G. Borland and A. N. Nicholson, *Brit. J. Clin. Pharmacol.* **2**, 215 (1975).
229. F. W. Kerner, *FDA Bylines* No. 1, 15 (1973).
230. J. B. Ragland and V. J. Kniross-Wright, *Anal. Chem.* **36**, 1356 (1964).
231. V. R. White, Ch. S. Frings, J. E. Villafranca, and J. M. Fitzgerald, *Anal. Chem.* **48**, 1314 (1976).
232. G. Sticht and H. Kaeferstein, *Z. Rechtsmed.* **82**, 105 (1978).
233. D. W. Mendenhall, H. Kobayaski, M. M. L. Shih, L. A. Sternson, T. Higuchi, and C. Fabian, *Clin. Chem.* **24**, 1518 (1978).
234. M. P. Neary, R. Seitz, and D. M. Hercules, *Anal. Lett.* **7**, 583 (1974).
235. W. R. Seitz and D. R. Hercules, in *Chemiluminescence and Bioluminescence*, M. J. Cormier, J. Lee, and D. M. Hercules, eds., Plenum Press, New York (1972) p. 427.
236. W. R. Seitz, W. W. Suydam, and D. M. Hercules, *Anal. Chem.* **44**, 957 (1972).
237. J. K. Baker and Cheng-Yu Ma, *IARC Sci. Publ.* **16**, 19 (1978).
238. D. H. Fine, D. Lieb, and F. Rufeh, *J. Chromatogr.* **107**, 351 (1975).
239. D. H. Fine and D. P. Roundbehler, *J. Chromatogr.* **109**, 271 (1975).
240. D. H. Fine, F. Huffman, D. P. Roundbehler, and N. M. Belcher, Analysis of *N*-nitroso compounds by combined high performance liquid chromatography and thermal energy analysis, in *Environmental N-nitrosa Compounds-Analysis and Formation*, E. A. Walker, P. Bogovski, and L. Griciute, eds., Lyon, International Agency for Research on Cancer, IARC Scientific Publications No. 14 (1976) pp. 43–50.

Index